Conservation in the 21st Century: Gorillas as a Case Study

DEVELOPMENTS IN PRIMATOLOGY: PROGRESS AND PROSPECTS
Series Editor: Russell H. Tuttle, University of Chicago, Chicago, Illinois

This peer-reviewed book series melds the facts of organic diversity with the continuity of the evolutionary process. The volumes in this series exemplify the diversity of theoretical perspectives and methodological approaches currently employed by primatologists and physical anthropologists. Specific coverage includes: primate behavior in natural habitats and captive settings; primate ecology and conservation; functional morphology and developmental biology of primates; primate systematics; genetic and phenotypic differences among living primates; and paleoprimatology.

BEHAVIORAL FLEXIBILITY IN PRIMATES:
CAUSES AND CONSEQUENCES
By Clara B. Jones

NURSERY REARING OF NONHUMAN PRIMATES IN THE
21ST CENTURY
Edited by Gene P. Sackett, Gerald C. Ruppenthal and Kate Elias

NEW PERSPECTIVES IN THE STUDY OF MESOAMERICAN
PRIMATES: DISTRIBUTION, ECOLOGY, BEHAVIOR,
AND CONSERVATION
Edited by Paul Garber, Alejandro Estrada, Mary Pavelka
and LeAndra Luecke

HUMAN ORIGINS AND ENVIRONMENTAL BACKGROUNDS
Edited by Hidemi Ishida, Russell H. Tuttle, Martin Pickford,
Naomichi Ogihara and Masato Nakatsukasa

PRIMATE BIOGEOGRAPHY
Edited by Shawn M. Lehman and John Fleagle

REPRODUCTION AND FITNESS IN BABOONS: BEHAVIORAL,
ECOLOGICAL, AND LIFE HISTORY PERSPECTIVES
Edited By Larissa Swedell and Steven R. Leigh

RINGAILED LEMUR BIOLOGY: *LEMUR CATTA* IN MADAGASCAR
Edited by Alison Jolly, Robert W. Sussman, Naoki Koyama
and Hantanirina Rasamimanana

PRIMATE ORIGINS: ADAPTATIONS AND EVOLUTION
Edited by Matthew J. Ravosa and Marian Dagosto

LEMURS: ECOLOGY AND ADAPTATION
Edited by Lisa Gould and Michelle L. Sauther

PRIMATE ANTI-PREDATOR STRATEGIES
Edited by Sharon L. Gursky and K.A.I. Nekaris

CONSERVATION IN THE 21ST CENTURY:
GORILLAS AS A CASE STUDY
Edited by T.S. Stoinski, H.D. Steklis and P.T. Mehlman

Conservation in the 21st Century: Gorillas as a Case Study

Edited by

T.S. Stoinski
The Dian Fossey Gorilla Fund International
Atlanta, Georgia, USA
and
Zoo Atlanta
Atlanta, Georgia, USA

H.D. Steklis
The Dian Fossey Gorilla Fund International
Atlanta, Georgia, USA
and
Department of Anthropology
Rutgers University
New Brunswick, New Jersey, USA

P.T. Mehlman
The Dian Fossey Gorilla Fund International
Atlanta, Georgia, USA

 Springer

T.S. Stoinski
The Dian Fossey Gorilla Fund
  International
800 Cherokee Avenue, SE
Atlanta, Georgia 30315
USA
tstoinski@zooatlanta.org

H.D. Steklis
HC1 Box 576
23 Wildlife Lane
Elgin, AZ 85611

P.T. Mehlman
Conservation International
2011 Crystal Drive, Suite 500
Arlington, VA 22202

Library of Congress Control Number: 2007922443

ISBN 978-0-387-70720-4   e-ISBN 978-0-387-70721-1

Printed on acid-free paper.

9 8 7 6 5 4 3 2 1

springer.com

Introduction

It is a sign of the times that it need hardly be said that the African apes are in a state of crisis. Dire predictions abound about their chances of survival in the wild into the next century, or even to the end of this one. Not only are African apes threatened with extinction, but also the many other species that share their habitats, as today we are witnessing a loss of species and the degradation or disappearance of entire ecosystems at a rate unprecedented in human history. Some authorities calculate that, around the world, we are already losing more than 100 species per day. At this rate, an estimated 25% of the world's species present in the mid-1980s may be extinct by the year 2015 or soon thereafter.

The reasons for this mounting conservation crisis are manifold and complex, but principally concern some combination of growing human economic needs, unsustainable hunting, natural resource exploitation, and, in the case of the African apes, cross-transmission of disease. In Africa, there is immense pressure to unsustainably exploit the tropical forests; there is also a serious lack of conservation monies and conservation expertise. As such, the loss of species is expected to be particularly high in the tropical forests of Africa unless appropriate actions are taken soon.

Given this alarming state of affairs, many will be attracted to this volume because, like us, they are active conservationists searching for both a current assessment of the gorilla's conservation status and, importantly, for ideas and tools that show promise of halting or reversing population declines and putting us on a path to achieving a stable, long-term co-existence of human and ape populations. Many others, who simply have an interest in gorillas, great apes, or wildlife conservation in general, also may find this book appealing, because they will recognize both that the problems and challenges facing gorilla conservation are broadly familiar to those encountered with other wildlife populations and that the ideas and methods described herein may thus be more widely applicable.

In putting this volume together, and in treating gorillas as a case study, it is indeed our objective to reach a very broad audience, surely including but reaching well beyond those active in gorilla conservation. As the ideas for this collection came together, on the heels of a symposium celebrating the 100th anniversary of the 1902 scientific discovery of the mountain gorilla, we were squarely focused on novel approaches to gorilla conservation, both theoretical and methodological. We intended for such a sampling to convey the necessarily changing landscape of conservation practice, along the way providing realistic hope for resolving the grave problems faced by most wild gorilla populations. Accordingly, in preparing this book, we drew on the expertise of field scientists in a variety of disciplines to discuss current conservation threats, novel approaches to conservation, and potential solutions. Our hope is that the book, while focused on gorillas, can serve as a "conservation handbook" for a variety of species, as well as providing specific information on current conservation issues faced by gorillas in the wild.

The book's 17 chapters are grouped into four thematic sections. The first presents an in-depth assessment of the current status of wild gorilla populations, the second and third sections present several novel approaches to conservation that have been explored at several field sites, including new conceptual and technological tools, and also examine the pros and cons of some generally accepted "solutions" (e.g., ecotourism) to conservation issues. Chapters in the final section take a broader view by exploring the role international and national political entities, and nongovernmental organizations, including zoos, can and must play in gorilla conservation. We hope that, on reading these chapters, all readers will take away the message that the conservation community—particularly local communities in African ape habitat countries—are bringing to bear unprecedented energies, commitment, and novel solutions to protect and preserve Africa's remaining great ape populations. Although most of the chapters focus on gorillas, there are several that discuss conservation programs with other species and even on continents other than Africa (e.g., Chapters 5, 7, 9); these chapters were selected because of the relevance of the approaches discussed to issues relating to gorilla conservation and more broadly conservation as a whole.

Finally, we would like to extend our gratitude to those who made the book possible. First, we thank Zoo Atlanta for agreeing to host the symposium to celebrate the 100th anniversary of the discovery of the mountain gorilla to western science. Second, we thank all of the authors who contributed to this volume; we greatly appreciate their efforts and patience on our behalf. Third, we thank all the individuals who improved the individual chapters and book as a whole through their reviews. And last, but certainly not least, many thanks to Elizabeth Price and Angela Legg, who were instrumental in preparing the chapters for publication.

List of Contributors

Francis Akindes
Laboratoire d'Économie et de Sociologie Rurale
Université de Bouaké
Bouaké, Côte d'Ivoire

Keith Alger
Conservation International
Washington, D.C., USA

Natalie D. Bailey
Bushmeat Crisis Task Force
Washington, D.C., USA

Christophe Boesch
Wild Chimpanzee Foundation
Geneva, Switzerland;
Max Planck Institute for Evolutionary Anthropology
Leipzig, Germany

Hedwige Boesch
Wild Chimpanzee Foundation
Geneva, Switzerland

Jean Martial Bonis-Charancle
Innovative Resources Management
Washington, D.C., USA

Charlotte Boyd
Conservation International
Washington, D.C., USA

Michael Brown
Innovative Resources Management
Washington, D.C., USA

Ann P. Byers
Conservation Breeding Specialist Group (SSC/IUCN)
Minneapolis, Minnesota, USA

Amos Courage
John Aspinall Foundation
Port Lympne, Kent, England

Heather E. Eves
Bushmeat Crisis Task Force
Washington, D.C., USA

Kay H. Farmer
Scottish Primate Research Group
Department of Psychology
University of Stirling
Stirling, Scotland, UK

Nick Faust
Georgia Tech Research Institute
Georgia Institute of Technology
Atlanta, Georgia, USA

Claude Gnakouri
Ymako Teatri
Abidjan, Côte d'Ivoire

Gali Hagel
The Law Office of Gali L. Hagel, P.C.
Marietta, Georgia, USA

Ilka Herbinger
Wild Chimpanzee Foundation
Geneva, Switzerland;
Centre Suisse de Recherches Scientifiques
Abidjan, Côte d'Ivoire

Miroslav Honzák
Conservation International
Washington, D.C., USA

Michael Hutchins, Ph.D
The Wildlife Society
Bethesda, Maryland, USA;
Graduate Program in Conservation Biology
    and Sustainable Development
University of Maryland
College Park, Maryland, USA

Eugene Kayijamahe
Department of Geography
National University of Rwanda
Butare, Rwanda

Séverin Kouamé
Laboratoire D'Économie et de Sociologie Rurale
Université de Bouaké
Bouaké, Côte d'Ivoire

Robert C. Lacy
Conservation Breeding Specialist Group (SSC/IUCN)
Minneapolis, Minnesota, USA;
Chicago Zoological Society
Brookfield, Illinois, USA

Annette Lanjouw
International Gorilla Conservation Program
Nairobi, Kenya

Francis Lauginie
Wild Chimpanzee Foundation
Geneva, Switzerland

Cathi Lehn
Biodiversity Alliance
Cleveland Metroparks Zoo
Cleveland, Ohio, USA

Mark Leighton
Department of Anthropology
Harvard University
Cambridge, Massachusetts, USA

Carla A. Litchfield
University of South Australia
Adelaide, SA, Australia

Kristen E. Lukas
Cleveland Metroparks Zoo
Cleveland, Ohio, USA

Scott Madry
Department of Anthropology
University of North Carolina at Chapel Hill
Chapel Hill, North Carolina, USA;
Informatics International, Inc.
Chapel Hill, North Carolina, USA

Luis Marques
Ymako Teatri
Abidjan, Côte d'Ivoire

Andrew J. Marshall
Department of Anthropology and Graduate Group in Ecology,
University of California at Davis,
Davis, California, USA

Patrick T. Mehlman
Conservation International
Washington, D.C., USA

Philip S. Miller
Conservation Breeding Specialist Group (SSC/IUCN)
Minneapolis, Minnesota, USA

Zephyrin Mogba
Innovative Resources Management
Washington, D.C., USA

Karl Morrison
Conservation International
Washington, D.C., USA

Mountain Gorilla Veterinary Project
The Maryland Zoo in Baltimore
Baltimore, Maryland, USA

Toshisada Nishida
Japan Monkey Centre,
Inuyama
Aichi, Japan

Grégoire Nohon
Wild Chimpanzee Foundation
Geneva, Switzerland

Jim Sanderson
Center for Applied Biodiversity Science
Conservation International
Washington, D.C., USA

H. Dieter Steklis
Department of Anthropology
Rutgers University
New Brunswick, New Jersey, USA;
The Dian Fossey Gorilla Fund International
Atlanta, Georgia, USA

Netzin Gerald Steklis
The Dian Fossey Gorilla Fund International
Atlanta, Georgia, USA

Emma J. Stokes
Wildlife Conservation Society
Nouabalé-Ndoki Project
Brazzaville, Republic of Congo

Tara S. Stoinski
The Dian Fossey Gorilla Fund International
Atlanta, Georgia, USA;
Zoo Atlanta
Atlanta, Georgia, USA

Rachna Sundararajan
Innovative Resources Management
Washington, D.C., USA

Moustapha Traoré
Laboratoire d'Économie et de Sociologie Rurale
Université de Bouaké
Bouaké, Côte d'Ivoire

Paul Waldau
Center for Animals and Public Policy
Cummings School of Veterinary Medicine at Tufts University
North Grafton, Massachusetts, USA

Rees Warne
Innovative Resources Management
Washington, D.C., USA

Frances R. Westley
Gaylord Nelson Institute for Environmental Studies
University of Wisconsin
Madison, Wisconsin, USA

Wildlife Conservation Society
Field Veterinary Program
Bronx, New York, USA

Richard W. Wrangham
Department of Anthropology
Peabody Museum
Harvard University
Cambridge, Massachusetts, USA

Contents

Section 1
Current Status of Gorillas

Chapter 1
Current Status of Wild Gorilla Populations and Strategies for Their Conservation

Patrick T. Mehlman

1. Introduction

Gorilla populations and their habitats throughout Africa are severely threatened and in decline. To understand and respond to this crisis, conservationists need up-to-date information on gorilla distributions, their relative population sizes, and the rates at which these populations are changing through time. Unfortunately, there are enormous gaps in our current knowledge about distribution and abundance measures for gorillas. This chapter describes and examines these gaps, discusses why they exist, and considers how they might be addressed. This chapter also reviews and compares the specific regional and local threats to wild gorillas, which, if left to continue at current trends, will substantially reduce or extirpate most wild populations within decades. Hopefully, this review will aid a new generation of field workers to plan and execute census work in the 21st century. Such census work will be critical for understanding where and how these animals are disappearing by human hands, and will be the principal method for developing strategic and systematic conservation approaches in the 21st century. Without new and rapid monitoring approaches linked to strategic conservation interventions, the vast majority of gorillas and the ecosystems they inhabit will not survive into the 22nd century.

2. Gorilla Nomenclature and Systematics

Extant western and eastern gorilla populations are divided by a distance of approximately 900 km (Figure 1.1). Proposed taxonomic approaches to this geographic, phenotypic, and genetic divide vary between considering these populations as three subspecies of one species of gorilla (e.g., Groves, 1966, 1967, 1970), or, more recently, as several subspecies of two separate species, as suggested by Butynski (2001) and Grubb et al. (2003) relying on several sources (Ruvulo et al., 1994; Sarmiento and Butynski, 1996; Ryder et al., 1999; IUCN/SSC PSG 2000; Sarmiento and Oates, 2000; Groves, 2001).

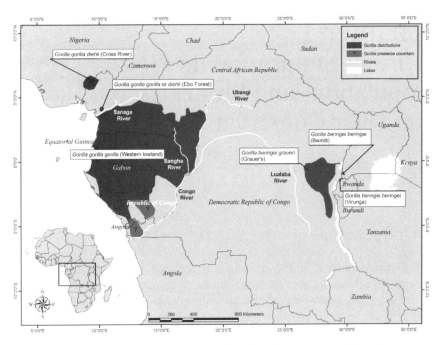

FIGURE 1.1. Approximate gorilla distributions in Africa, with current provisional taxonomy.

Figure 1.1 displays a provisional taxonomy representing two species and four subspecies, following recent findings on mtDNA phylogenies (Ruvulo *et al.*, 1994; Garner and Ryder, 1996; Seaman *et al.*, 1998, 2001). Note, however, that the validity of all mtDNA studies of gorillas has recently been challenged (Thalmann *et al.*, 2004).

Western gorilla populations may not be as cladistically uniform as once thought or as might be imagined from their distributional range. First, the Cross River clade can be distinguished from the rest of the western lowland gorilla populations by its mtDNA (Oates *et al.*, 2003; although see Ryder, 2003; Clifford *et al.*, 2004) and in some aspects of its cranial morphology (Albrecht *et al.*, 2003; Oates *et al.*, 2003; Stumpf *et al.*, 2003). Second, western lowland gorillas themselves may be split into at least two mtDNA clades, east and west of the Sangha river (Clifford *et al.*, 2003; Figure 1.1), perhaps corresponding with "demic" craniomorphological variation sometimes detected in western lowland populations (e.g., Groves, 1970; Leigh *et al.*, 2003). It is unknown where a small population recently observed in the Ebo area of Cameroon (Morgan *et al.*, 2003; Figure 1.1) fits with respect to the Cross River and western lowland gorilla groups.

It is now thought that Grauer's gorilla populations (sometimes referred to as "eastern lowland gorillas," see below) represent a single clade that may have only recently expanded after the last glacial maximum (~18,000 b.p.) from a bottlenecked population in a refuge area. They have low mtDNA heterogeneity (Jensen-Seaman *et al.*, 2001), as well as low craniometric diversity when compared with western lowland populations (Albrecht *et al.*, 2003; Leigh *et al.*, 2003). Although some studies have detected variation in morphological traits (Groves, 1970; Casimir, 1975; Groves and Stott, 1979), this most likely represents a cline associated with altitude and concomitant dietary habits (Groves and Stott, 1979) and will not likely warrant any further subspecific "splitting," as suggested by some authors (Sarmiento and Butynski, 1996; Groves, 2003). Note, however, that the morphological and genetic studies to date are based on very small sample sizes that are not representative of the full distribution and possible variation for Grauer's gorillas.

The Bwindi and Virunga populations appear to exhibit some morphological differences (Sarmiento *et al.*, 1996; but note that this study is based on small sample size). These differences, like those of Grauer's gorillas, most probably reflect a cline in diet and altitude, since Bwindi and Virunga populations appear to be the same clade based on mitochondrial DNA studies (Garner and Ryder, 1996; Jensen-Seaman *et al.*, 2001; Clifford *et al.*, 2004; but see Leigh *et al.*, 2003, and Thalmann *et al.*, 2004, concerning difficulties in using mtDNA and DNA-Y-chromosome studies).

Given that gorilla taxonomy is clearly a "work in progress" (Grubb *et al.*, 2003), this chapter will identify gorilla populations by their biogeographical distributions and their currently suggested mtDNA clades: eastern populations are referred to as *Grauer's, Virunga,* and *Bwindi* populations, or all together as *Eastern gorillas*; western populations are referred to as *western lowland* populations, the *Cross River* population, and the *Ebo* population, or altogether as *Western gorillas*. The designation "eastern lowland" gorilla is not used here since it is misleading; the altitudinal range for Grauer's gorillas varies from 550 m to more than 3000 m (Mts. Kahuzi and Tshiaberimu), nearly as high as for Virunga gorillas (4000 m, Harcourt and Fossey, 1981). For a review of the rich and colorful history of gorilla taxonomy, the reader is referred to Groves (2003).

3. Geographical Ranges for All Gorilla Populations

The approximate geographic range of all Western and Eastern gorillas, or their extent of occurrence (IUCN, 2000), is displayed in Figure 1.1. In west central Africa, western lowland gorillas are found from the Ubangi River, Central African Republic, in the east, north to the Sanaga River, Cameroon, and west to its Atlantic coasts, Gabon, Equatorial Guinea, Republic of Congo towards the southeast in an irregular distribution following tropical forest blocks, and, finally, east to the Congo River. The Cross River population isolates are found near the Cross River on the border of Cameroon and Nigeria. A gorilla population has recently been observed in the Ebo Forest area of Cameroon (Morgan *et al.*, 2003; Figure 1.1), and recent field reports suggest a surviving gorilla population in Angola.

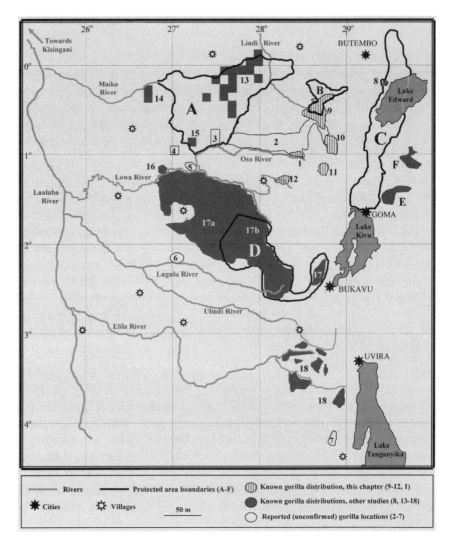

FIGURE 1.2. Known distributions for Eastern gorillas, 1996. Protected areas: A—Maiko National Park, DRC; B—Tayna Gorilla Reserve, DRC; C—Virunga National Park, DRC; D—Kahuzi-Biega National Park, DRC; E—Virunga Conservation Area of DRC, Rwanda, and Uganda (see text), where Virunga mountain gorillas are located; F—Bwindi Impenetrable National Park, Uganda, where Bwindi gorillas are located. Areas 2–7 are areas where gorillas have been reported, but not verified: 2—UGADEC staff; 3–4 —Hart and Sikubwabo (1994); 5–6 —Hall et al. (1998a); 7—Omari et al., (1999). Areas 1, 9–12 are areas where gorilla populations have been reported in 2005 by UGADEC and Tayna Gorilla Reserve staff. Areas 13–15 are areas reported to contain gorillas by Hart and Sikubwabo (1994) from surveys in the early 1990s. Areas 17a and b represents gorilla distribution as reported by Hall et al. (1998a) for surveys in 1994–1995; they also reported a small population north of the Lowa River (area 16). Area 17 represents the distribution of Grauer's gorillas for the highland sector of Kahuzi-Biega (Hall et al., 1998b). Area 18 represents gorilla distribution for the Itombwe mountains as reported by Omari and colleagues (1999).

TABLE 1.1. Distributional and occupancy ranges for gorillas.

	Geographical range km^2	Occupancy area km^2	References for range (R) and occupancy (O)
Western Gorillas			
Western lowland gorillas	709,000	?	R: Butynski, 2001; also see Harcourt, 1996
Cross River gorillas	2,000	~250	R and O: Oates, personal communication
Ebo Forest gorillas	n/a	n/a	Sightings reported in Morgan *et al.*, 2003
Eastern Gorillas			
Grauer's gorillas	52,000	21,595	R and O: this chapter, Table 1.5
Bwindi gorillas	330	~300	R and O: Plumptre *et al.*, 2003
Virunga mountain gorillas	447	~380	Virunga Census 2003 (Census Report., 2005: ICCN, ORTPN, UWA, IGCP, DFGFI)

In east Africa, gorillas are found in three blocks of dissimilar sizes. Grauer's populations are located in the eastern Democratic Republic of Congo (DRC), and their irregular and patchy distribution stretches from the escarpment of the Albertine Rift in the east, to the Lindi River in the north, to about 26° 45″ E in the west, and then to the Itombwe mountains in the south (Figure 1.2). The Bwindi population is found in Uganda, in and near the Bwindi Impenetrable National Park (with a few excursions into neighboring DRC), and the Virunga population is found in the Virunga Volcanoes region[1] (Figure 1.2). The approximate sizes of these geographical ranges and areas of occupancy (IUCN, 2000) are displayed in Table 1.1. Distribution ranges and areas of occupancy are discussed further in sections below.

4. Methods for Determining Distribution and Abundance and Sources of Error

Current knowledge of gorilla distribution and abundance in the wild derives from a few common field methodologies. These are reviewed below along with their potential sources of error.

---

[1] The Virunga Volcanoes region encompasses the Volcanoes National Park of Rwanda, the Mgahinga National Park of Uganda, and the Mikeno sector of the Virunga National Park, South Sector, Democratic Republic of Congo.

4.1. Distribution

Distribution is determined by surveying areas of reported or known gorilla presence, conducting reconnaissance walks along abandoned roads, existing human trails or forest trajectories that follow a path of least resistance (ridge tops, stream beds, etc.), and recording all gorilla sign (trail, nests, sightings) on maps (e.g., Schaller, 1963). Although the meandering routes taken during these reconnaissance walks do not provide representative habitat sampling (and therefore are of limited use in generating abundance/density estimates), they remain the most simple, rapid, and effective way of assessing distribution and survival threats over large and often remote areas. Although this technique is sometimes referred to as "footpath or forest reconnaissance," we suggest this term is probably best avoided to not confuse it with the "reconnaissance transect method" (see below), and suggest instead the term *prospecting* or *prospection*. Prospecting is now made more precise by obtaining GPS points for gorilla sign, which can be transformed into GIS maps overlaid on satellite images. One important potential source of error is the level of confidence with respect to observations of gorilla sign; it is, therefore, obligatory to differentiate between observations made by trained observers and any other sources of data, such as local informants or hunters (not devaluing their often important contributions).

An important technique for recording gorilla (and other faunal) distribution is the use of a *quadrant grid methodology*. This approach divides a target area into sample squares for which presence or absence can be noted (e.g., for gorillas, Tutin and Fernandez, 1984; Hart and Hall, 1996). A useful grid size for large mammals is one-tenth degree latitude by one-tenth degree longitude (i.e., near the equator, approximately 11.2 km by 11.2 km, producing a quadrant that is 125 km² in size). This has the advantage of aiding field workers, since it corresponds with latitude-longitude locations (100 quadrants per one degree latitude and longitude) and the use of GPS units and GIS-produced maps in the field; it can be further broken down into four equal-size subquadrants as necessary, and can easily be designed and mapped using GIS methods.[2] Quadrants (or subquadrants) can then be surveyed by any of the methods mentioned here and identified by a simple absence/presence score to build a clearer picture of occupancy range.

---

[2] Before the advent of GIS techniques, quadrants were often created as 10×10 km grids (Tutin and Fernandez, 1984) producing an obvious advantage of using 100 km² quadrants for calculations. However, today, these 100 km² quadrants are less convenient to use in more modern GIS latitude and longitude projections and can result in miscalculations and distortions. For example, what appears to be an accidental assumption that one-tenth degree latitude by longitude was 100 km² in size (Hart and Sikubwabo, 1994) produced significant map distortions and has rendered their previous quadrant analysis of gorilla and other faunal distributions quite difficult to use by current fieldworkers using GIS methods.

The distribution of small isolated populations, such as the Virunga, Bwindi, and Cross River populations, can also be more precisely determined by *long-term observations of the ranging behavior of gorilla groups* (e.g., isolated populations of Barbary macaques, Mehlman, 1989). Further, it is important to evaluate whether small, isolated gorilla groups (or even local populations) separated by 20–50+ km might be connected by gene flow through single males traveling these distances (as in Barbary macaques, Mehlman, 1986). Such data for gorillas have never been published with these objectives in mind, but will be a fruitful avenue for future analyses of gorilla metapopulation dynamics related to the distribution of suitable habitat types. (Analyses like these are currently being conducted on Cross River gorillas, Sunderland-Groves et al., 2005; Oates, personal communication.)

4.2. Abundance

There are two types of survey methods that have been employed to estimate gorilla abundance for a specific target area. The first type is a *complete census*, which in theory would entail a count of all animals in the population, but in practice is modified for small, isolated populations of gorillas, such as for Bwindi and Virunga, to incorporate nest counts for some members of the population over a short period of time. For this method, teams of rangers sweep through sectors of the target area and obtain nest counts for all groups and solitary males. These counts are then supplemented and made more precise by up-to-date observations of the numbers of known individuals in habituated groups (Weber and Vedder, 1983; Sholley, 1990, 1991; Yamawiga et al., 1993). This method provides an overall abundance number, which, when added to known ranging behavior, can provide estimates of density.

A second group of survey methods utilizes *transect-based density estimations*. These provide an estimate of gorilla abundance by subsampling smaller, prestratified zones (usually on the basis of broad habitat differences or levels of presumed human disturbance, e.g., Walsh et al., 2003) to obtain density estimates thought to be representative of a larger occupancy area. The subsample density estimates are sometimes obtained by labor-intensive *line-transect* census techniques (e.g., Yapp, 1956; Eberhardt, 1978; Tutin and Fernandez, 1984; Fay et al., 1989; Plumptre, 2000). In this technique, observers cut and walk a series of randomly or systematically placed, highly regulated linear trajectories, and count gorilla nests encountered as an indirect representation of individual animals. Using a variety of techniques for extrapolating individual gorilla numbers from the nesting evidence along transects, the numbers of estimated gorillas per surface areas for transects representing each subsample zone are calculated as local densities. These are then applied to the various strata for the entire occupancy area to derive an estimate of abundance.

In surveys in Kahuzi–Biega National Park, Hall et al. (1998a) found a significant positive correlation between gorilla nest encounter rates derived from reconnaissance walks and encounter rates from associated line transect surveys. Using linear regression, they transformed encounter rates (nest sites/km) from their reconnaissance walks to yield density estimates. This is known as a "*reconnaissance-transect*" methodology or "*recce*" (see Barnes and Jensen, 1987; White and Edwards, 2000) and is formalized by walking a path of least resistance across a large survey area and conducting short line transect/recce pairs (200–500 m) cut perpendicular to the principal path of travel at regularly placed intervals. Comparisons of encounter rates from transect/recce pairs are used to calibrate a density estimate using nest encounter rates on the principal reconnaissance path. These density estimates are then extrapolated to give density estimates for entire regions in much the same way as line transect estimates.[3]

4.3. Abundance Estimates and Associated Errors

Complete censuses can be highly accurate (especially when they are conducted on populations for which many groups are already habituated and observed daily). They have only a limited number of biases or sources of error, such as the number of unweaned individuals, or the potential for some nest sites to be missed during the field "sweeps." In contrast, density estimations based on line and reconnaissance transecting and the abundance measures they provide are plagued by multiple and cumulative sources of bias (Table 1.2), many related to extrapolating numbers of gorillas by using the indirect measure of nests observed (e.g., Plumptre and Reynolds, 1997; Sarmiento, 2003). Thus, abundance numbers derived from density estimation methods can easily be biased by tens of percent (Walsh *et al.*, unpublished manuscript). Chimpanzee nests can often be confused with gorilla nests where these two great apes are sympatric; in one study, it was assumed that gorilla nest counts were 18% higher than the observed counts because of misidentification of gorilla nests as chimpanzee nests (Hall *et al.*, 1998a). These kinds of bias, when added together and introduced into a calculation of abundance, can easily produce error of more than 40%, even before sampling error confidence intervals are applied to transect methods (and confidence intervals themselves have been called into question, Sarmiento, 2003).

---

[3] Sometimes abundance estimates for large areas are also generated by a "forest cover model" applying a density estimate to the amount of surface area still covered by primary forest as a proxy for occupancy range (Harcourt, 1966). For example, if we use Harcourt's model (overall gorilla density estimated to be 0.25 individuals/km²), the Minkebe forest block in northern Gabon (approximately 30,000 km²), would contain about 7,500 western lowland gorillas. However, we now know that the nest encounter rate at Minkebe has declined by about 98% (Walsh *et al.*, 2003) since the Tutin and Fernandez (1984) survey (it was this survey that Harcourt used to calibrate density in his forest cover model). This kind of approach can thus generate enormous error (by a factor of about 20 times too high) and should be rejected as a reliable method for estimating abundance for large regions.

TABLE 1.2. Potential sources of error in line and reconnaissance transect sampling.

Subsamples are not representational for occupancy area	Reconnaissance and line transects may not represent the occupancy area, leading to over- or underestimations. Transects rarely sample more than 1% of a target area.
There are interobserver differences in detecting nests along transects	Observer skills are quite variable, depending on experience, fatigue, weather, season, and even time of day, resulting in nest sites being missed; when overall sample numbers are quite small, this can lead to underestimations.
Nest decay and degradation rates vary by site and season	The increase between a decay rate of 78 days and 106 days can lower a density estimate by 27%; at 120 days, the estimate is 35% lower (Figure 1.5). These decay rates vary by site, by season, and by nest materials used in construction.
Misidentification of gorilla and chimpanzee nests	Some studies suggest that between 18% (Hall et al., 1998a) and 33% of gorilla nests might be miscounted as chimpanzee nests (Tutin et al., 1995); adjusting findings upward to compensate for this may lead to large overestimations if misidentifications are not as frequent as believed.
Not all gorillas build nests every night	At some sights, as many as 45% of gorillas in a group do not build nests each night (Mehlman and Doran, 2002); this can lead to significant underestimations.
Gorillas can build more than one nest per night	On occasion, gorilla groups build two sets of nests during the night and move their site (personal observations); with small sample sizes along transects, this can lead to significant overestimations.
Unweaned gorillas do not make nests; juveniles sometimes do not make nests	Density estimates need to be adjusted upwards to account for those immatures that do not make nests; in many areas, demographic information and immature nest building behavior data are lacking to accurately determine these adjustments, leading to over- or underestimations that can easily exceed 10%.

Traditional line transect surveys are extremely labor intensive and time consuming and are impractical in areas with complex mountainous topography. Although the reconnaissance transect technique has many advantages (greater coverage per unit time, smaller teams, etc.), including the use of short line transect segments, it is important to recognize that density estimates generated from these are plagued by the *same* multiple biases associated with line transect surveys. Current density and abundance estimates based on line and reconnaissance transect surveys are, therefore, at best, approximations, and, at worst, inaccurate by orders of magnitude and not at all comparable between sites. However, for western lowland and Grauer's gorillas that are distributed (often patchily) in vast occupancy ranges, these are the only techniques available.

5. Grauer's Gorillas: A Case Study on Distribution and Abundance

To highlight current gaps in knowledge concerning gorilla distribution and abundance throughout Africa, we can examine as a "case study" the present knowledge concerning Grauer's gorillas. This has several advantages. First, it will reveal gaps and weaknesses in how we are currently conducting scientific studies as field conservationists; these lessons learned are equally applicable in defining the limits of our current knowledge for Western gorillas. Two, it will allow us to introduce into the literature recent information collected on abundance and distribution for Grauer's gorillas. Three, by examining current gaps and weaknesses in this case study, we can improve how we plan and execute future studies investigating the distribution and abundance of *all* gorillas—a critical exercise if we have any hopes of succeeding in our conservation efforts.

5.1. Status of Grauer's Gorillas: 1959

Schaller and Emlen conducted the first comprehensive survey of Eastern gorillas in 1959 using prospecting techniques, combined with interviews of local inhabitants (Emlen and Schaller, 1960; Schaller, 1963). Combining the Graueri, Bwindi, and Virunga populations, they suggested a total population size of between 3,000 and 15,000 individuals distributed in a series of isolated and small populations, with a central area of continuous distribution (see Figure 1.6).

5.2. Status of Grauer's Gorillas: 1996

From 1989 through 1995, there were three large surveys that provided abundance and density estimates via transect methods for Grauer's gorillas in the Itombwe plateau (Omari *et al.*, 1999), in the lowland sector of Kahuzi-Biega, in a region northwest of that called "Kasese" (Hall *et al.*, 1998a), and in Maiko National Park (Hart and Sikubwabo, 1994; Hart and Hall 1996).[4] Figure 1.2 displays these 1996 population distributions overlaid on a GIS-based map of more recently discovered (2002–2005) Graueri populations by the staff of the Tayna Gorilla Reserve and other community conservation projects supported by the Dian Fossey Gorilla Fund International (Kakule and Mehlman, 2004; Sivalingana-Matsitsi *et al.*, 2004).

---

[4] For a comprehensive review of all other minor surveys in East Africa until 2001, see Sarmiento (2003, Table 18.2).

5.3. *Challenges with Determining Distribution*

In an overview article by Hall and colleagues (1999b), a number of local populations east of Maiko National Park did not appear on their distribution map, and are briefly discussed in their text as being small and unknown (Figure 1.2, areas 9–12, 1). Recent work indicates that this area contains at least one local population now known to be distributed over more than 600 km², and, more likely, a metapopulation that inhabits an area of about 1000 km² (Sivalinga-Matsitsi *et al.*, 2004; Figure 1.4, areas 9 and10, probably connected by movements of solitary males or small groups). These appear to have been omitted by Hall and colleagues (1999b) due to inadequate knowledge of the area.

In the western Maiko area (Figure 1.2, areas 3–5, 15 and 16) Hall and colleagues (Hart and Sikubwabo, 1994; Hall, 1999b; Hart *et al.*, 1999b) indicated only one known 100 km² quadrant in Maiko National Park to be occupied by gorillas, and further suggested, "the westernmost population [in Maiko] is believed extinct" (Hart *et al.*, 1999b: 124). They are probably referring to the Peneluta site (Figure 1.2, area 15), although it does not appear on their distribution map (Hall *et al.*, 1999b: Figure 1.1).

The 2005 distribution status for this western area of Maiko, however, is quite different from that portrayed in Hall and colleagues (1999b, referring to Hart and Sikubwabo, 1994). A just-completed survey of the area by ICCN[5] Maiko, UGADEC, and DFGFI staff (Nixon *et al.*, 2005; see Figure 1.3) indicates that the gorilla distribution in this area was significantly underestimated in the early 1990s (compare Figures 1.2 and 1.3). Unless Grauer's gorillas have substantially increased their range between the 1990s and 2005, an unlikely proposition that is not supported by knowledgeable local informants and ICCN staff (Nixon, personal communication), the distribution surveys conducted in the early 1990s (Hart and Sikubwabo, 1994; Hart and Hall, 1966) were probably inadequate with respect to detecting gorilla presence.

In the Kahuza-Biega National Park and adjacent area northwest extending to the Lowa River (Figure 1.2, areas 17a and b), Hall and colleagues (1999a, 1999b) indicated (as did Schaller, 1963) that this vast area of more than 12,000 km² represented a continuous gorilla distribution, albeit at varying local densities. The 1990s surveys (Hall *et al.*, 1999a) did not cover about a third of the quadrants for this area (Hart and Hall, 1996), however, and, given that local population fragmentation might have already occurred by the time of their surveys, it is not so clear that this population was continuously distributed. Future studies, therefore, should seek to conduct more complete survey coverage for this area and focus on metapopulation distribution and dynamics related to anthropogenic factors and habitat types.

---

[5] ICCN is the *Institut Congolais pour la Conservation de la Nature*, the DRC Wildlife Authority.

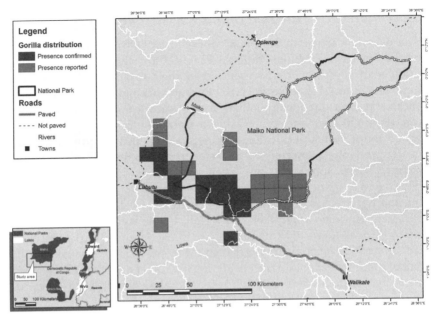

FIGURE 1.3. Known distribution for Grauer's gorillas in the southern Maiko area between the Lowa and Maiko Rivers, 2005. This map indicates the location of gorillas using a 125 km² quadrant system based on latitude and longitude (see text), based on a survey from February to May 2005, conducted by the ICCN (DRC Wildlife Authority), research staff of the Dian Fossey Gorilla Fund International, and staff from UGADEC (Table 1.5, footnote a). Darker quadrants indicate the presence of gorilla (feeding sign, trail, nests) detected by trained members of the survey team. Lighter quadrants indicate gorilla presence as reported by other ICCN staff stationed in this area and knowledgeable locals (hunters, community conservation workers). (Further information in this survey can be found in Nixon et al., 2005.)

5.4. Abundance Estimates for Grauer's Gorillas: 1996

Hall et al. (1998b), incorporating their census work (Hall et al., 1998a) and referring to other recent surveys at that time, provided a total estimate of Grauer's gorillas at 16,902. This well-quoted number (in reality, a central tendency) of 17,000 Grauer's gorillas as of 1996, may, however, have been a significant overestimate at that time. First, it reported an estimate of 1,155 gorillas for the Itombwe area, 298 more than the Itombwe surveyors published in the next year (Omari et al., 1999). Second, it reported an estimate of 859 gorillas for the Maiko area, when the published census results from Maiko at that time only appear to estimate at best between 170 and 571 gorillas.[6] Third, the estimate included 14,659 gorillas from their survey of Kahuzi-Biega and Kasese (Figure 1.2, areas 17a and b), which may have been an overestimate for several reasons: 1) it was derived by adding an additional 18% to baseline estimates to account for gorilla nests that were misidentified as chimpanzee nests; this adjustment was applied to all weaned nests, which included an approximate 2,500 solitary male nests (which because of their size are unlikely to be misidentified); 2) the estimate was derived by adding 0.33 infants per adult female and another 0.16 per all weaned individuals to represent juveniles who were assumed to not make nests but rather sleep with their mothers (this assumption results in an increase of 1,300 individuals, since the 16% increase appears to be compounded on the 18% increase for misidentification of chimpanzee nests); and 3) the estimates for unweaned infants and the ratio of juveniles to weaned individuals were derived from mountain gorilla data (Vedder *et al.*, 1986; Hall *et al.*, 1998b quoting Watts, 1991), which may not be applicable to Grauer's gorillas. If we correct for these possible sources of error and rework the original estimates of Hall *et al.* (1998b) with different assumptions, we see that a more conservative estimate would have been 21% lower (Table 1.3). Combining these figures with the other estimates available at the time reveals that, rather than 17,000 Grauer's gorillas, the central tendency may have been closer to 13,000 individuals; the former is 29% larger than the latter (Table 1.4).

5.5. Distribution and Abundance for Grauer's Gorillas in the Tayna Sector: 2002–2004

The Tayna Gorilla Reserve and nearby areas (Figure 1.4) consist of a mix of closed-canopy primary mixed forest with about 15% abandoned and active agricultural clearings, with altitudes from 900 to 2000 m. Between 2001 and 2004, field staff from the Tayna Gorilla Reserve and from the project for the Primate Reserve for Bakambale (ReCoPriBa) collected GPS points during prospecting patrols for the location of all gorilla sign; they discovered a contiguous gorilla distribution that encompasses a surface area of approximately 675 km² (Figure 1.4, area 9). From late 2001 through early 2003, field staff from

---

[6] In Maiko, 51 quadrants (each at 100 km²) were surveyed, representing about half the Park's surface area. From these total quadrants, 17 were found to contain gorilla sign, and another seven were reported by locals to contain gorilla, but the survey teams were unable to find sign (Hart and Sikubwabo 1994). In the original report, the authors determine density to be 0.10 individuals/km² (Hart and Sikubwabo, 1994); later, with the same data, density is reported to be 0.25 individuals/km² (Hart and Hall, 1996). Applying these two densities as a range for the 17 quadrants, this would provide an abundance measure of 170–425 gorillas, or if the additional seven questionable quadrants are included, a range of 240–600 gorillas, neither of which approach the figure of 859 reported in Hall *et al.* (1998) or the range of 350–1,000 gorillas for Maiko reported in Hart and Hall (1996). Using the average of the two densities reported (0.17 individuals/km²), and applying it the 17 quadrants, this produces 289 weaned individuals, and adjusting this upward for unweaned and misidentified nests (using same calculations as in Hall *et al.,* 1998b), the abundance number for Maiko would have been 395 gorillas. It is precisely these kind of estimations, already based on wide amounts of error inherent in transect methods (Table 1.2) that call into question the whole exercise of providing abundance estimates.

TABLE 1.3. Examining the calculations used for density and abundance estimates: case study on Grauer's gorillas.[a]

Stratum	Sampling zones	Stratum area (km²)	Weaned density (km²)	Nest site rate per km of transect	Total no. transect (km)	No. nest sites	Area of transect as per DISTANCE (km²)	Est. no. solitary nests encountered	Est. no. group nest sites encountered	Density of solitary males (km²)	Density of inds. found in group nests (km²)	Estimate of solitary males	Estimate of inds. found in groups	Estimate of unweaned at 15% group size	Total gorillas with unweaned
1	KB 1	1,110	1.73	0.45	76	34.2	19.8	7.2	27.0	0.36	1.37	404	1,516	227	2,148
2	KB 2	260	3.21	0.85	68	57.8	18.0	12.2	45.6	0.68	2.53	176	659	99	933
3	KB 3	1,555	1.43	0.39	75	29.3	20.5	6.2	23.1	0.30	1.13	468	1,756	263	2,487
4	KB 4	1,719	0.75	0.22	100	22.0	29.3	4.6	17.4	0.16	0.59	271	1,018	153	1442
6	K 1	2,367	0.31	0.19	53	10.1	25.2	3.9	6.2	0.16	0.24	368	579	87	1,034
6	K 2	2,367	0.50	0.31	54	16.7	41.9	6.5	10.2	0.16	0.24	368	579	87	1,034
6	K 3	2,367	0.39	0.24	54	13.0	32.4	5.0	7.9	0.16	0.24	368	579	87	1,034
5	KB	1,026										80	125	19	224
Subtotals (or means) for Kahuzi-Biega		5,670	1.26	0.45	319	143.3		30.2	113.1	0.34	1.29	1,400	5,073	761	7,234
Subtotals (or means) for Kasese		7,101	0.40	0.25	161	39.8		15.5	24.3	0.16	0.24	1,105	1,736	260	3,101
Totals		12,771			480	183		46	137			2,504	6,809	1021	10,335
18% increase for misidentified nests												2,504	8,035	1021	11,560

[a] Hall and colleagues (1998a) published a total abundance estimate of 14,659 individuals (11,020 weaned individuals) for the lowland sector of Kahuzi-Biega and the Kasese area (Figure 1.2, areas 17a and b). The assumptions they used for these calculations are discussed in the text. In this simulation, their same data are used but with the following assumptions: 1) 18% chimpanzee nest misidentification is only applied to group nest sites; 2) an increase of 15% for unweaned gorillas is added to only group nest sites. This simulation thus produces a total abundance estimate including unweaned gorillas of 11,560, 21% lower than their total estimate of 14,659. Note that without the chimpanzee nest misidentification adjustment, the total would be 10,335.

16

TABLE 1.4. Two different estimates for Grauer's gorillas censuses completed in 1996.

	Hall et al. (1998b)	Other estimates	References
Kahuzi Biega Highland	262	262	Hall et al. (1998b)
Kahuzi-Beiga Lowland and Kasese	14,659	11,560	Table 1.3, and text
Maiko	859	395[a]	Note #6 and this chapter
Itombwe	1,155	857	Omari et al. 1999
Lowa	13	13	Hall et al. (1998b)
Tshiaberimu	16	16	Hall et al. (1998b)
Masisi	28	28	Hall et al. (1998b)
Totals	16,992	13,131	

[a] This would have been the estimate for Maiko with known information in 1996. By 2005, new information (this chapter, Table 1.6) suggests that this area contains more than 558—3,177 individuals.

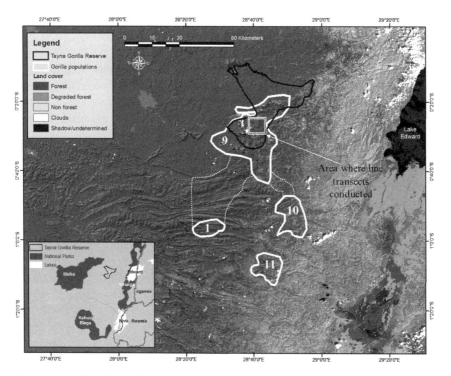

FIGURE 1.4. The Tayna Gorilla Reserve and distribution of Grauer's gorillas. White polygons indicate gorilla populations (GPS located) identified by staff of the Tayna Gorilla Reserve and UGADEC (Table 1.5, footnote a). From local reports and remote imaging, it is likely that areas 9, 10, and 1 are connected by other gorilla groups and/or the ranging of solitary males. Area 11 (Masisi) is totally isolated from other gorilla populations to the north (author's personal field observations).

the Tayna Gorilla Reserve also conducted a series of line transects (68.9 km) in which gorilla and chimpanzee nest sites were counted to estimate great ape densities (Figure 1.4). They utilized a variable-width line transect methodology and the formula of Ghigleri (1984), as modified by Tutin and Fernandez (1984), with an estimated decay rate of 108 days (Hall *et al.*, 1998a). For gorillas, they found no difference between mean nest site size for fresh sites (8.0 ± 4.3) and older sites (9.5 ± 3.5) and thus multiplied the overall mean nest site size (9 ± 4.0) by the number of nest sites ($n = 18$) on the transects (effective total strip width was 24 m, DISTANCE, Laake *et al.*, 1994) to derive an estimated density of weaned gorillas of 0.92 individuals/km^2. This density was then multiplied by the effective occupancy zone for gorillas in the Tayna Reserve and environs, a total area of 675 km^2 (Figure 1.4, area 9), producing an estimate of 624 *weaned* gorillas for the entire contiguous sector in this area encompassing the Tayna Reserve.

5.6. Challenges with Determining Abundance

The estimate for Tayna and nearby areas, like all estimates from transect methods, has a wide range of error. First, the estimation of abundance is an exercise in only two dimensions, when in fact these gorillas are distributed over a mountainous area with slopes that often approach 45°, distorting calculations of true surface area. Second, the decay rate of nests is highly variable; for Western gorillas, decay rate can be anywhere between 78 and 120 days (herbaceous *versus* arboreal nests: Tutin and Fernandez, 1984). Third, the detection width of transects is an estimate that can only be calculated for the perpendicular distance of nest sites to transects on which nests were detected (DISTANCE, Laake *et al.*, 1994); it does not represent accurately the detection width for transects where no nests were detected. Given the possible variation both in decay rate and transect width, we can model the Tayna results using a series of possible transect widths from 16 to 36 m and a series of possible decay rates that varied between 78 days and 120 days (Figure 1.5). This reveals that the estimate of abundance of weaned gorillas for the Tayna Sector (Figure 1.4, area 9) could vary by more than 300%, from 367 to 1,269 individuals. These sources of bias are present in all transect methodologies even before we consider adjustments based on misidentifying chimpanzee nests or the inability to detect immatures not making nests.

5.7. Summary of Case Study

As this case study of Grauer's gorillas demonstrates, there are many sources of error associated with estimates of gorilla distribution and abundance. With respect to gorilla distribution, we have seen that studies have a tendency to focus on areas where their surveys occurred, but in combining their findings with other studies in review-oriented articles, there is a tendency to over-extrapolate distribution in their study zone, while under-representing distributions outside their study zone, probably due to incomplete knowledge of other zones or a reliance on questionable results from other surveys.

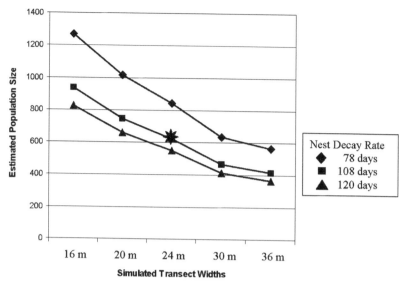

FIGURE 1.5. Range in abundance estimates related to various possible transect widths and nest decay rates. These graphs show a series of possible abundance estimates for area 9 (Figure 1.4) for the Tayna Gorilla Reserve, which had an overall mean nest site size of 9 ± 4.0 and 18 nest sites for 68.9 km of linear transects with total estimated strip width of 24 m. Using a nest decay rate of 108 days produces an abundance estimate of 624 weaned gorillas for this area (asterisk). Note that if nest decay rate were quicker (78 days), this abundance estimate would rise to 844 individuals; if the detection strip width was 20 m rather than 24 m, the estimate rises to 1,017. This figure thus illustrates that small errors in estimating decay rate or the calculation of strip width can produce large variations in the abundance estimate.

We thus conclude that caution should be exercised in interpreting any "reviews" of gorilla distribution, and we urge field workers to always consult original sources when planning and executing new censuses.

For abundance estimations, we have also seen that there are far too many inherent sources of error in transect methods to provide reliable central tendencies. First, we are sampling nests, not individuals, and our results are thus compromised by the fact that gorillas do not always make nests (immatures as well as adults, see Mehlman and Doran, 2004), sometimes make two or more nests per evening, and sometimes make nests that are difficult to differentiate from those of chimpanzees. Moreover, these nests remain in different sample areas for far different rates of time (decay rate) depending on site and habitat characteristics, seasonal factors, climate, and types of nest construction (Table 1.2). Results can also be compromised by whether transects are representational across habitat types, as well as by errors in detection rate, including interobserver differences (experience, fatigue, type of habitat, etc.) and estimated transect widths (Figure 1.5).

Importantly, many of these sources of error creep into our calculations even *before* statistical confidence interval limits are calculated. Thus, no amount of statistical "bootstrapping," or adjustments based on "uniform key or hazard rate cosine adjustments" (see Laake *et al.*, 1994; Hall *et al.*, 1998a) can make abundance estimates and their confidence intervals any more precise when the original counts are modified upward by tens of percent to represent misidentified chimpanzee nests or to estimate the number of uncounted juveniles and infants that do not make nests.

In gorilla conservation, we seem to be confronted with the problem that our western culture demands numbers, and our constituencies and the public need a "sound bite": a number of gorillas that we believe exist in an area. This apparently causes us to engage in a "numbers game" in which we as scientists provide a number that we know is derived from: 1) a statistical central tendency with very wide confidence limits; and 2) a field methodology that is very imprecise and has multiple sources of error. Our science also seems to fall into this cultural trap, as our "numbers" begin to take on a life of their own when we publish reviews that cite abundance estimates that are sometimes decades old (see "Current Status of Western Gorillas," below).

6. Current Status of Eastern Gorillas

6.1. The Virunga and Bwindi Populations

The distribution of the Virunga mountain gorillas is limited to an area of 447 km^2 in the Virunga Volcanoes region (see note #1). Detailed *complete censuses* between 1971 and 1986 showed a population that was estimated to vary by about 13%, between 252 and 285 individuals (Harcourt and Groom, 1972; Groom, 1973; Weber and Vedder, 1983; Vedder *et al.*, 1986). In 1960, Schaller first estimated this population to be between 400 and 500, using a combination of group counts, prospecting, and extrapolations (Schaller, 1963). Given the differences in methodologies between Schaller and subsequent field workers, a cautious interpretation is that we cannot really be certain that this represented a decline, nor, if so, can we assess the magnitude of the decline between 1959 and 1971, when complete census methods began to be employed (Harcourt and Groom, 1972). There are, however, good reasons to suspect some decline during this period, since it is estimated that the total protected surface area of the Virunga Volcanoes was reduced by more than 20% between 1958 and 1973 (Harcourt and Curry-Lindahl, 1979; Weber, 1987).

Although it is tempting to analyze the various complete census fluctuations between 1971 and 1986 as representing either small decreases or increases every few years, each census during this period included a range of between 20 and 30 individuals, precisely the range of all the censuses (252–282) for the entire 15 years, suggesting a relatively stable population experiencing fluctuations of approximately 10% around a central tendency. However, a subsequent census in 1989 (Sholley, 1991) estimated there to be 324 individuals, clearly indicating an increase of between 15 and 28% from the years 1971–1986 to 1989. A more recent census[7] in 2003 now estimates 380 individuals, another 17% increase over the most recent 14-year period.

The Bwindi gorillas are found in and around the Bwindi Impenetrable National Park of Uganda and occupy a range of approximately 300 km^2 (Sarmiento *et al.*, 1996; Butynski and Kalina, 1998; McNeilage *et al.*, 1998). More than seven years ago, their numbers were estimated to be approximately 300 individuals using a complete census methodology (McNeilage *et al.*, 1998); this number has likely increased today.

6.2. Threats to the Bwindi and Virunga Gorillas

These are isolated island populations in upland areas surrounded by some of the highest human densities found on the African continent with extremely poor, agricultural-based local economies. As such, these gorillas will continue to be severely threatened by anthropogenic disturbance, such as agricultural conversion and illegal extraction of resources (snare setting for smaller mammals that entrap gorillas, cattle grazing, etc.). While these gorillas are no longer hunted for their meat in this region, they are, however, the focus of illegal animal trafficking. This threat, in which members of a group are killed and wounded (with the group sometimes disintegrating as a result) in an effort to trap an infant for the black market, is particularly severe for the Virunga population. In 2002, three separate incidents accounted for the death of at least six adults and three infants. In 2004, another infant mountain gorilla was confiscated, suggesting that one of the Virunga groups suffered at the hands of poachers.

Despite anthropogenic threats, the direct poaching, and the insecurity in both areas related to the 1994 Rwandan genocide and the subsequent Congolese civil wars, the known increase in the Virunga population and the good health of the Bwindi population must both be attributed to one important factor. These populations are located in national parks with well-developed protection and enforcement programs supported by many international conservation organizations, whose programs originated with Dian Fossey's efforts to focus world attention on the plight of mountain gorillas. This is surely a conservation success story. However, there may be unintended negative consequences. For example, in Rwanda and Uganda, conservation efforts are partially supported by mature, revenue-generating ecotourism industries, with a high percentage of gorilla groups experiencing tourist visits on a daily basis. This brings with it another significant threat: cross-transmission of disease from human tourists

---

[7] This census was carried out jointly in the Virunga Volcanoes Range by the Institut Congolais pour la Conservation de la Nature (ICCN), the Office Rwandaise du Tourisme et Parcs Nationaux (ORTPN) and the Ugandan Wildlife Authority (UWA), with the support of the International Gorilla Conservation Programme (IGCP), Dian Fossey Gorilla Fund International (DFGFI), Institute of Tropical Forest Conservation (ITFC), Max Plank Institute of Evolutionary Anthropology (MPIEA), Wildlife Conservation Society (WCS), Dian Fossey Gorilla Fund Europe (DFGFE), and Berggorilla & Regenwald Direkthilfe (BRD).

(reviewed in Butynski 2001; Woodford *et al.*, 2002). This combined with cross-transmission threats from local inhabitants (e.g., Lilly *et al.*, 2002), as well as their livestock, probably poses the greatest conservation threat for Bwindi and Virunga populations in the 21st Century (MGVP/WCS, Chapter 2, this volume). Today, these populations are highly managed through various types of veterinary interventions; their future may thus be one of becoming a managed rather than a truly wild population. Further, future studies will need to determine whether their small population sizes and potential for inbreeding depression could produce deleterious effects that might contribute to their demise.

6.3. Grauer's Gorilla

Schaller's (1963) first surveys of this gorilla in 1959 indicated there may have been as many as 15,000 individuals distributed as several fragmented local populations. By 2005, various surveys (e.g., Hall *et al.*, 1999a; Omari *et al.*, 1999), and new information presented in this chapter, reveal that many blocks of the distribution reported by Schaller no longer contain gorillas. Figure 1.6 displays the Eastern gorilla distributions as Schaller and Emlen reported them for 1959, with an overlay generated from Figures 1.2 and 1.3. It thus provides an indication of which gorilla populations have been extirpated between 1959 and 1996; these are described in Table 1.5. Thus, depending on the accuracy of Schaller's first surveys, it appears that at least 24% of the occupancy range of Grauer's gorillas has disappeared during the last 36 years, representing approximately 6700 km^2.

At present, there are four large populations of Grauer's gorilla, each still probably maintaining internal gene flow, but subject to increasing amounts of fragmentation due to anthropogenic disturbance: 1) a Tayna population; 2) a Maiko population; 3) a Kasese-Kahuzi-Biega population; and 4) an Itombwe population (Figure 1.6). Gene flow between these larger populations is doubtful, with the exception of Maiko and Tayna, and even that remains to be researched (Table 1.5). The Itombwe population appears to be experiencing the most fragmentation. There are also several small, isolated populations with between 18 and 50 individuals remaining (determined by nest counts): 1) Tshiaberimu (Figure 1.6, area A); 2) Masisi (area C); and 3) Walikale (area D). The highland sector of Kahuzi-Biega (Figure 1.6, between area K and Lake Kivu) is also a small isolated population, with only 170 gorillas remaining (Hart and Liengola, 2005; Hart, personal communication). Clearly, our current knowledge of the distribution of Grauer's gorilla is only large scale at best, and quite incomplete with respect to the distribution of local populations relative to fragmentation and loss of suitable habitats.

Our knowledge concerning the abundance of Grauer's gorilla is also quite imperfect.[8] A previous review estimated a range of abundance between 8,660 and 25,499 individuals (Hall *et al.*, 1998b); given the new information

---

[8] Deriving estimates of abundance by applying estimated densities to estimated range areas (Harcourt, 1966) should be viewed with a healthy skepticism. See note #3.

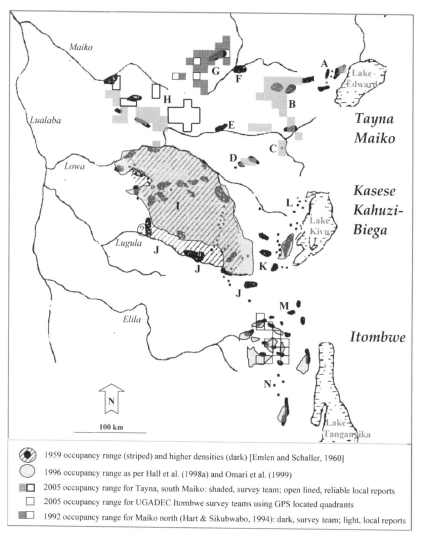

FIGURE 1.6. Loss of occupancy range for Grauer's gorillas, 1960–2005. Occupancy ranges from recent studies (see Table 1.5; G—100 km² quadrants, Hart and Sikubwabo, 1994; H—125 km² quadrants, this chapter and Nixon et al., 2005; M—125 km² quadrants, UGADEC, 2005) are indicated by lighter shaded polygons and quadrants, and are overlaid on a distribution map of Grauer's gorillas (black polygons) published by Emlen and Schaller (1960). Occupancy range loss is therefore represented by black polygons and shaded area with no overlays (Areas A, F, J, K, L, M); occupancy loss for Grauer's gorillas from 1960 to 1994–2005 is approximately 24% of their entire range as it must have been in 1960 (Table 1.5).

presented in this chapter, we should enlarge these limits to 5,500 and 28,000 (Table 1.6). There are no available data to make any quantitative change detection estimates for what we suspect has been a decline in the population since it was first surveyed by Schaller in 1959.

TABLE 1.5. Loss of occupancy range for Grauer's gorilla between 1959 and 1996–2005.

| | Schaller | | Range loss | | | |
Figure 1.6 and Emlen	1959[a]	2005	1959–2005	Locations	Population notes	Sources[b]
A 1,050	1,050	25	1,025	Tshiaberimu and Lubero	Only 20 gorillas reported for Tshiaberimu today, **totally isolated** from all other populations. Lubero populations extirpated due to forest conversion and subsequent subsistence hunting.	Author (pers. comm. with ICCN staff and travels in the Lubero region)
B 750	2,500	2,500	—	Tayna and south	Tayna population is most likely interconnected (Fig. 1.4) through small groups and solitary males; may be connected to Maiko south through mountainous corridor (Fig. 1.2, area 2; Fig. 1.6, area E); not likely connected to Maiko by the area to the northwest.	This chapter, UGADEC staff, DFGFI staff.
C 25	25	25	—	Masisi	About 26 gorillas reported from nest counts, **totally isolated** from populations to the North.	UGADEC staff, DFGFI staff.
D 150	150	75	75	Walikale	Small population **totally isolated** from all others.	UGADEC staff, DFGFI staff.
E 200	200	200	?	Lowa River in Walikale Territory	Has not been investigated, but could form part of corridor between Tayna and Maiko (Fig. 1.2, area 2; locals report presence of gorillas.	UGADEC staff, DFGFI staff.
F 200	—	0	—	Oninga	There may still be an isolated local population here, but not likely; Schaller may have misidentified location with those further north in Maiko (Fig. 1.6, area G).	UGADEC staff, reports from locals
G 275	2,100	2,100	—	North Maiko	Local interconnected population that most likely is in contact with Maiko south (Fig. 1.6, area G), but not likely connected to Tayna (area B).	Hart and Sikubwabo, 1994; UGADEC staff
H 700	2,000	2,000	—	South Maiko	Likely interconnected population with known occupancy of 2000 km², but ICCN and locals report another 2000 km² that is occupied (Fig. 1.6, area H). This population is most likely connected to Maiko North, and probably not	This chapter, recent census by ICCN and DFGFI staff (Figure 1.3)

Occupancy ranges and loss through time (km²)

				Location	Description	Reference
I J	15,870	15,870	12,770	3,100	Kasese Kahuzi-Biega (lowland sector)	connected to Kasese-Kahuzi populations, since individuals would have to cross the Lowa River barrier. ~3100 km^2 of occupancy range has been lost between Schaller's surveys in 1959 and those of Hall and colleagues – in the southern part of the range north of the Lugala River. Not clear whether portions of the main occupancy range have become fragmented (see text).	Hall et al., 1998a
K	1,700	1,700	500	1,200	Kahuzi-Biega highland sector and corridor to lowland sector	Large losses of occupancy range surrounding the high land sector and the corridor between highland and lowland sectors; this area is now degraded and it is unlikely that solitary males could travel between highland and lowland sectors. Highland sector is totally isolated from other populations (Hall et al., 1998b).	Hall et al., 1998b
L	300	300	0	300	Lake Kivu mountains	Schaller found small fragmented populations here; the area is now totally degraded and does not contain gorillas.	Hall et al., 1998b; personal observations from aerial flyovers.
M N	2,400	2,400	1,400	1,000	Itombwe	In 1996, Omari and colleagues recorded small fragmented populations that were probably partially interconnected (exceptions: Nundu and Fizi, local populations to the south); it is suspected that small populations north of the Ulindi River are now extinct. There are reports of gorillas in the southernmost area (near Fizi); it is unknown whether they are now extinct. **The Itombwe population is isolated from any populations to the north**, and is itself undergoing severe fragmentation.	Omari et al. 1999; UGADEC staff

(Continued)

TABLE 1.5. Loss of occupancy range for Grauer's gorilla between 1959 and 1996–2005—Cont'd.

Occupancy ranges and loss through time (km²)					
Schaller and Emlen	1959ᵃ	2005	Range loss 1959–2005	Locations	Population notes	Sourcesᵇ
Figure 1.6						
Totals	23,620	28,295	21,595	6,700	At least 24% of occupancy range has been lost from 1960 through 1996, possibly more in areas in Kasese-Kahuzi-Biega and Itombwe from 1996 through 2005.	

ᵃ These figures include occupancy areas from 1994 to 2005 that Schaller did not detect for 1959: it assumes therefore that he underestimated some areas, rather than gorillas having increased their range in the last several decades. This cannot be ruled out however.
ᵇ ICCN, see note #3 of this chapter. UGADEC (*Union des Associations de Conservation des Gorilles pour le Développement Communautaire à l'Est de la République Démocratique de Congo*) is a federation of community-managed projects seeking to establish a corridor of community-managed nature reserves between Maiko and Kahuzi-Biega National Parks, and in the Itombwe mountains. It is supported technically and financially by Dian Fossey Gorilla Fund International (DFGFI); the Tayna Gorilla Reserve (Figure 1.4) is the first government-sanctioned community-managed nature reserve created under this initiative. All UGADEC projects have field staff conducting gorilla monitoring.

TABLE 1.6. Current estimates of upper and lower estimates for abundance of Grauer's gorilla.

Areas as in table 1.5, figure 1.6	Lower abundance estimate	Upper abundance estimate	Source
Tshiaberimu	20	20	ICCN pers. comm.
Tayna and south	367	1,129	Figure 1.5
Masisi	26	26	Nest counts, UGADEC
Walikale	25	25	Estimate from local population
Lowa River in Walikale Territory	13	13	Hall et al. (1998b)
North Maiko	160	1,440	1600 km² quadrants with gorilla presence at 0.10–0.90 ind./km²
South Maiko	200	1,880	2000 km² quadrants with gorilla presence (observed and reported) at 0.10–0.90 ind./km²
Kasese Kahuzi-Biega (lowland sector)	4,000	22,203	Lower—adjusting downward using lower 95% confidence interval from Hall et al., 1998a and lower estimates from Table 1.3; Upper—Hall et al., 1998b.
Kahuzi-Biega highland sector and corridor to lowland sector	170	170	(Hart, WCS survey, pers. comm.)
Itombwe	400	831	Approximated from Omari et al., 1999 and Hall et al., 1998b
Total	**5,588**	**27,623**	

6.4. Threats to Grauer's Gorillas

The most severe threat to Grauer's gorillas is habitat loss and fragmentation through agricultural and pastoral expansion related to the westerly movement of people from high-density areas in the east to the forests of the west. Villages in forest zones practice slash-and-burn (or shifting) agriculture. As fields fatigue and hunting resources are depleted, these settlers move farther (west) into the forest, often selling their previous fields and land to local commercial interests, who then clear-cut large areas for pasturage of dairy and meat cattle, sheep, and goats (a process quite similar to that found in Central and South America). In North Kivu, for example, this process has resulted in the loss of large expanses of primary forest and the extinction of local fauna through subsistence hunting, including gorilla populations (e.g., Lubero: Figure 1.6, Table 1.5). In Kahuzi-Biega, a similar invasion process has effectively degraded the corridor zone between the highland and lowland sectors (Draulans and Van Krunkoisven, 2002). In Itombwe, the process of agricultural/pastoral expansion has been even more intense (Doumenge, 1998).

Mining activity is also a threat to Grauer's gorillas. The exploitation of coltan (columbite-tantalite used in micro-capacitors for electronic equipment) created a well-publicized late 20th century conservation crisis throughout this area. A boom-and-bust cycle of pricing lured thousands of people away from agriculture into mining camps, where anarchy was common and bushmeat hunting sustained the miners; it has also been identified as fueling the civil war in Democratic Republic of Congo (Tegera, 2002). Recently, however, a fall in coltan prices has caused many coltan miners to turn to the animal trafficking trade or to other mining activities. The newest mining threat is tin mining (cassiterite, SnO_2), which is now quite intense throughout the range, especially in the area between Maiko and Kahuzi-Biega National Parks, with an estimated 40 million U.S. dollars of cassiterite leaving the area each year (cassiterite is used in lead-free solder, important in the electronics industry, and is sometimes traded as "coltan"). Gold mining is present throughout the Grauer's gorilla range, with a cline of decreasing intensity from north (Maiko—most intense) to south (present in Itombwe). Diamond mining is common in the north, specifically within Maiko National Park. Mining techniques for all these ores are nearly identical; miners divert small and medium watercourses through damming (or digging large craters in the streambeds) and then practice sluice and pan techniques, or extract rock from pits and most often use water from nearby diverted streams to separate out the ore. Although these activities can be described as small scale, their heavy presence throughout these forests has a cumulative large-scale effect. They cause both direct and indirect environmental damage, such as forest clearance, stream pollution, erosion, firewood cutting, tree debarking (panning trays), liana cutting, disturbance to freshwater ecology, and bushmeat hunting (Redmond, 2001).

There is a total absence of commercial logging through the range of Grauer's gorilla, and since there are no commercial logging roads, the extraction and commercial transport of bushmeat, while still a persistent threat, exists at far lower levels than, for example, in Cameroon or other areas of Central West Africa.[9] Mining camps (many as large as 1,000 individuals) are probably the most important factor in bushmeat extraction, which is especially intense, for example, in Maiko and Kahuzi-Biega National Parks. While many of us working in this area believe that this has severely reduced elephant populations (Bishikwabo, 2001, Nixon *et al.*, 2005), monkeys, and other fauna, the impact on gorilla populations is not so clear, since gorillas do not appear to be targeted for bushmeat *per se* (although see Redmond, 2001). Layered onto hunting threats from mining camps is the ongoing and

---

[9] Deriving estimates of abundance by applying estimated densities to estimated range areas (Harcourt, 1966) should be viewed with a healthy skepticism. See note #3.

persistent threat of subsistence hunting by agriculturalists expanding throughout the region. Gorillas (and chimpanzees) are severely threatened by an illegal animal trafficking for the pet trade, with field prices for young apes varying between $1,000 and $5,000. The author is aware of nine gorillas and five chimpanzees confiscated by wildlife authorities in 2004–2005, indicating a brisk trade in the area; this is further complicated by a lack of wildlife sanctuaries, decreasing the motivation of law enforcement officials to engage in confiscation.

A number of socioeconomic impediments, common to most developing nations, have been exacerbated in the DRC by the cruel exploitation policies of the Belgian colonials, the Mobutu regime, and finally the recent civil wars of the last decade: poverty, disease, malnutrition, unemployment, lack of education, lack of medical care, and corruption. The regional and civil wars have added genocide, massacres, rape, starvation, child soldiers, destruction of property, and the disintegration of civil society to produce little more than anarchy. This has had profound negative consequences for conservation work throughout the range of Grauer's gorilla. This is changing. There is now a political unification process underway in DRC, with increasing integration of all armed groups into a national army, a reduction in armed conflicts, and improvements in governance and civil security. With these changes, however, new "peace threats" will certainly emerge. As a unified DRC government begins to function and attempts to solve the complex socioeconomic needs of the country, it may inevitably turn to the exploitation of natural resources, such as future lumber, mining, and petroleum concessions.

7. Current Status of Western Gorillas

7.1. The Cross River and Ebo Populations

The Cross River gorillas have a range area that encompasses 2,000 km^2, with an approximate occupancy area of 300 km^2 (Oates *et al.*, 2003; Oates, personal communication, 2005). Their range straddles the Nigerian and Cameroonian border in a mountainous area (200–1700 m) composed of a series of government and forest reserves, wildlife sanctuaries, and a national park. Repeated censuses indicate a population of between 250 and 300 individuals (Oates, personal communication). The population is reported as highly fragmented, with groups (or small clusters of groups) inhabiting isolated areas separated by distances of 5–30 km, with perhaps lone males providing gene flow between them. Threats to this population appear to be hunting pressures, combined with increasing habitat loss via agricultural conversion and forest fires due to a long dry season (Oates *et al.*, 2003).

Until recently, western lowland gorillas were only observed south of the Sanaga River, Cameroon. However, in late 2002, a small population of gorillas was discovered north of the river, in the Ebo Forest, Littoral Province,

Cameroon (Morgan *et al.*, 2003). Approximate numbers and area of occupancy are still unknown at present, as is the relationship of this population to either Cross River or western lowland gorillas.

7.2. Western Lowland Gorillas

These gorillas are found in Cameroon, Central African Republic, Gabon, Republic of Congo, and Equatorial Guinea. The fate of small populations inhabiting western DRC and northern Angola is unknown (although recent reports indicate they may still be present in Angola). Three recent reviews (Harcourt, 1996; Butynski, 2001; Sarmiento, 2003) provide abundance estimates of 111,000, 94,700, and 85,000 individuals, respectively. Each of these reviews, however, relied heavily on four important country-wide line transect census studies, three of which are now one to two decades old (Gabon: Tutin and Fernandez, 1984; Republic of Congo: Fay and Agnagna, 1992; Caroll, 1988; Equatorial Guinea: Gonzalez-Kirchner, 1997). They also utilize abundance measures derived from Harcourt's (1996) method of multiplying *estimated* occupancy areas (*from closed forest cover estimates*) times a mean density of gorillas of 0.25 individuals/km^2 (or twice this density for protected areas).[3] These kinds of reviews perpetuate "the numbers game" by providing abundance estimations that appear to have somewhat narrow error limits. As our earlier case study on Grauer's gorillas and the discussion of census methods (above) have shown, these estimations can be inaccurate by tens of percent. This is further compounded by using estimates for Gabon generated from a census in 1980–83, when, by the time of the above reviews from 1966 to 2003, there was good reason to suspect that habitat loss and the bushmeat trade must certainly have produced declines.

The decline in fact happened, and appears to be devastating ape populations. In a novel study led by Peter Walsh (Walsh *et al.*, 2003), investigators compared the combined encounter rate of both gorilla and chimpanzee nests from the country-wide Gabon survey in 1980–1983 (Tutin and Fernandez, 1984) with encounter rates derived from censuses conducted from 1998 to 2002 in areas containing high densities for gorillas and chimpanzees (261 km of line transects and 4793 km of reconnaissance surveys in existing and potential protected areas). They found an ape decline of 56% (95% confidence interval, 30–70%), and indicate that their methods may in fact underestimate the true decline. Moreover, they argue that the threats responsible for this decline (below) are quite similar in all the remaining host countries for western lowland gorillas, and that overall catastrophic declines are probably occurring there as well.

Currently, we can summarize what little we know about western gorilla abundance by providing a range of between 27,000 and 67,000 individuals (Table 1.7), with the following two assumptions: 1) the original estimates for ranges of abundance were relatively accurate (and as we might suspect, this may not be the case); and 2) the arguments presented by Walsh *et al.* (2003)

TABLE 1.7. 2005 Range in abundance estimates for the Western Gorilla: 27,000–66,000 individuals.[a]

Year	Totals	CAM	GAB	PRC	EG	CAR
1983			**35,000**[c]			
1984			33,355			
1985			31,787			
1986			30,293			
1987	104,166	25,470	28,870	39,281	1,545	**9,000**[f]
1988	99,270	24,273	27,513	37,436	1,471	8,577
1989	94,604	23,132	26,220	35,677	1,402	8,174
1990	90,158	22,045	24,987	**34,000**[d]	1,336	7,790
1991	85,920	21,009	23,813	32,402	1,273	7,424
1992	81,882	20,021	22,694	30,879	1,213	7,075
1993	78,034	19,080	21,627	29,428	1,156	6,742
1994	74,366	18,184	20,611	28,045	1,102	6,425
1995	70,871	17,329	19,642	26,727	1,050	6,123
1996	67,539	16,515	18,719	25,470	**1,000**[e]	5,836
1997	64,365	15,738	17,839	24,273	953	5,561
1998	61,341	**15,000**[b]	17,000	23,132	908	5,300
1999	58,458	14,295	16,201	22,045	866	5,051
2000	55,710	13,623	15,440	21,009	825	4,813
2001	53,092	12,983	14,714	20,022	786	4,587
2002	50,597	12,373	14,023	19,081	749	4,372
2003	48,219	11,791	13,364	18,184	714	4,166
2004	45,952	11,237	12,736	17,329	680	3,970
2005	*43,793*	10,709	12,137	16,515	648	3,784
2005a	*66,226*	12,586	21,943	24,730	826	6,142
2005b	*31,483*	9,162	7,434	11,823	531	2,534
2005c	*27,000*	9,162	5.948	9,390	531	1,971

[a] This table assumes a linear decrease of 4.7% per annum for all range countries, derived from a recent Gabon census (Walsh *et al.*, 2003) demonstrating a great ape population loss of 56% between 1983 and 2000. Walsh and colleagues indicate 95% confidence intervals of between 30% and 70% decline; thus the range for a current abundance estimate falls between 31,483 and 66,226 individuals (2005a and 2005b). If, for example, we take the minimum abundance estimate for the Gabon study in 1983 (e.g., 27,000, Tutin and Fernandez, 1984) and apply the highest rate of decline of 6.8% per annum (70% decline, Walsh *et al.*, 2003), a total minimum abundance estimate for Western lowland gorillas in 2005 could be as low as 27,000 individuals (2005c). In this table, some country populations are extrapolated in reverse (e.g., Cameroon, using the figure of 4.7% per annum loss) to show that the oft-quoted abundance estimate of 100,000 might have been applicable in 1987–1988, but is certainly not the case today.
[b] Using a figure from Butynski (2001) quoting personal communication from L. Usango.
[c] Tutin and Fernandez (1984).
[d] Fay and Agnagna (1992), see text.
[e] Gonzalez-Kirchner (1997).
[f] Caroll (1988).

that the decline observed in Gabon for apes can represent gorillas as well and is also applicable to the other range countries. We have little up-to-date information on distribution, although we can assume with recent and current threats that many areas of distribution that may have once been continuous are now quite fragmented.

7.3. Threats to Western Lowland Gorillas

Bushmeat consumption in the Congo Basin (recently estimated to exceed 1,000,000 tons per annum, Walsh *et al.*, 2003) is the major source of protein for Basin inhabitants, and, although gorillas are not the most pursued animal in this trade, their lower densities and relative slow rates of reproduction cause them to be particularly susceptible to this threat (Wilkie and Carpenter, 1999; Eves *et al.*, Chapter 17, and Stokes, Chapter 15, this volume). The bushmeat trade is thought to be devastating western lowland gorilla populations, with what is by now the well-known litany of its link to logging activities and infrastructure, its upscale market value in urban areas, and its lower price than domestic-based protein such as poultry, beef, or fish (e.g., reviews in Ammann, 2001; Butynski, 2001; Wilkie, 2001, Walsh *et al.*, 2003). Because this trade is so surreptitious, quantifying rates of extraction for gorillas is quite difficult. In northeastern Republic of Congo, Kan and Asato (1994) estimated that 5% of gorillas were being killed for bushmeat each year. In the mid-1990s, Rose (1997) made qualitative estimates that 3,000–6,000 apes were being killed for the bushmeat trade each year (off-take estimates are also reviewed in Butynski, 2001).

Habitat loss, due both to human population expansion and agricultural clearing, as well as the commercial logging industry, is also a significant threat to western lowland gorillas. For example, the percentage loss of forest from 1990 to 2000 for Cameroon, Central African Republic, Republic of Congo, Equatorial Guinea, and Gabon together was 3%, representing close to 3 million hectares of mature forest,[10] with Cameroon and Equatorial Guinea leading the list with 11% and 6%, respectively. As Harcourt (1966) indicates, these rates of decline do not even capture the effects of selective logging but only losses of the larger tracts of forest. These rapid rates of deforestation appear to be linked to increases in human density and host-country economies, particularly their foreign debt (Harcourt, 1966; Barnes, 1990). Selective logging is sometimes suggested to be a sustainable approach in Congo basin forests, but some argue that Africa's tropical forests cannot be sustainably logged (Reitbergen, 1992; Struhsaker 1996), given corruption, logging in protected areas (Horta, 1992; Williams, 2000), and the massive, negative side-effects associated with logging roads (Wilkie *et al.*, 2001).

Cross-transmission of infectious disease between humans and closely related gorillas is a well-known and documented threat for Virunga gorillas, and has always been seen as a looming threat for western gorillas as their habitats become smaller and more fragmented and contacts with human populations increase (reviewed in Butynski, 2001).

---

[10] Calculated as a ratio of net change of total forest area (1990–2000) to extent of forest with >50% canopy cover (2000) for these countries (World Resources Institute, www.wri.org).

As was the case for chimpanzees in the Taï Forest of Ivory Coast (Boesch and Boesch-Achermann, 1995), it is now clear that Ebola hemorrhagic fever has struck western lowland gorillas and is significantly implicated in the dramatic decline of gorilla populations in Gabon and the Republic of Congo (Walsh *et al.*, 2003; Stokes, Chapter 15, this volume). The vector(s) for Ebola, its transmission routes, and its total impact on gorilla populations is unknown at present. Many researchers suggest that Ebola is the primary cause for significant declines in gorillas in some areas such as Minkebe in Gabon (95% decline since the mid-1990s) and Odzala in Republic of Congo (Leroy et al., 2004; Walsh, personal communication).

The direct threats to western lowland gorillas, such as Ebola fever, bushmeat consumption, and habitat loss related to agricultural expansion and logging, are inexorably intertwined, and, despite what may appear to be still remaining large areas of gorilla distribution with concomitant large population sizes, these factors make for pessimistic predictions about their future. Contrary to conventional wisdom in this field, Western gorillas may be more at-risk of extirpation throughout most of their range than Eastern gorillas (see below).

8. Which Gorilla Populations are Most at Risk?

8.1. Cross River, Virunga, and Bwindi Gorillas

These three gorilla populations, by virtue of being small (200–400 individuals each), completely isolated, and undergoing continual anthropogenic and disease threats, as well as having the potential deleterious effects of inbreeding depression, are the most at-risk gorilla populations on the continent. This is reflected by their current World Conservation Union (IUCN) Red List status as *Critically Endangered Subspecies*.

The Virunga and Bwindi populations benefit from numerous conservation interventions by National Wildlife Authorities and several international conservation organizations (see note #5). Further, because these two populations are small, isolated, and well studied, their numbers, demographics, and health status are monitored relatively accurately by complete censuses, ongoing studies of habituated groups, and veterinary scientists. In large measure, as a result of these intense conservation efforts, the Virunga population in a recent census was found to have increased in size by 17% since 1989 (above). The Bwindi population is also thought to be in good health, with some of its conservation and research supported by the Mgahinga and Bwindi-Impenetrable Forest Conservation Trust Fund (Plumptre *et al.*, 2003).

Thus, among these three isolated and at-risk populations, the Cross River gorillas are the most critically endangered, having the smallest population of the three, while still being exposed to critical threats such as habitat loss, fragmentation, and bushmeat hunting. Conservation interventions for the Cross

River population are not as mature as those for the Virunga and Bwindi populations, but are now gaining momentum (reviewed in Oates *et al.*, 2003); recently, a new conservation partnership for the Cross River gorillas was implemented between the Cameroon Ministry of Environment and Forests, WCS and WWF.[11]

8.2. Western Lowland and Grauer's Gorillas

The IUCN Red List status for these two groups is *Endangered* (although there has been a call to reclassify them as Critically Endangered, Walsh *et al.*, 2003). Evidence for significant rates of decline in western lowland gorillas has been presented above: over a period of 18 years, Walsh and colleagues estimated a 56% decline in great ape populations in Gabon and suggest these rates of loss typify Western Central Africa. Over a period twice as long (36 years), Grauer's gorillas are estimated to have disappeared from 24% of their occupancy range (Table 1.5). If density was uniform throughout the range, *this would represent a rate of decline for Grauer's gorillas that was less than half that for western lowland gorillas.* The rate of decline might even be less given that many areas of occupancy that were lost were already marginalized and fragmented populations even in 1959 (especially those in the east, see Figure 1.6) and would not have represented the highest densities and abundances of Grauer's gorillas at that time.

Obviously, there are numerous caveats to this oversimplified comparison of two different measures of loss. First, Walsh and colleagues measured great apes together; there could easily be a differential decline between gorillas and chimpanzees. Second, and even more importantly, loss of occupancy range is not a good proxy measure for declines in abundance, since, in the remaining occupancy areas, the density of Grauer's gorillas may have also declined significantly. There is some limited evidence, however, to suggest that this is not the case. Even though we have large gaps in our data sets, we can model "estimates" of average density throughout the entire range of Grauer's and western lowland gorillas based on the information presented in this chapter (Figure 1.7). If we take the upper and lower abundance figures for Grauer's gorillas today (Table 1.6) and calculate a range of overall density estimates from their known occupancy range (Table 1.5), this falls between 0.25 and 1.30 individuals/km² (Figure 1.7). This is higher than the range of current

---

[11] The WCS program in the area has also organized an endorsement of the Environment Ministers of Cameroon and Nigeria for Cross River gorilla conservation with a workshop in Limbe, 2003, and WCS continues to work closely with state and federal governments, local communities, other NGOs and international development agencies in Nigeria and Cameroon on behalf of the conservation of these gorillas (Oates, personal communication.). Also see WWF – World Wildlife Fund, WCS – Wildlife Conservation Society (http://www.panda.org/downloads/africa/apeupdateno1(english).pdf.

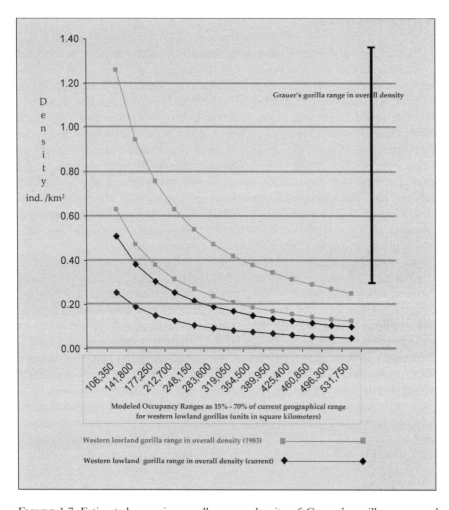

FIGURE 1.7. Estimated range in overall average density of Grauer's gorillas compared
with range in western lowland gorillas. X-axis displays a range of possible percentages
(15–70%) of occupancy range/geographical range with vertical numbers indicating total
area of occupancy range for each percentage. Y-axis plots upper and lower limits of
density (individuals per km²) based on upper and lower estimates for total abundance
(Table 1.6) for: 1) dark bar at far right—Grauer's gorillas (5,500–30,000 individuals, see
text) with current occupancy range estimated at 21,595 km² (Table 1.5); 2) dark lines
and diamonds—current upper and lower ranges in overall density for western lowland
gorillas modeled on increasing percentages of occupancy range (15–70%, intervals of
5%) to total estimated geographical range of 709,000 km² (Table 1); and 3) gray lines
and diamonds—Western lowland gorilla ranges estimated for 1983, that is twice the current overall density to represent the 56% loss of great apes described by Walsh et al.
(2003). Note that in this modeling exercise, the current range of possible densities for
western lowland gorillas only approaches the lower estimated range for Grauer's gorilla
when the occupancy range for the former is modeled at 15–20% of the total geographical range (709,000 km²), that is 106,350–141,800 km². Similarly, if we take a point equidistant between the upper and lower ranges to model a central tendency (i.e., 0.87
gorillas per km² for Grauer's gorillas), the simulated density for western lowland gorillas 18 years ago only reaches that of Grauer's gorillas if their occupancy range 18 years
ago was as small as 106,350 km², or 15% of the total geographical range.

overall estimated densities we might similarly model for western lowland gorillas and higher than the majority of the range in overall density for western lowland gorillas 18 years ago, using an average decline figure of 56% from Walsh and colleagues' (2003) study (Figure 1.7). If we rule out the possibility that Grauer's gorillas several decades ago existed at overall densities much higher than today (a reasonable exclusion, given Schaller's survey results, and the modeling in Figure 1.7), we are led to suspect that the loss of occupancy range for Grauer's gorillas may correlate reasonably well with presumed losses in abundance (or at least, using loss of occupancy range as a proxy does not lead to underestimates in abundance declines). If so, and we accept the findings for western gorillas by Walsh and colleagues (2003), it appears that western lowland gorillas have disappeared at much faster rates throughout their range than has been the case for Grauer's gorillas.

Although this suggestion needs far more quantitative data to support it, it is perhaps not as surprising as it might first seem. The bushmeat trade appears to be a much more severe threat to western lowland gorillas than for Grauer's gorillas.[12] In western lowland gorillas, the bushmeat trade is intense and amplified by logging infrastructure and a regional-cultural tendency to prefer wild over domestic meats (or fish). Further, bushmeat consumption in Western Central Africa has significantly increased in the last decade (e.g., Walsh *et al.*, 2003). For Grauer's gorilla, we have anecdotal evidence indicating that, in the last decade, certain military and rebel groups, as well as miners in their camps, did hunt and eat gorilla (Yamagiwa, 2003), but that there does not appear to have been any *systematic or long-distance trade* in gorilla meat (notwithstanding that, where gorillas were extirpated, this was obviously linked to subsistence hunting in areas under pastoral and agricultural conversion). The most significant threats to Grauer's gorillas are (and were in recent decades): 1) deforestation linked to agricultural conversion and 2) the illegal animal trade. It is difficult to imagine that these factors together could produce a steeper decline in Grauer's than in western lowland gorillas, especially when agricultural conversation of forest is also occurring in West Central Africa,[13] combined with the history and continued impact of commercial logging in many areas (especially in Cameroon and Equatorial Guinea).

Ebola, linked to the significant decline in numbers of western lowland gorillas (Walsh *et al.*, 2003), is not known to be present in Grauer's gorillas in Eastern DRC. From this, we can suggest as a working hypothesis the following: In Western Central Africa, the bushmeat trade and Ebola are removing western lowland gorillas from forests (the "empty forest phenomenon") much more rapidly than is the case for Grauer's gorillas. This exercise thus provides us a counterintuitive caution: despite the larger overall range and higher abundance of western lowland gorillas, they appear to be more at-risk and disappearing at faster rates than Grauer's gorillas. Clearly, we need new

---

[12] This needs to be quantified in future studies.
[13] This needs to be quantified in future studies.

and up-to-date data on occupancy ranges and abundance measures (and their rate of change) for both western lowland and Grauer's gorillas to test this hypothesis and set our conservation priorities.

At present, all of the gorilla populations on the African continent are facing a conservation crisis. The Cross River gorillas by virtue of their small size and ongoing threats are the most critically endangered. Western lowland gorilla populations may likely be at higher risk than Grauer's gorillas for extirpation of many large, local populations, *although present threats and circumstances could change at any time*. The Virunga and Bwindi gorillas are somewhat more secure as a result of long-term, intensive conservation interventions, but both are at-risk by virtue of their small size and from threats due to human population pressure and the potential for local- and tourist-based cross-transmission of disease (MGVP/WCS, Chapter 2, this volume).

9. Formulating a Strategy for Gorilla Conservation

The *direct or proximate factors* that threaten the survival of gorillas in this century are well known: bushmeat hunting, human encroachment, including agricultural and grazing conversion, commercial logging, disease cross-transmission, and a new complication, Ebola. The *indirect or ultimate factors* that underlie and contribute to these direct threats in Africa, the poorest continent on our planet, seem even more insurmountable: poverty, disease, hunger and malnutrition, war, lack of education, lack of adequate governance, corruption, and perhaps underlying all of these, economic underdevelopment. Further, funding for conservation (and development) activities is a limited resource, especially with the current poorly performing economies of developed countries, the primary sources for conservation funding. Given this crisis mode, how can conservationists formulate the important elements of a strategy for gorilla conservation?

9.1. Address the Gaps in Knowledge on Distribution and Abundance of Gorillas

As should be clear from this chapter, we need a wide array of new data to understand current distribution and relative abundance (and rates of change for both variables), especially for western lowland and Grauer's gorillas. How can we address this need for data and prioritize our activities, given the expense, time, and manpower necessary to conduct surveys and prospections? First, any gorilla censuses, be they prospective or transect-based, must also include data collection on all large and medium-sized mammals; we cannot afford to conduct gorilla-centric or great ape surveys (see below).

Second, we must abandon the notion that we can provide accurate or reliable central tendencies about abundance and focus instead, as did Walsh and colleagues (1999), on detecting rates of change. In their methodology, they avoided many of the intervening confounds by directly comparing nest encounter rates of gorillas and chimpanzees combined together over a period of time. This method has great promise and could be applied wherever previous surveys have collected nest encounter rate data. For example, for Grauer's gorillas in the lowland sector of Kahuzi-Biega and the Kasese region, we could employ the same reconnaissance transect methodologies as did Hall and colleagues (1998a) in the mid-1990s (i.e., conduct as close a repetition of methods as possible) and with new data, calculate rates of change. This will enable us to determine whether relative ape density has declined for at least one core area of Grauer's gorillas during the last decade. Similarly, for western lowland gorillas, a repeat survey should be conducted in the Republic of Congo, replicating the methods and areas sampled by Fay and Agnagna (1992) in 1989–1990. This would have the advantage of producing more information on rates of (presumed) decline for western lowland gorillas and could also test the assumption of Walsh and colleagues (2003) that their findings from Gabon are applicable to neighboring countries. These suggested repeat surveys, based on transect methodologies, would be labor intensive and require substantial funding. Given the time to conduct them, they would also not likely produce results for at least another two or more years even if started today (bearing in mind the desire to publish results in scientific journals).

Despite the high priority we must assign to conducting more transect surveys, we also need other more rapid information to determine the scale of loss over large tracts of occupancy range. If we distill abundance and density measures to their simplest forms (i.e., they are present at some density, or they are absent with zero density), however, we can more rapidly collect absence-presence information using a standardized quadrant methodology based on 1/10th degrees latitude longitude (above) over large range areas for both western lowland and Grauer's gorillas. At present, we do not even have baseline data on the overall occupancy ranges for these gorilla forms. In several years' time, this information could help us at least determine whether we are losing or gaining occupancy range, and where this is occurring spatially. With only small investments in training (use of GPS units and interpretation of GIS-generated maps), baseline data on absence-presence distribution could be collected by park rangers, community conservation workers, and local stakeholders (Brown *et al.*, Chapter 10, this volume). We need only provide a rapid training program that ensures standardization of methods by trained observers who record gorilla trail, feeding, feces, and nests (absence-presence should also be recorded for large and medium-sized mammals and anthropogenic disturbance) and a protocol for prospection that ensures that areas patrolled have maximum coverage.

This is not at all a suggestion to abandon transect surveys. Where funding and well-trained participants are available, transect-based surveys should be conducted in tandem with distribution surveys; these will address current knowledge gaps and lay down important baselines for future change detection analyses. In the interim, however, we must admit that our knowledge of gorilla occupancy range is very poor (as is probably our knowledge for most large and medium-sized mammals), and the best proxy, given our current crisis situation, would be to create baseline databases on occupancy ranges and then to repeatedly look for losses and gains with the most basic and inexpensive change detection function available: absence-presence measures. These databases could be analyzed spatially, using functions related to topography and distance to cities and bushmeat markets, as well as a number of other spatial variables, such as Ebola outbreaks (Walsh *et al.*, 2003). An initiative such as this would require an umbrella organization or agency partnering with host-country wildlife agencies and international conservation NGOs to both standardize data collection, verify incoming data, clean the data, and enter it into easily accessible on-line databases for rapid access (e.g., GRASP, see below). At present, we may not be able to afford the delays associated with scientific publication; perhaps a process must be developed on-line that can acknowledge the sources of information, conduct reviews, accept the findings, and make them rapidly available.

With respect to threats, it is also clear that we lack *quantitative* comparative data on the basic threats facing each of the gorilla populations (as well as other fauna), such as spatially located rates of deforestation and rates and types of bushmeat consumption. There are a number of ongoing and new GIS and remote sensing initiatives currently measuring rates of deforestation in the overall Congo basin (e.g., Global Forest Watch from the World Resources Institute, www.globalforestwatch.org), as well as in all of the CBFP Landscapes (University of Maryland, supported by CARPE, http://carpe.umd.edu/); many of these data are now available or will be produced in the next few years. The standardization of protocols and collection of Congo Basin-wide data on bushmeat consumption are also a CARPE goal for the 11 Landscapes targeted in the CBFP[14] initiative; presumably data from this program will also be available within the next several years. Other consortia and joint programs, such as the *Bushmeat Crisis Task Force* and *Traffic,* are facilitating these efforts, and hopefully will aid conservationists to determine local "hotspots" for gorilla hunting and trafficking so that more targeted interventions (awareness raising and law enforcement efforts) can be designed and rapidly executed.

9.2. Gorilla Conservation in the Context of "Landscape Conservation"

Gorilla populations cannot be taken out of their ecological context; a conservation strategy for them must therefore preserve *intact ecosystems* sizeable enough to support linked populations of larger mammals that must maintain

---

[14] CARPE is an acronym for USAID's Central African Regional Program for the Environment; CBFP represents the Congo Basin Forest Partnership.

connectivity over thousands, even hundreds of thousands, of hectares (e.g., forest elephant, buffalo, chimpanzees, and gorillas). This may necessitate an approach that provides conservation interventions on a large scale, engaging multiple sectors of society, and one that combines conservation support both for protected areas (PAs) and conservation and development support for areas outside PAs, such as buffer zones and potential ecological corridors.

A recent example of this *Landscape* approach (e.g., also known as a "Corridor approach," Conservation International (Morrison *et al.*, Chapter 9, this volume) or a "Heartlands approach," African Wildlife Foundation) was launched by the United States and South Africa, along with 27 public and private partners, at the World Summit on Sustainable Development in Johannesburg in September, 2002. This program, entitled the Congo Basin Forest Partnership (CBFP) was provided approximately $45 million (from 2003 to 2006) by the U.S. government to encourage public-private partnerships and leverage more funding from other government partners, as well as matching funding from nonprofit, international conservation organizations, and the private sector. The U.S. administers its funding support through the USAID CARPE II program (Central African Regional Program for the Environment) and has as its goals the:

"promotion of economic development, poverty alleviation, improved governance, and natural resources conservation through support for a network of national parks and protected areas, well-managed forestry concessions, and assistance to communities who depend upon the conservation of the outstanding forest and wildlife resources of eleven key landscapes in six Central African countries." (http://carpe.umd.edu).

Some conservationists, however, urge caution with respect to combining sustainable development with conservation initiatives. As Walsh and colleagues recently commented:

"The solution to the hunting crisis will not result in investment in poverty alleviation or sustainable development. Rural poverty is too intractable to be solved before hunting further demolishes ape populations. . . . The bulk of conservation investment should, instead, be focused on formally protected areas that still have enough apes to be viable in the long run. The immediate priority is a massive investment in law enforcement, which has long been under funded in WEA [West Equatorial Africa]. . . . *We must also provide* [host] *national governments with economic incentives, linking aid and debt relief to verifiable measures of conservation performance."* (Walsh *et al.*, 2003)

The two quotes above therefore perhaps represent two different paradigms for 21st century conservation: 1) a "protected area reinforcement conservation strategy" (PARCS) that suggests a more concentrated focus on formally protected areas, and the use of available (limited) conservation funding for protection and law enforcement (and capacity building of host countries to execute these activities); or 2) a Landscape strategy (LANDSCAPE) that attempts to widen conservation approaches by integrating them into larger, multi-sector, multi-partner sustainable development initiatives, while simultaneously supporting protected areas.

Given the current challenges in conservation and limitations in funding, there is a certain appeal to PARCS. It is an approach that primarily focuses on *addressing direct threats* and it suggests that conservationists should do what they do best, that is, concentrate on purely conservation-oriented activities *in strictly protected areas*, and unbundle any development interventions from the former. Clearly, an increased emphasis on law enforcement of protected area rules and regulations is one major component of a comprehensive strategy to improve biodiversity protection (Gibson *et al.*, 2005). There is no question that protected areas, even including those that are not even completely functioning, do function to preserve biodiversity. For example, evidence gathered by a questionnaire methodology from 93 protected areas in 22 tropical countries revealed that the presence of protected areas reduces land clearing and, to some degree, lessens logging, hunting, grazing incursions, and fire within their boundaries (Bruner *et al.*, 2001).

Despite its appeal, there may be some weaknesses with a strictly PARCS approach. First, it could lead us to abandon, and therefore lose, high-biodiversity areas outside of PAs. It presumes that the number, spatial distribution, and connectivity of *existing* protected areas are sufficiently large and strategically located to ensure preservation of biodiversity in all its forms. We currently do not have sufficient information to suggest this is true. For example, many of the current protected areas in Africa were former hunting reserves and/or areas of low human population density and were gazetted long before any biodiversity studies identified them as centers of endangered and endemic flora. This approach can also inadvertently shift to become "ape-centric" or "species-centric" approaches, as, for example in the statement, *"the bulk of conservation investment should, instead, be focused on formally protected areas that still have enough apes to be viable in the long run"* (Walsh *et al.*, 2003).

Second, PARCS does little to address the ultimate causes creating the current conservation crisis in Africa: primarily poverty and underdevelopment. Despite any gains in law enforcement (and meaningful prosecution) in PAs, that is, no matter how many poachers are arrested and imprisoned, the ultimate factors of extreme poverty and hunger (as well as Africa's current demographics) will continue to provide an inexhaustible supply of new (young) people willing to risk prosecution and imprisonment in order to survive. Investing heavily in law enforcement recalls the "War on Drugs" approach in the United States; this certainly filled U.S. prisons, but concentrating on law enforcement and prosecution while ignoring the root causes of the problem did very little to stem the tide of drug abuse (Gray, 2001). This approach can also result in a situation where people living outside PAs perceive the lack of attention to their livelihoods needs as insensitive. This generates ill will and alienation, and increases exponentially the lack of respect for PAs, as well as the continued illegal extraction activities within their borders.

Third, a PARCS perspective is not at all synchronized with African host-country politics. To many African government authorities, this strategy may appear to embody policies of international conservation NGOs from the developed world, which are interpreted as blind to the reality of the human needs in their countries. African policy-makers have political constituencies and many are attempting to solve quite difficult economic and political problems; to issue decrees oriented on law enforcement and prosecution, without addressing the needs of local populations, would be political suicide (whether in democracies or the various alternative forms of government). The related concept of providing economic incentives to national governments achieving conservation performance (Walsh *et al.*, 2003) is a top-down approach and, while it should be tested, may suffer from bureaucratic inertia related to issues of measurement and compliance. It will also take political will and quick action on the part of developed nations, not an easy formula.

In a LANDSCAPE approach, incentives can be directly targeted on local populations, which can contribute to overall conservation goals. Conservationists taking a LANDSCAPE approach are interested in focal regions, larger tracts of intact forest with remaining high levels of biodiversity. By providing development interventions for local populations living in and around these focal areas, they increase the chances of building stakeholder constituencies that support conservation. In this approach, conservation activities are linked to interventions that 1) address critical needs for food, nutrition, medical care; 2) improve livelihoods and educational opportunities (with an added component of conservation-oriented curricula); and 3) establish economic alternatives that may replace bushmeat extraction and other exploitative activities. These targeted development initiatives can secure the good will of local populations (and host governments), and can attempt to diminish the root causes of nonsustainable, exploitative activities.

In some instances, the LANDSCAPE approach can empower local people to protect biodiversity by creating new and novel types of protected areas. For example, in one of the CBFP Landscapes in Eastern DRC (#10), a community-managed, government-sanctioned nature reserve, the Tayna Gorilla Reserve, is significantly adding to conservation efforts in the region.[15] Further, it is serving as a model for six other community groups wishing to establish similar reserves that will, by virtue of their connectivity, provide an ecological corridor between two National Parks (Figure 1.8). This creates a conservation-minded stakeholder constituency in the buffer zones of these two parks that adds to and synergizes all conservation efforts, and is addressing the ultimate causes creating the conservation crisis for this region.

The LANDSCAPE approach does not preclude interventions for national parks and other protected areas, nor does it preclude a heavier investment

---

[15] The Tayna Gorilla Reserve is a Nature Reserve created under the existing DRC Forestry Code (revised in August, 2002) and a Environment Ministerial Decree (#274); it is managed by two traditional communities, the Batangi and Bamate Nations, and has a 900 km^2 integral zone that sanctions complete protection of all flora and fauna within its boundaries (also see Figure 1.8)

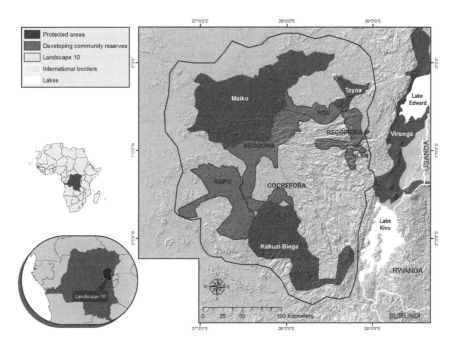

FIGURE 1.8. The location of the integral zone of the Tayna Gorilla Reserve (900 km^2) in Landscape #10, the Maiko Tayna Kahuzi-Biega Landscape of the CBFP. This integral zone is sanctioned by local stakeholders to provide complete protection of biodiversity, identical to the level of protection of National Parks in DRC. Also shown are six other community-based projects and their projected integral zones as they model their efforts after the Tayna Reserve, and attempt to create a biological corridor of community-managed nature reserves between Maiko and Kahuzi-Biega National Parks.

in law enforcement and prosecution; these are critical interventions that address the direct threats to biodiversity in protected areas. A LANDSCAPE approach, however, does necessitate additional foci on:1) forests outside of (and sometimes linking) protected areas; and 2) even more importantly, the local stakeholders residing in and adjacent to these areas. Given this, the most critical question, therefore, is how funding should be partitioned between direct support for protected areas and more development-linked conservation initiatives. This important question will generate continued debate by host countries, local stakeholders, and international and national NGOs wishing to conduct conservation. We suggest that the solution is not likely a formulaic one, such as whether the PARCS or LANDSCAPE solution is best throughout the Congo Basin. Funding decisions need to be context-specific with respect to target areas and linked to both biodiversity and threats assessments, such as measures of the distribution and density of human populations relative to important biodiversity forest tracts, their resource extraction patterns, their socioeconomic parameters and critical needs, the existence and types of commercial logging activities, as well as other commercial resource use such as mining and oil extraction.

At this time, we suggest that it would be far too radical to retreat to a PARCS approach and abandon the LANDSCAPE approach. The latter needs to be tested and then critically evaluated, with a focus on measurable indicators of conservation success (and decisions about continued funding need to be performance-based). One of the challenges of 21st-century conservation will be to determine, through empirical means, how to strike a proper balance between a strictly protected area– and law enforcement–oriented approach to conservation, and those approaches that successfully incorporate development interventions and new models for conservation and resource management.

9.3. Gorilla Conservation and Multidisciplinary Training and Teamwork

Given that much of gorilla conservation will be occurring in the context of landscape planning and a multi-sectorial approach, the gorilla conservationist of the 21st century must work within large, multidisciplinary teams. These are often composed of a variety of specialists in conservation planning, GIS science, socioeconomic science, political science, and law, etc. In the last century, gorilla conservationists often began as site researchers trained in biology or primatology. Today, the complexity of conservation demands that biologists and primatologists be at least exposed to the rudimentary aspects of these disciplines in their university training before they go into the field. This will prepare them to communicate efficiently with other specialists in a multidisciplinary context. There will always be a need for site-based gorilla research on behavior, demography, local population dynamics, etc., by specialists in primatology and biology. Indeed, the presence of long-term research sites plays an important role in conservation (e.g., see Oates *et al.*, 2003; Steklis and Steklis, Chapter 6, this volume). However, in this century, many of these site-based research programs are joining forces and becoming integrated into more regional and global conservation efforts, as for example, with the "Western Gorilla Network" a consortium of site-based gorilla researchers who have joined forces to aid gorilla conservation.

Development of Action Plans

In 2002, a group of conservation and great ape experts assembled in Abijan, Cote d'Ivoire, and developed the *Regional Action Plan for Conservation of Chimpanzees in West Africa* (Kormos and Boesch, 2002). This approach has now been replicated for great apes in West Central Africa, with a recent workshop held in Brazzaville, Republic of Congo, in May 2005. This has now produced the *Regional Action Plan for Conservation of Chimpanzees and Gorillas in Western Equatorial Africa* (Tutin *et al.*, 2005). In this endeavor, participants not only identified and summarized threats to the great apes but developed a set of priority areas and area-specific priority actions with an associated five-year budget. A consensus-building effort such as this, which brings authorities, experts, funding agencies, and, importantly, host country representatives together, represents the best efforts of conservationists to solidify their goals and objectives, and to coordinate and plan their efforts for the future. A similar approach for Eastern gorillas (and chimpanzees) is most certainly a next step in great ape conservation.

9.5. Gorillas as a Focal Species

Gorillas, being the largest of primates with spectacular male displays, figure large as an icon in western culture. Similarly, for African cultures, gorillas are the stuff of lore: sometimes they are a protected totem animal, and sometimes they represent a "kill" signifying a hunter's prowess and bravery (which does not always mean that they are hunted regularly for protein). Thus, gorillas are perhaps one of the best charismatic megafauna to provide a pathway into people's relationship with nature and animals and to sensitize them about preservation of biodiversity. In western countries, great apes certainly generate conservation funding, and they are also central to a subcultural movement that sees them as not only closely genetically related to people, but grants them the ethical status of personhood, with basic "human rights" (Teleki, 1997; *Declaration on Great Apes and the Great Ape Project*, reviewed in Butynski, 2001).

In Africa, they are being used as a focal species at the local level to successfully interest local stakeholders in conservation activities and community-based conservation management (e.g., the Tayna Gorilla Reserve, see above). Along with chimpanzees and bonobos, they also form the basis of an international-level initiative by the United Nations Environmental Programme (UNEP) and the United Nations Educational, Scientific and Cultural Organization (UNESCO) called *GRASP* (Great Ape Survival Project), launched in 2001, which seeks to use great apes as focal species, and to develop a global strategy for their survival, or as UNEP states:

"to provide a framework into which all the individual conservation efforts of governments, wildlife departments, academics, non-governmental organisations (NGOs), UN agencies and others can be layered to ensure maximum efficiency, effective communication and successful targeting of resources. GRASP is an ambitious initiative, which while recognising the autonomy and independence of existing initiatives, seeks to create a harmonious and coordinated network, in order to halt the decline of the great apes." (http://www.unep.org/grasp)

Among some conservationists, there is now a move to lobby the United Nations to classify great apes as a "World Heritage Species," akin to the UNESCO program of World Heritage Sites. Using the gorilla and other great apes as a focal species to ignite interest in conservation at host-country and international levels is an ideal strategy, but it is only a small, first step towards appreciating the inherent value of entire, intact ecosystems and the preservation of natural resources. We must be careful to link these messages about great apes to larger conservation topics, such as global climate change, carbon sequestration, massive deforestation and related loss of evapotranspirative water cycling, and preservation of freshwater ecosystems, etc. Although gorillas are an ideal focal species, we must be continually mindful of our overall conservation goals. If not, we perhaps risk great expenditures in energy, time, and funding that are solely focused on saving gorillas and great apes, and lose precious time and opportunities to foster approaches that preserve entire ecosystems. To do otherwise mistakes the real level of biodiversity crisis the planet is currently experiencing. An inexact analogy in marine conservation is the oft-repeated emphasis on saving dolphins, while in reality the majority of the ocean's fisheries are nearing collapse (e.g., Clover, 2004). People are quite content to eat tuna that are "dolphin safe" thinking that they are contributing to conservation, when they unknowingly consume a number of endangered fish species in high-end seafood restaurants that is contributing to the collapse of fisheries and the world's marine ecosystems (Clover, 2004).

In the 21st century, gorilla conservationists, through their knowledge and passion for gorillas, can serve as important spokespersons to introduce global conservation topics by using a charismatic focal species in rapid decline. But to do so, they must be trained and prepared to provide the next step in communicating a conservation message above and beyond the important activity of protecting gorillas: entire ecosystems are at risk of collapse, and conservation can often only address these issues using a landscape (or larger) scale approach that considers people to be integral components of the environment. People are both the problem and the only real hope for solutions.

10. Conclusions and Recommendations

Those of us practicing gorilla conservation and research, or those of us preparing to do so, have much to do in the 21st century. In the arena of site-related research activities, we can provide more information on feeding behavior and habitat characteristics, with a more focused emphasis on understanding carrying capacity. We can contribute to modeling demographic parameters at the group and local population levels, and understanding the important factors that shape group size, as well as fission and fusion processes. We can conduct research that will provide a clearer understanding of local population dynamics, and we can begin to examine metapopulation processes, such as the distribution and local sizes of gorilla groups related to variations in habitat types, as well as the genetic connectivity of local populations through emigration and immigration. All of these research activities will provide important information that can shape the underlying science of gorilla conservation in its context of conservation at large.

We also know that our very presence on a *long-term* basis at research sites aids in protection efforts, and, by active participation with local stakeholders, we can do much for awareness-raising and conservation education. Site-based researchers need to link their conservation and research efforts with landscape, regional, and global initiatives, and be prepared to continually train and work in a multidisciplinary environment and to enter into partnerships and cooperative agreements with other agencies and organizations.

For those of us conducting more general gorilla conservation activities, there is no doubt that their use as an iconic focal species can catalyze both fundraising and conservation efforts, and we must continue to engage donors as well as stakeholders by increasing an awareness of gorillas, their life-ways, and their relatedness to ourselves. However, in executing our conservation activities, be they scientific or political, we cannot afford to concentrate solely on gorillas. Our conservation activities must be broad-based and relate to the preservation of all fauna and flora within "gorillascapes," that is, they must focus on conservation efforts for habitats and ecosystems in which gorillas play an integral role. Fortunately, these efforts often intersect. For example, an excellent set of recommendations specific to great ape conservation was published in 2001 (Butynski, 2001). Even though they were recommended for great ape conservation, they wholly intersect with goals for overall conservation efforts in Africa.

The present chapter does not intend to be exhaustive in its review. Clearly, it has reinforced what Butynski concluded in 2001, *"there is an alarming lack of reliable information on the distribution and numbers of chimpanzees and gorillas"* (2001:41). Moreover, in examining the case study for Grauer's gorillas and discussing the sources of error in abundance measures, this chapter has re-emphasized the question of the reliability of our current data sets and further suggested that, in future surveys, we attempt to increase our knowledge sets for distribution and occupancy ranges, and that we re-orient our analyses to detecting changes in density and abundance, rather than engaging in a static "numbers game."

It would be incorrect to state that nothing much has changed since Butynski's review in 2001. For example, Walsh and colleagues (2003) have published new information on western lowland gorillas and chimpanzees that can help to refine our methods and which adds to our knowledge base. Unfortunately, their findings are dire. To counterbalance this, perhaps, we can add that, since Butynski's review, African conservationists and stakeholders have both launched a community-based conservation approach (Tayna Gorilla Reserve) and discovered through their own scientific research that another large local population of Grauer's gorillas has gone virtually undetected by western scientists. In Eastern DRC, Maiko National Park and environs appear to contain more gorillas than previously thought. As this chapter tentatively suggests, Grauer's gorillas, although no doubt facing severe threats, may not be disappearing at rates as fast as previously suggested; more data will tell.

Since Butynski's review, there is a new emphasis in gorilla conservation, which is examining and developing preventative measures for disease cross-transmission (Lilly *et al.*, 2002; MGVP/WCS, Chapter 2, this volume). This good news is dampened by the finding that we must focus our efforts on understanding why Ebola is now a significant threat for great apes in Western and Central Western Africa (Walsh *et al.*, 2003).

Since Butynski's review, conservation funding has also increased. Examples of this abound: the Gorilla Directive sponsored by the U.S. Congress and administered by USAID provided about $4.5 million for gorilla-based conservation; the CBFP initiative, launched by the United States via USAID CARPE funding of more than $45 million for three years, has leveraged matching funds from international NGOs, and will leverage more funding from Japan and other nations of the European Union; GRASP funding is now making its way to the field. In DRC, for example, a new initiative sponsored by UNDP is targeted at rehabilitating protected areas in DRC; and the World Bank, while currently providing conservation and related funding, is preparing a substantial funding intervention to aid several DRC national parks.

Recently, the U.S. Congress reauthorized the Great Ape Conservation Act to be administered by the U.S. Fish and Wildlife Service. Even more funding support is on the way, especially from the European Commission, much of it directed toward overall conservation in those areas still inhabited by gorillas. An optimistic view of these events is that the global community and especially the developed nations have become aware of the conservation crisis in the Congo Basin. This has been facilitated by several large U.S.-based conservation organizations that cooperated together to fast track the CBFP, as well as other important consortia, like the Bushmeat Crisis Task Force and the IUCN. The Congo Basin, and the gorillas living therein, now seem to be on the world's radar screen. It remains to be seen whether there is enough time to substantially reduce the threats and stem the tide of disappearing biodiversity. Fortunately, there is a new generation of African conservationists who have now become intensely involved in this effort, from stakeholders on the ground, to conservationists employed by international and national conservation NGOS, to political policy-makers in African national governments and international organizations. Let us suggest that this is their century. Gorillas are their biological heritage.

Acknowledgments. The author would like to thank the efforts of Peter Walsh and John Oates, who through comprehensive and thorough reviews, substantially improved this chapter. Stuart Nixon, who with his wife, Francine Nixon, led the Maiko South Survey, also reviewed and improved this work. The author would also like to thank Tara Stoinski and Alexia Lewnes for review and commentary. Bradley Mulley kindly provided several of the maps in this work: Figures 1.1, 1.3, 1.4, and 1.8. A portion of the author's salary and all support for DFGFI fieldworkers and partners is provided by Conservation

International, through USAID CARPE funding and through Conservation International's Global Conservation Fund. Further, the surveys described in this chapter, both in Maiko and in Tayna, were supported by funds from the USAID Gorilla Directive, Conservation International's Global Conservation Fund, and USAID CARPE funding from Conservation International, the Louisville Zoo, and the Thorne Foundation. This study would not have been possible without the support of the author's organization, the Dian Fossey Gorilla Fund International. The author also gratefully acknowledges the contributions of the field staff who contributed heavily to data presented in this chapter and are members of UGADEC, the Tayna Gorilla Reserve, and Maiko National Park, ICCN. The 21st century is theirs indeed.

References

Albrecht, G.H., Gelvin, B.R., and Miller, J.M.A. (2003). The hierarchy of intraspecific craniometric variation in gorillas: A population-thinking approach with implications for fossil species recognition studies. In: Taylor, A.B., and Goldsmith, M.L. (eds.), *Gorilla Biology: A Multidisciplinary Perspective.* Cambridge University Press, New York, pp. 62–103.

Ammann, K. (2001). Bushmeat hunting and the great apes. In: Beck, B.B., Stoinski, T.S., Hutchins, M., Maple, T.L., Norton, B., Rowan, A., Stevens, E.F., and Arluke, A. (eds.), *Great Apes and Humans: The Ethics of Coexistence*, Smithsonian Institute Press, Washington D.C., pp. 71–85.

Barnes, R.F.W. (1990). Deforestation trends in tropical Africa. *Journal of Ecology* 28:161-173.

Barnes, R.F.W. & Jensen, K.L. (1987). How to count elephants in the forest. African Elephant & Rhino Specialist Group Technical Bulletin 1:1–6.

Bishikwabo, K. (2001). Gorillas without elephants? A query for research. *Gorilla Journal* 22:17.

Boesch, C., and Boesch-Achermann, H. (1995). Taï chimpanzees confronted with a fatal Ebola virus. *Pan Africa News* 2:2–3.

Butynski, T.M. (2001). Africa's great apes. In: Beck, B.B., Stoinski, T.S., Hutchins, M., Maple, T.L., Norton, B., Rowan, A., Stevens, E.F., and Arluke, A. (eds.), *Great Apes and Humans: The Ethics of coexistence*, Smithsonian Institute Press, Washington D.C., pp. 3–56.

Butynski, T.M., and Kalina, J. (1998). Gorilla tourism: A critical look. In: Milner-Gulland, E.J., and Mace, R. (eds.) *Conservation of Biological Resources,* Blackwell Science, Oxford, pp. 294–313.

Caroll, R. (1988) Relative density, range extension, and conservation potential of the lowland gorilla (*Gorilla gorilla gorilla*) in the Dzanga-Sangha region of southwestern Central African Republic. *Mammalia* 52:309–323.

Casimir, M.J. (1975). Feeding ecology and nutrition of an eastern gorilla group in the Mt. Kahuzi region (Republique du Zaire). *Folia Primatologica* 24:81–136.

Clifford, S.L., Abernethy, K.A., White, L.J.T., Tutin, C.E.G., Bruford, M.W., and Wickings, E.J. (2003). Genetic studies of western gorillas. In: Taylor, A.B., and Goldsmith, M.L. (eds.), *Gorilla Biology: A Multidisciplinary Perspective.* Cambridge University Press, New York, pp. 269–292.

Clifford, S.L., Anthony, N.M., Bawe-Johnson, M., Abernethy, K.A., Tutin, C.E., White, L.J.T., Bermejo, M., Goldsmith, M.L., McFarland, K., Jeffery, K.J., Bruford, M.W., and Wickings, E.J. (2004). Mitochondrial DNA phylogeography of western lowland gorillas (*Gorilla gorilla gorilla*). *Molecular Ecology* 13(6):1551–1565.

Clover, G. (2004). *The End of the Line: How Over-fishing Is Changing the World and What We Eat.* Ebury Press, London.

Doumenge, C. (1998). Forest diversity, distribution, and dynamics in the Itombwe mountains, South Kivu, Congo Democratic Republic. *Mountain Research and Development* 18(3):249–264.

Draulans, D., and Van Krunkoisven (2001) The impact of war on forest areas in the Democratic Republic of Congo. *Oryx* 36(1):35–40.

Eberhardt, L.L. (1978). Transect methods for population studies. *Journal of Wildlife Management* 42: 1-31.

Emlen, J.T., and Schaller, G.B. (1960). Distribution and status of the mountain gorilla (*Gorilla gorilla beringei*), 1959. *Zoologica* 45:41–52.

Fay, J.M. (1989). Partial completion of a census of the western lowland gorilla (*Gorilla g. gorilla* (Savage and Wymand)) in southwestern Central African Republic. *Mammalia* 53(2):203–215.

Fay, J.M., and Agnagna, M. (1992). Census of gorillas in northern Republic of Congo. *American Journal of Primatology* 27(4):275–284.

Garner, K.J., and Ryder, O.A. (1996). Mitochondrial DNA diversity in gorillas. *Molecular Phylogenetics and Evolution* 6(1):39–48.

Ghiglieri, M.P. 1984. The chimpanzees of Kibale Forest: a field study of ecology and social structure. Columbia University Press, New York.

Gonzalez-Kirchner, J.P. (1997). Census of western lowland gorilla population in Rio Muni region, Equatorial Guinea. *Folia Zoologica* 46(1):15–22.

Gray, J.P. (2001) Why our drug laws have failed and what we can do about it: A judicial indictment on the war on the drugs. Temple University Press, Philadelphia.

Groom, A.F.G. (1973). Squeezing out the mountain gorilla. *Oryx* 12:207–215.

Groves, C.P. (1967). Ecology and taxonomy of the gorilla. *Nature* 213:890–893.

Groves, C.P. (1970). Population systematics of the gorilla. *Journal of Zoology* 161: 287–300.

Groves, C.P. (2001). *Primate Taxonomy,* Smithsonian Institution Press, Washington, D.C.

Groves, C.P. (2003). A history of gorilla taxonomy. In: Taylor, A.B., and Goldsmith, M.L. (eds.), *Gorilla Biology: A Multidisciplinary Perspective.* Cambridge University Press, New York, pp. 15–34.

Groves, C.P., and Napier, J.R. (1966). Skulls and skeletons of gorilla in British collections. *Journal of Zoology* 148:153–161.

Groves, C.P., and Stott, K.W. (1979). Systematic relationships of gorillas from Kahuzi, Tshiaberimu and Kayonza. *Folia Primatologica* 32:161–179.

Grubb, P., Butynski, T.M., Oates, J.F., Bearder, S.K., Disotell, T.R., Groves, C.P., and Struhsaker, T.T. (2003). Assessment of the diversity of African primates. *International Journal of Primatology* 24(6):1301–1357.

Hall, J.S., Saltonstall, K., Inogwabini, B.I., and Omari, I. (1998a). Distribution, abundance and conservation status of Grauer's gorilla. *Oryx* 32(2):122–130.

Hall, J.S., White, L.J.T., Inogwabini, B.I., Omari, I., Morland, H.S., Williamson, E.A., Saltonstall, K., Walsh, P., Sikubwabo, C., Bonny, D., Kiswele, K.P., Vedder, A., and Freeman, K. (1998). Survey of Grauer's gorillas (*Gorilla gorilla graueri*) and eastern chimpanzees (*Pan troglodytes schweinfurthi*) in the Kahuzi-Biega National park lowland sector and adjacent forest in eastern Democratic Republic of Congo. *International Journal of Primatology* 19(2):207–235.

Harcourt, A.H. (1996). Is the gorilla a threatened species? How should we judge? *Biological Conservation* 75(2):165–176.

Harcourt, A.H., and Curry-Lindahl, K. (1979). Conservation of the mountain gorilla and its habitat in Rwanda. *Environmental Conservation* 6:143–147.

Harcourt, A.H., and Fossey, D. (1981). The Virunga gorillas: Decline of an "island" population. *African Journal of Ecology* 19:83–97.

Harcourt, A.H., and Groom, A.F.G. (1972). Gorilla census. *Oryx* 11:355–363.

Hart, J.A., and Hall, J.S. (1996). Status of eastern Zaire's forest parks and reserves. *Conservation Biology* 10(2):316–324.

Hart, J., and Liengola, I. (2005). Post conflict inventory of Kahuzi Biega National Park. *Gorilla Journal* 30: 3.

Hart, J.A., and Sikubwabo, C. (1994). Exploration of the Maiko National park of Zaire 1989-1992. Working Paper No. 2. Wildlife Conservation Society, New York.

Horta, K. (1992). Logging in the Congo: Massive fraud threatens the forests. *World Rainforest Report* No. 24.

IUCN/SSC Primate Specialist Group (PSG). (2000). Threatened African Primates.

IUCN/SSC PSG Workshop on Primate Taxonomy. Conservation International, Washington, D.C., Unpublished report.

Jensen-Seaman, M.I., Altheide, T.K., Deinard, A.S., Hammer, M.F., and Kidd, K.K. (2001). Nucleotide diversity at autosomal, Y-chromosomal, and mitochondrial loci of African apes and humans. *American Journal of Physical Anthropology* 32:86.

Kakule, P., and Mehlman, P.T. (2004). Community-based conservation and the creation of protected areas in eastern Democratic Republic of Congo. *Folia Primatologica* 75(S1):286.

Kano, T., and Asato, R. (1994). Hunting pressure on chimpanzees and gorillas in the Motaba River area, northeastern Congo. *African Study Monographs* 15(3):143–162.

Kormos, R., and Boesch, C. (2002). *Regional Action Plan for the Conservation of Chimpanzees in West Africa (Pan troglodytes versus and Pan troglodytes vellerosus).* Conservation International. Washington, D.C.

Laake, J.L., Buckland, S.T., Anderson, D.R., and Burnham, K.P. (1994). *DISTANCE User's Guide V2.1.* Colorado Cooperative Fish and Wildlife Research Unit, Colorado State University, Fort Collins, CO.

Leigh, S.R., Relethford, J.H., Park, P.B., and Konigsberg, L.W. (2003). Morphological differentiation of gorilla subspecies. In: Taylor, A.B., and Goldsmith, M.L. (eds.), *Gorilla Biology: A Multidisciplinary Perspective.* Cambridge University Press, New York, pp. 104–131.

Leroy, E.M., Rouquet, P., Formenty, P., Souquière, S., Kilbourne, A., Froment, J.M., Bermejo, M., Smit, S., Karesh, W., Swanepoel, R., Zaki, S.R., and Rollin, P.E. (2004). Multiple Ebola virus transmission events and rapid decline of central African wildlife. *Science* 303 (5656):387–390.

Lilly, A.A., Mehlamn, P.T., and Doran, D. (2002). Intestinal parasites in gorillas, chimpanzees, and humans at Mondika Research site, Dzanga-Ndoki National Park, Central African Republic. *International Journal of Primatology* 23(3):555–573.

McNeilage, A., Plumptry, A., Doyle, A.B., and Vedder, A. (1998). Bwindi Impenetrable National Park, Uganda Gorilla and Large Mammal Census, 1997. Wildlife Conservation Society, New York, pp. 52.

Mehlman, P. (1986). Male intergroup mobility in a wild population of the Barbary macaque (*Macaca sylvanus*), Ghomaran Rif Mountains, Morocco. *American Journal of Primatology* 10(1):67–81.

Mehlman, P.T. (1989). Comparative density, demography, and ranging behavior of Barbary macaques (*Macaca sylvanus*) in marginal and prime conifer habitats. *International Journal of Primatology* 10(4):269–292.

Mehlman, P.T., and Doran, D.M. (2002). Nest building in western lowland gorillas (*Gorilla gorilla*), Mondika, Central African Republic. *International Journal of Primatology* 23(6):1257–1285.

Morgan, B.J., Wild, C., and Ekobo, A. (2003). Newly discovered gorilla population in the Ebo Forest, Littoral Province, Cameroon. *International Journal of Primatology* 24(5):1129–1137.

Nixon, S.C., Ngwe' E.E., Mufabule, K., Nixon F., Bolamba, D., and Mehlmam, P.T. (2005). The status of Grauer's gorilla in the Maiko South Region: A Preliminary Report. *Gorilla Journal* November.

Oates, J.F., McFarland, K.L., Groves, J.L., Bergl, R.A., Linder, J.M., and Disotell, T.R. (2003). The Cross River gorilla: Natural history and status of a neglected and critically endangered subspecies. In: Taylor, A.B., and Goldsmith, M.L. (eds.) *Gorilla Biology: A Multidisciplinary Perspective*. Cambridge University Press, New York, pp. 472–497.

Omari, I., Hart, J.A., Butynski, T.M., Birhashirwa, N.R., Upoki, A., M'Keyo, Y., Bengana, F., Bashonga, M., and Bagurubumwe, N. (1999). The Itombwe Massif, Democratic Republic of Congo: Biological surveys and conservation, with an emphasis on Grauer's gorilla and birds endemic to the Albertine Rift. *Oryx* 33(4):301–322.

Plumptre, A.J. (2000) Monitoring mammal populations with line transect techniques in African forests. *Journal of Applied Ecology* 37:356–368.

Plumptre, A.J., and Reynolds, V. (1997). Nesting behavior of chimpanzees: Implications for censuses. *International Journal of Primatology* 18(4):475–485.

Redmond, I. (2001). Coltan boom, gorilla bust. *Gorilla Journal* 22:13–16.

Reitbergen, S. (1992). Forest management. In: Sayer, J.A., Harcourt, C.S., and Collins, N.M. (eds.) *Africa: The Conservation Atlas of Tropical Forest*. Macmillan, London, pp. 62–68.

Rose, A.L. (1997). Growing commerce in bushmeat destroys great apes and threatens humanity. *African Primates* 3(1–2):6–12.

Ruvulo, M., Pan, D., Zehr, S., Goldberg, T., Disotell, T.R., and von Durnom, M. (1994). Gene trees and hominid phylogeny. *Proceedings of the National Academy of Sciences USA* 91:8900–8904.

Ryder, O.A. (2003). An introductory perspective: Gorilla systematics, taxonomy, and conservation in the era of genomics. In: Taylor, A.B., and Goldsmith, M.L. (eds.) *Gorilla Biology: A Multidisciplinary Perspective*. Cambridge University Press, New York, pp. 239–246.

Ryder, O.A., Garner, K.J., and Burrows, W. (1999). Non-invasive molecular genetic studies of gorillas: Evolutionary and systematic implications. *American Journal of Physical Anthropology* 28:238.

Sarmiento, E.E. (2003). Distribution, taxonomy, genetics, ecology, and causal links of gorilla survival: The need to develop practical knowledge for gorilla conservation. In: Taylor, A.B., and Goldsmith, M.L. (eds.), *Gorilla Biology: A Multidisciplinary Perspective*. Cambridge University Press, New York, pp. 432–471

Sarmineto, E.E., and Butynski, T.M. (1996). Present problems in gorilla taxonomy. *Gorilla Journal* 12:5–7.

Sarmiento, E.E., and Oates, J.F. (2000). The Cross River gorillas: A distinct subspecies, *Gorilla gorilla diehli* Matschie 1904. *American Museum Novitates* 3304:1–55.

Sarmiento, E.E., Butynski, T.M., and Kalina, J. (1996). Gorillas of Bwindi-Impenetrable Forest and the Virunga Volcannoes: Taxonomic implications of morphological and ecological differences. *American Journal of Primatology* 40(1):1–21.

Schaller, G.B. (1963). *The Mountain Gorilla: Ecology and Behavior.* University of Chicago Press, Chicago, pp. xvii, 431.

Seaman, M.I., Saltonstall, K., and Kidd, K.K. (1998). Mitochondrial DNA diversity and biogeography of Eastern gorillas. *American Journal of Physical Anthropology* 26:199.

Sholley, C.R. (1991). Conserving gorillas in the midst of guerrillas. *AAZPA Annual Conference Proceedings*, pp. 30–37.

Sivalingana-Matsitsi, D., Kamal-Piliplili, M., Eliba-Omba, B., N'Sasi-Soki, M., Mutakilo-Ngoy, E., Mutsorwa, B., Kambale-Kioma, G., Mufabule, K., Paluku-Mukakirwa, A., Kakule, P., and Mehlman, P.T. (2004). Redefining the geographical distribution of Grauer's gorilla in eastern Democratic Republic of Congo. *Folia Primatologica* 75(S1):334.

Sunderland-Groves, J.L., Oates, J.F., and Bergl, R. (2005) Box 7.1 The Cross River gorilla (*Gorilla gorilla diehli*). In: Caldecott, J., and Miles, L. (2005) *World Atlas of Great Apes and Their Conservation.* Prepared at UNEP-WCMC. University of California Press, Berkeley.

Struhsaker, T.T. (1996). A biologist's perspective on the role of sustainable harvest in conservation. *African Primates* 2:72–75.

Tegera, A. (2002) The coltan phenomenon: How a rare mineral has changed the life of the population of war-torn North Kivu province in the East of the Democratic Republic of Congo. Pole Institute, Goma.

Teleki, G. (1997). Human relations with chimpanzees: A proposed code of conduct. *African Primates* 3:35–38.

Thalmann, O., Hebler, J., Poinar, H.N., Paabo, S., and Vigilant, L. (2004). Unreliable mtDNA data due to nuclear insertions: a cautionary tale from analysis of humans and other great apes. *Molecular Ecology* 13(2):321–335.

Tutin, C.E.G., and Fernandez, M. (1984). Nationwide census of gorilla (*Gorilla g. gorilla*) and chimpanzee (*Pan t. troglodytes*) populations in Gabon. *American Journal of Primatology* 6(4):313–336.

Tutin, C., Stokes, E., Boesch, C., Morgan, D., Sanz, C., Reed, T., Blom, A., Walsh, P., Blake, S., and Kormos, R. (2005). *Regional Action Plan for the Conservation of Chimpanzees and Gorillas in Western Equatorial Africa.* Conservation International, Washington D.C.

Vedder, A., Aveling, C., and Condiotti, M. (1986). Census update on Virunga gorilla population (*Gorilla gorilla beringei*): 1986. *Primate Report* 14:199.

Walsh, P.D., Lahm, S.A., White, L.J.T., and Blake, S. (unpublished manuscript). Methods for monitoring African apes: some cold, hard facts about bias.

Walsh, P.D., Abernethy, K.A., Bermefo, M., Beyers, R., De Wachter, P., Akou, M.E., Huijbregts, B., Mambounga, D.I., Toham, A.K., Kilbourn, A.M., Lahm, S.A., Latour, S., Maisels, F., Mbina, C., Mihindou, Y., Obiang, S.N., Effa, E.N., Starkey, M.O., Telfer, P., Thibault, M., Tutin, C.E.G., White, L.J.T., and Wilkie, D.S. (2003). Catastrophic ape decline in western equatorial Africa. *Nature* 422(6932):611–614.

Watts, D.P. (1991). Mountain gorilla reproduction and sexual behavior. *American Journal of Primatology* 24(3/4):211–225.

Weber, A.W. (1987). Socioecologic factors in the conservation of afromontane forest reserves. In: Marsh, C.W., and Mittermeier, R.A. (eds.) *Primate Conservation in the Tropical Rain Forest*. Alan R. Liss, New York, pp. 205–229.

Weber, A.W., and Vedder, A. (1983). Population dynamics of the Virunga gorillas: 1959–1978. *Biological Conservation* 26(4):341–366.

White, L., and Edwards, A. (2000). *Conservation Research in the African Rain Forests: A Technical Handbook*. Wildlife Conservation Society, New York, pp. xvi, 454.

Wilkie, D.S. (2001). Bushmeat trade in the Congo Basin. In: Beck, B.B., Stoinski, T.S., Hutchins, M., Maple, T.L., Norton, B., Rowan, A., Stevens, E.F., and Arluke, A. (eds.), *Great Apes and Humans: The Ethics of Coexistence*. Smithsonian Institute Press, Washington D.C., pp. 86–109.

Williams, J. (2000). The lost continent: Africa's shrinking forests. In: Ammann, K. (ed.) *Bushmeat: Africa's Conservation Crisis*. World Society for the Protection of Animals, London, pp. 8–15.

Woodford, M.H., Butynski, T.M., and Karesh, W.B. (2002). Habituating the great apes: The disease risks. *Oryx* 36(2):153–160.

Yamagiwa, J. (2003). Bushmeat poaching and the conservation crisis in Kahuzi-Biega National Park, Democratic Republic of the Congo. *Journal of Sustainable Forestry*, 16:115–135.

Yamawiga, J.N., Mwanza, N. Spangenberg, A., Maruhashi. T., Yumoto, T, Fisher, A., and Steinhauer-Burkart, B. (1993). A census of the eastern lowland gorillas (*Gorilla gorilla graueri*) in Kahuzi-Biega National Park with reference to mountain gorillas in the Virunga Region, Zaire. *Biological Conservation* 64:83–89.

Yapp, W.B. (1956) The theory of line transects. *Bird Study* 3:93–104.

Section 2
Approaches—On the Ground

Chapter 2
Conservation Medicine for Gorilla Conservation

Mountain Gorilla Veterinary Project (MGVP, Inc.)
Wildlife Conservation Society (WCS)

1. Introduction

This chapter will discuss historical, present, and future approaches of veterinary science and medicine in the context of gorilla conservation and conservation medicine. The emphasis is placed on the mountain gorilla due to the intensity of study this subspecies has received. Disease is now ranked the third most serious threat to the sustainability of gorilla populations in general, and in areas of protected habitat it is considered the primary threat.

There is greater interaction, contact, and confrontation with wildlife as human populations, with their associated agricultural practices and domestic animals, grow and consume and/or utilize more natural resources. Global environmental changes and this human expansion have led to the emergence of new diseases and new host susceptibility to old pathogens as species that have never come into close contact before are forced into novel relationships (Wolfe *et al.*, 1998). This phenomenon presents increased challenges to health management (conservation medicine) of wildlife and their habitats.

Conservation medicine can be defined as the medical practice that seeks to ensure ecological health and well-being of a defined habitat. In terms of veterinary science, and medicine in a broad sense, it is the study of the pathogen flow and interventions to reduce pathogen exchange among wildlife, humans, and domestic animals. Since disease pathogens play important ecological roles, a particular medical problem in a species of interest must be viewed in the context of conspecifics and habitat quality in order to define health for an ecosystem. Veterinary input into such conservation efforts has expanded greatly in the last decade (Karesh and Cook, 1995).

Several key aspects of medicine with respect to gorilla conservation are discussed in this chapter, including:

A) The different philosophical approaches to clinical and preventive veterinary management of large gorilla populations such as *Gorilla gorilla gorilla* and *Gorilla beringei graueri*, and the small populations of *Gorilla gorilla delhi* and *Gorilla beringei beringei*.

B) The wildlife/domestic, animal/human interface and the need for collaboration between the wildlife veterinarian, local practitioners, and human health experts.

C) The consequences of recent human interactions with gorilla populations, including habituation, poaching and resulting orphans, and ecotourism.

D) The modern approaches to collection, storage, and analysis of biological samples.

E) The development of information systems, including electronic databases, that will allow the rapidly expanding medical knowledge of gorillas to be analyzed and integrated into computer modeling to aid with sound management and evidence-based policy formation.

2. Health Objectives Common to all Gorilla Subspecies

Health objectives common to all gorilla subspecies include:

A) The development of contingency plans involving all the conservation partners to reduce the potential devastating effects of disease epidemics in gorillas and other wildlife populations.

B) The methodical and consistent collection of baseline health data and information (such as the prevalence and incidence of diseases present in a population) so future health changes, due to human activities or other factors, can be evaluated.

C) The development of research programs and data analysis to gain a clear understanding of the relationship between human and domestic animal health and the health of gorillas and other wildlife in different habitats.

D) The formation of a professional and holistic collaborative approach among conservation partners and government agencies to address health concerns.

3. Health Management and Conservation Medicine Practices for Gorillas

3.1. *Large Versus Small Gorilla Population Health Management*

The health management of large populations of wildlife, such as gorillas, requires a baseline understanding of diseases and health abnormalities present in populations that appear to be viable and demographically healthy (Karesh *et al.*, 1998; Kilbourn *et al.*, 2003). In the larger populations, such as the western lowland gorilla, where management focuses on the population more than the individual, data is gathered via both opportunistically and proactively scheduled sample collections that may be either noninvasive or invasive in nature. The benefit of acquiring the samples is worth the risk to the population. Invasive samples are gathered from clinically normal individuals for the sole purpose of data that reflect the health status of, and allow for informed management of, the overall population. Preventive health programs are based on this information and directed to protect the health and viability of the populations as well as the ecosystem. Conversely, the management of small populations, such as the mountain gorilla, where each individual is an integral part of the population's sustainability, proactive sampling is very restricted with respect to invasive collection of samples (Cranfield *et al.*, 2002). Limited data on healthy individuals is gathered only opportunistically. Most of the data from invasive sampling are from sick or injured individuals that may not be representative of the overall population. With the mountain gorillas, in contrast to the western lowlands, a large percentage of time and resources is spent monitoring and treating individual animals to enhance the population numbers.

3.2. Health Management of the Western Lowland Gorilla

For western lowland gorillas, a regional, standardized approach is being implemented in the Central African Republic, the Republic of Congo, and Gabon. To date, dozens of park guards, park managers, and researchers have been trained in concepts of infectious diseases, basic health surveillance methods, record keeping, and preventive medicine. Extensive protocols standardizing invasive and noninvasive testing procedures, record keeping, monitoring techniques and procedures, and human health-related issues have been distributed to sites with western lowland gorillas (full protocols are available at http://www.fieldvet.org/). Community outreach programs, including education on the risks of disease transmission between animals and humans, have begun at some sites and are being expanded to other areas. Several months following these educational programs, surveys conducted in these villages suggested that the hunting and consumption of non-human primates declined (Karesh, personal communication). Interventions involving immobilizations and physical examinations of "normal" western lowland gorillas have provided initial baseline information on their exposure, or lack of exposure, to common infectious diseases. "Apparently normal" gorillas had a wide variety of exposure to infectious agents, including yellow fever virus, *Treponema* sp. (Yaws), influenza virus, etc. (WCS, unpublished data). Knowledge of exposure to seemingly less significant viruses can have very meaningful implications. If an outbreak develops, knowing which infectious agents healthy gorilla populations have been exposed to prior to the outbreak may help to rule out these agents as a cause of concern. All western lowland gorillas tested had antibody titers to adenovirus. This information from banked serum samples has been critical in evaluating the potential for using an Ebola hemorrhagic fever virus (EHFV) vaccine in development that is on adenovirus vector (i.e., the gorilla's immune systems may neutralize the vaccine, making it ineffective).

3.3. Health Management of the Mountain Gorilla

The small populations of mountain gorillas in the Virungas and the Bwindi Impenetrable Forest, number approximately 300–350 individuals in each habitat. They are the most intensely studied and habituated gorilla populations, with the highest density of humans and associated land use around their very limited habitats. Studies have shown that each individual's genetic makeup is extremely important to the viability of the population (Garner and Ryder, 1996). The approach to conservation medicine with the mountain gorillas is different from other gorilla populations because of the closely monitored activities and health of gorilla individuals, and the fact that habituated individuals can be clinically managed with less risk to the veterinarian or animal than is the nonhabituated gorillas. The Mountain Gorilla Veterinary Project (MGVP, Inc.) approaches health management from both a population management level as well as individual clinical cases.

MGVP, Inc. was started in 1986 at the request of Dian Fossey, who had seen the mountain gorilla population decline in the Virungas to an estimated 260 individuals, in part due to human-induced snare wounds, trauma, and illness. MGVP, Inc., is thought to be the first health care initiative responsible for *in situ* clinical care of individuals of a wild population through its clinical medicine and pathology programs. It has been credited as one of the efforts responsible for the increase in the number of gorillas in recent years (Butynski and Kalina, 1998). Veterinarians routinely visit habituated groups to evaluate clinical signs and occasionally treat animals when life-threatening or human-induced health problems occur. Biological samples are collected invasively only during interventions for health problems, and therefore represent a biased subset of the population, making it difficult to establish normal baseline health parameters.

3.4. Clinical Medicine of Mountain Gorillas

Few, if any, wildlife species are given the degree of clinical veterinary care that habituated mountain gorillas receive. While clinical medicine is not without technical, logistical, and political difficulties, it is generally feasible with habituated gorillas. The interests of the multiple stakeholders involved in gorilla conservation, combined with the widely accepted tenet of not interfering with natural processes, often complicate or even abrogate possible clinical procedures. However, well-designed policies and protocols, proper contingency planning, and excellent communication allow for successful clinical interventions when deemed necessary.

Once decisions are agreed upon, the inherent difficulties with obtaining and organizing equipment, transportation, and personnel, which are true of any clinical wildlife procedure, must be overcome. In addition, the rugged habitat, unpredictable climate, and political insecurity in most gorilla range countries make field procedures more difficult. For established projects like the MGVP, Inc., most of this preparation has become routine, but individual procedural preferences and assistant personnel instructions should always be clearly communicated or reinforced before every procedure. Moreover, it is impossible to fully predict or anticipate the exact situation an intervention team will encounter with wild gorillas, and all parties must, therefore, maintain a fair degree of flexibility.

There are two broad categories of interventions: those involving immobilization of a gorilla or gorillas, and those in which gorillas are not immobilized. The first category obviously encompasses procedures requiring hands-on veterinary care, like snare removals or surgical wound treatment, as well as rare events, like obtaining samples to confirm certain diagnoses. Occasionally, female gorillas have to be immobilized for procedures involving their infants. This category would also include routine examinations and sample collection for background data and sentinel health monitoring that have been discussed for other gorilla populations. Anesthetic procedures usually carry a higher number of risks to both the gorillas and the people involved and are therefore avoided when non-immobilization methods are possible. Situations that might allow nonimmobilization interventions include treatment of various infections (e.g., bacterial upper respiratory infections or parasitic skin infections) or very rarely for prophylactic protection from certain diseases by vaccination (e.g., measles). These procedures are dependent on the ability to deliver antibiotics, antiparasitics, or vaccines by remote injection with darting equipment. It must be kept in mind that these situations usually require a measure of certainty of the diagnosis as well as the safety of the agents used and are therefore commonly follow-up procedures after gorillas have been previously immobilized and samples have been analyzed.

Regardless of the exact nature of the procedure, virtually all interventions involve darting at least one gorilla, either with an immobilization agent or a treatment drug. Proper and safe darting requires a fair degree of experience with gorilla behavior and should always involve the human personnel who best know the individual gorillas involved. It is usually preferable to dart the target gorilla when it is at least somewhat isolated from the others, though this is unfortunately not always possible. Most of the agents chosen for use are designed for intramuscular injection, wide margin of safety, quick inductions, and recovery. Because doses can be large and dart accuracy variable, darts are usually aimed at the large muscle groups in the upper arms and legs or sometimes the lateral back muscles if the gorillas are large enough. It cannot be overstated that many gorillas can and will react aggressively when seeing even a small dart gun barrel, and all parties must be for such reactions. Likewise, gorillas will defend group members, especially when they are ailing, falling under the effects of immobilization agents, or being approached or handled by people. Therefore, part of the immobilization team will be dedicated solely to driving and keeping away any uninvolved gorillas.

Because immobilizations are rare events and there is little historical data on gorilla baseline health, these procedures are usually used to the fullest extent for diagnostic specimen collection. Routine protocols involve collection of blood, fecal, urine, and hair samples. In some cases, skin scrapings and/or biopsies are obtained along with any apparent ectoparasites. When timely analysis is possible, bacterial culture swabs are collected from wounds or abnormal discharges and exudates and sometimes from throats or noses to document normal flora. Portable anesthetic blood monitors, blood chemistry analyzers, and field microscopes can provide some analysis during interventions, and advances in immunogenic and molecular detection systems show promise for detection for limited disease agents. In general, however, complete diagnostics cannot be performed in the field. In the cases of snare removals or wound treatments this information is usually ancillary. Biological samples from individual gorillas help in three ways: 1) to diagnose the immediate clinical problem, 2) to assess the health status of the group from which the individual's samples have been taken, and 3) to bank for future research to monitor the health status of the population over time and events (Lehn, Chapter 6, this volume).

Bacterial cultures and antibiotic sensitivities would be beneficial for antibiotic treatment decisions on severely infected wounds or respiratory infections, but since multiple treatments per day or even daily treatments are impractical, the logistics dictate the use of long-acting antibiotics where possible and limit the selection to broad-spectrum drugs anyway.

If bacterial cultures reveal agents that might not be susceptible to these drugs, tough decisions must be made on the appropriate use of antibiotic therapy. It is generally possible and practical to treat individual gorillas by darting once, or even every 3–5 days, which the long-acting drugs allow. If, however, drugs that necessitate daily dosing are required, it may be better to opt for not treating rather than risk giving improper treatment regimens that might increase bacterial resistance. In the management of previous scabies outbreaks, a similar philosophy has held true. After studying the safety of a long-acting antiparasitic in captive monkeys, it was used because it remained effective long enough to kill any parasites that hatched after 10 days. It allowed time to treat all members of the group before the effects wore off from those first treated. The gorillas were thus unable to reinfect each other, which is a potential complication of infections like mange.

From 2001 through 2005, MGVP, Inc. conducted more than 60 gorilla treatments without immobilizations (e.g., given antibiotic or antiparasitic drugs via dart without anesthetizing gorillas), 22 full immobilizations of 27 gorillas (six mothers had babies that were worked on and/or partially sedated), and 29 necropsies/autopsies. These numbers are relatively high with respect to historical activities of the project and probably account for at least a third of the hands-on medical procedures since 1986.

Illnesses and injuries are now comprehensively and systematically monitored. Generally speaking, MGVP, Inc. observes injuries (cuts, wounds, etc.) around once a week and respiratory "illnesses" (coughing, snotty noses, etc.) about once a month and usually at a group level (e.g., many gorillas being sick). Other illnesses (e.g., scratching, scabies, diarrhea) are seen much less often, and snares are observed around 3–5/year in all four parks. These numbers will probably rise as securing and monitoring improve in DRC. While it is obvious that monitoring is important for detecting ailments and for proper follow-up on treated or sick animals, routine health monitoring is also essential to know the regular cycles and patterns of illness.

3.5. Pathology

3.5.1. Clinical Pathology

Clinical pathology is defined as the methods and procedures of analyzing biological samples that pertain to the prevention or diagnosis of a disease and the care of patients. This analysis is particularly important when working with wildlife like gorillas because diagnoses are otherwise only possible by visual examination from a distance or physical examination under anesthesia, which can alter physiological parameters.

A very important aspect of clinical pathology is examining serum titers or antibody levels that help to define exposure to selected organisms. This may indicate the vulnerability of a population and help assess the level of risk if exposed to a particular disease. When these titers are compared with those of the human population with potential exposure to gorillas, diseases of special interest or high risk can be better defined.

The importance of serology became evident during a respiratory outbreak in 1989, when rising titers indicated measles as the pathogen and vaccination of affected populations with a measles vaccine slowed the spread of the disease (Hastings et al., 1991).

3.5.2. Gross Examination and Histopathology

One of the most important tools for managing a wild population, large or small, is the understanding of the causes of morbidity and mortality within that population. Data from post-mortem examinations provides much of this information. The post-mortem consists of a gross examination in the field where organs are evaluated in situ. With the advent of digital cameras, visual information can be easily distributed to specialized pathologists in distant laboratories for confirmation of lesions and diagnosis. Representative samples of organ systems as well as lesions are collected for histopathology and possible culture, and toxicology. In 30–40% of the cases, the cause of death is diagnosed on gross post-mortem. Post-mortem examinations of western lowland gorilla populations are conducted by trained local field teams (for protocols, see http://www.fieldvet.org/ and http://mgvp. cfr.msstate. edu/mgvp/mgvp.htm). These examinations have helped demonstrate for the first time that lowland gorillas and some duiker species were in fact dying from EHFV (Leroy, et al., 2004).

Because of the intense monitoring, a high proportion of mountain gorillas that die are usually found within a 48-hour time period, which allows veterinarians to perform more informative necropsies and recover quality tissue samples. Even when the cause of death is obvious, such as a gunshot wound or severe trauma from intergroup aggression, a post-mortem can reveal incidental findings, such as subclinical infections, that can lead to management decisions affecting the rest of the gorilla population.

MGVP's pathology program (100 complete cases as of this writing; Table 2.1) has revealed trauma as the major cause of mortality in every age group. For infants, the primary type of trauma is infanticide (13/15), while for juveniles (7/9) and adults (15/16) direct or indirect poaching is the main type of trauma. Respiratory disease is the second most common cause of death and, in this data set, affects all age classes equally. Respiratory problems are the most common infectious cause of mortality, which corresponds with respiratory disease outbreaks as the major clinical problem seen with mountain gorillas (MGVP, unpublished data). Historically, MGVP field veterinarians have found that the morbidity is high and mortality is moderate. Evidence supports the clinical impression that the majority of respiratory outbreaks have a primary viral etiology that predisposes gorillas to secondary bacterial infections. Mortality from these outbreaks can be reduced by the appropriate use of antibiotics and/or vaccines. For cases in which the cause of death is undetermined (the third most common category), infants are usually

TABLE 2.1. Causes of death by age class.

Cause	Infant (birth to < 3 years)	Juvenile (3 to <10♀ and 13♂)	Adult (≥10♀ and 13♂)	% of total
Trauma	15	9	16	40%
Respiratory	8	6	10	24%
Undetermined	9	1	7	17%
Multifactorial	1		4	5%
Gastrointestinal	1	1	2	4%
Metabolic	1	1	1	3%
Cardiac			3	3%
Infectious - other			1	1%
Developmental	1			1%
Neurologic	1			1%
Parasitic	1			1%
Total	38	18	44	100%

suffering from decomposition that hampers diagnosis, and adults, especially aged individuals, frequently have multiple subclinical and/or presumably chronic processes, the impacts of which are difficult to assess in the absence of clinical laboratory data. Incomplete histology and autolysis also contributed to this category.

One of the purposes of a pathology program is to find patterns of mortality in order to effectively concentrate resources on addressing significant problems. The deaths in the mountain gorilla population have had a bi-modal age distribution, with most of the deaths occurring in the very young and very old animals (Lowenstine, unpublished data). The deaths in the older animals are often due to natural age-related, chronic problems. These problems are usually untreatable in free-ranging animals and fall outside the mission of MGVP, Inc. As with most natural animal populations and human populations with little health care, infant mortality is significant. While some causes of infant mortality can be addressed through management at a population level (e.g., decreasing poaching, minimizing infectious respiratory outbreaks), treatment of individual cases is difficult. Infant gorillas that are too small to dart safely are generally in physical contact or within arm's reach of their mother, which means that mothers need to be immobilized to examine and treat infants. The potential for other gorillas to carry off infants of immobilized mothers further complicates potential interventions. Rapid disease progression leading to sudden or peracute death is more common in infants, which often means that clinical signs are not observed or that infants die before treatment can be rendered. The possibility that the majority of infant mortality is from natural causes such as infanticide raises management issues of whether or not it is appropriate to intervene and what level of management is acceptable.

Continued investigation of gorilla morbidity and mortality through detailed post mortem examinations will improve our knowledge of the disease processes and helps to correlate these findings with ante-mortem clinical signs so that we can more effectively intervene when problems occur.

4. Human Disease as a Threat to Gorilla Survival

Recent (post-1960) circumstances have increased the extent to which humans control the fate of mountain gorillas. As an example, human population growth rates in the areas in which mountain gorillas live are very high (2–3% per annum) and the population density surrounding some areas is among the highest in the world (reaching as high as 807/km^2 around the Virunga Mountains) (The World Bank, 2003). Animal husbandry practices of the ever-increasing local community also inevitably affect the wildlife inside the adjacent parks.

Because humans and gorillas are genetically similar, sharing over 97% of their genetic makeup, gorillas are susceptible to many human infectious diseases (Sibley and Ahlquist, 1984). Prior to relatively recent habituation for research and tourism purposes, and encroachment of human settlements on gorilla habitat, humans and gorillas rarely spent time in close *physical* proximity. Gorillas, therefore, may represent a potentially "naïve" population (without acquired immunity) for some infectious organisms found in human populations. Organisms introduced into susceptible, naïve populations cause higher morbidity and/or mortality than in populations that have built resistance to a disease (either through endemic disease or immunization) (McCallum and Dobson, 1995). Consequently, human beings are potential carriers of infectious diseases that could have devastating effects on gorillas. With reductions in poaching and some control over cattle grazing in the protected areas, infectious disease has since been identified as one of the *major* risks to the remaining populations of mountain gorillas (Foster, 1993; Werikhe *et al.*, 1998). Infectious diseases of concern that could be transmitted from humans to gorillas (anthropozoonosis) include those spread through respiratory modes, such as measles, tuberculosis, and influenza, and diseases spread by fomites (inanimate objects) or fecal-orally, like poliomyelitis, shigellosis, mange helminthiasis and viral hepatitis (Homsy, 1999; Whittier *et al.*, 2001). Evidence supporting interspecies pathogen transmission comes mainly from studies of captive animals (Ott-Joslin, 1993). Data on disease or pathogen transmission in the wild is scant, due to the inherent difficulties of collecting data in a rigorous way that would provide more convincing evidence.

4.1. Groups of Humans Posing Potential Disease Threat

Humans who could potentially pose a health threat to gorillas can be divided into four groups, defined by their differing levels of potential exposure and the types of health interventions possible to minimize such exposure: 1) gorilla conservation workers, 2) tourists, 3) locals from communities surrounding protected areas, and 4) illegal extant populations.

4.1.1. Conservation Workers

Protected area employees include park managers and office staff, trackers, porters, guides, researchers, and veterinarians. Office-based staff likely have very limited, if any, exposure to gorillas, but in areas with habituated populations, trackers, porters, guides, and researchers may come into close physical proximity to gorillas on a daily basis. Veterinarians who provide clinical care to sick or injured individuals have less frequent but direct contact with gorillas during clinical interventions.

Providing preventive, diagnostic, treatment, and referral health services to gorilla conservation workers is a logical strategy for minimizing the risk of infectious disease transmission between this group of humans and gorillas with which many workers have daily contact. Such an occupational health program also reduces the risk of disease among the workers, an equally important program objective. Protocols for the types of health services that should be provided to workers coming into close contact with animals are available for personnel working with captive animals in developed country settings (Silberman, 1993), and recommendations for which services should be provided to employees working with wild animals have been proposed (Nutter and Whittier, 2001).

MGVP, Inc. initiated an Employee Health Program (EHP) in 2001 in Rwanda. EHP interventions during 2001, 2002, and 2003 involved medical history taking, a clinical exam, lab tests (on sputum, feces, blood, urine), prophylactic immunization (for tetanus), prophylactic treatment with mebendazole and metronidazole (for intestinal worms), treatment for acute diseases, referral for employees with more complicated/chronic diseases, and clinical follow-up care for anyone diagnosed during one of the clinical exams.

Recent data from MGVP's Rwanda employee health program (MGVP Employee Health Group, 2004) and bio-banked mountain gorilla blood samples from Rwanda and Uganda demonstrated that there is an overlap in the viruses for which one or more individuals from the human and gorilla groups tested antibody positive. This could represent cross-reactivity for similar viruses or positivity for the same viruses in the two species. Analysis of fecal samples from employees (2001–2003) revealed that one or more employees tested positive for helminths and enteric parasites (e.g., *Ascaris, Trichuris, Cryptosporidium, Giardia lamblia*, and hookworm).

As part of the EHP, employees are asked questions regarding potential job-related risk factors for disease. Data from 2002 revealed that the main risk factor for testing positive for any fecal pathogenic organism was use of a pit latrine at home. Public health sanitation interventions such as ventilated pit latrines and hygiene education would likely go a long way toward reducing the potential risk of oral-fecal disease transmission between gorillas and conservation workers. Extending health program benefits to family members, especially their children, would potentially reduce risk of pathogen transmission even more as children are often the ones at highest risk for many of the infectious diseases potentially transmissible to mountain gorillas.

4.1.2. Tourists

Tourism to view the gorillas first began in Uganda in the 1960s on a small scale and in Rwanda in 1978. Tourism has since become an important source of foreign revenue for mountain gorilla range countries (Butynski and Kalina, 1998), with the potential of several thousands of tourists a year visiting the gorillas. The frequent close contact by groups of tourists may increase stress, disturb gorilla behavior, and/or pose direct health risks.

A study conducted by Adams and Sleeman (1999) provides insights into the risk that this human group could pose to the health of great apes, including mountain gorillas. In that study, a self-administered questionnaire was filled out by tourists who visited Kibale National Park, Uganda, to view chimpanzees habituated for tourism. Almost 30% (12/43) reported that they were diagnosed with one of the diseases listed recently enough to be considered infectious at the time of the chimpanzee visit. In another study involving 21 guides, trackers, and rangers in Uganda and Rwanda (Homsy, 1999), 50% of the respondents indicated that maintaining tourists at a distance of more than 5 m was the regulation most difficult to enforce (often due to the gorillas approaching the tourists). While these two studies are not representative of all tourists visiting mountain gorillas, they provide some evidence to suggest that this human group could potentially pose a health risk to the animals being visited.

A report by Homsy (1999) summarizes numerous strategies for minimizing disease risk, and the evidence base supporting these recommendations. While promotion of many of these regulations has been in effect for a while, enforcement is not always easy. Awareness-raising among tourist agents and tourists themselves via brochures is one of the various strategies being used to assist with regulation compliance.

4.1.3. Local Human Communities

Gorillas have survived until recently in areas relatively inaccessible to humans. However, human settlements in many places now extend right up to the boundary of protected gorilla habitats and are expanding in unpro-tected gorilla habitat. The lack of sanitation services in these areas means household refuse and human waste are disposed of on-site. Trash piles or open pits and latrines (sometimes covered and sometimes not) are usually in proximity to farmland. Some gorillas that are less fearful of humans are attracted to the crops and refuse as an alternative or supplemental food source, and interactions between humans and gorillas at this intersection are increasing. This situation poses health risks to the gorilla from direct and indirect pathogen transmission (e.g., through fomites and vectors such as rats), and also potential exposure to environmental toxins and bodily injury during physical encounters. It also poses health, social, and economic risks to humans, ranging from reduced income due to crop destruction, to potential for pathogen transmission from gorilla feces or bites, to school dropouts due to the need to stay home to protect the family's crops from being raided.

Various studies support the suggestion that organisms are being transmit-ted among gorillas and community members living in close proximity. For example, MGVP has shown that *Cryptosporidium, Microsporidium,* and *Giardia* from gorillas, people, and cattle in the Bwindi area are genetically identical in the DNA sequences studied to date (Graczyk *et al.,* 2002; Nizeyi *et al.,* 2002a,b). A review of fecal studies in the *Gorilla beringei beringei, Gorilla beringei graueri,* and *Gorilla gorilla gorilla* indicates that some helminths and protozoa have a higher prevalence in mountain gorillas, possibly as a result of higher local density of, and therefore exposure to humans. However, possible differences in sampling and lab testing methods must also be considered as explanatory factors. A study of scabies in gorillas in Bwindi Impenetrable National Park in Uganda (Kalema *et al.*, 1998; Graczyk *et al.*, 2000) is highly suggestive of transmission to the gorilla population from the local human community. This is further supported by the results of comparative genotyping of mites obtained from gorillas and local humans that suggest the parasites to be genetically the same (Graczyk, personal communication).

Hygiene, sanitation, and other behaviors of community members determine the degree of risk associated with this group. For long-term effects, targeted and coordinated efforts among government and NGO health, environment, and community development agencies working in the area are needed. A pilot project between MGVP, Inc. and DFGF-I in Ruhengeri, Rwanda, involving fecal testing and treatment for community members exemplifies this kind of collaborative effort.

4.1.4. Illegal Activities of Local Populations

Although the habitat in which many mountain gorillas live is now protected area, humans continue to make use of the land and its natural resources, including the wildlife. The human populations near these protected areas often have reduced access to basic services, such as health care and schooling, making them some of the country's most marginalized citizens. They often still depend upon local natural resources to meet their most basic needs for food, clothing, building material, medicine, and disposable income. Hunting, therefore, still goes on illegally in the protected areas and, in addition to direct poaching, hunters potentially expose gorillas to disease from open latrines or uncovered refuse. During times of political instability, military personnel, or rebels may take refuge in or near protected areas. Thousands of refugees camped close to and accessed the park after being displaced in Rwanda in the mid-1990s. The park's natural resources also lure people from other areas of the country or neighboring countries who then illegally hunt, log trees, or collect things like plants or honey. In the process, they also potentially expose gorillas to infectious disease.

No data exist in the literature on the transmission of disease from extant populations due to the problems of monitoring illegal activities. However, because of their existence in the forest, this population presents a potential disease source. Interventions to improve the health of the whole community will help, as many in this group come from local communities. Awareness-raising about the risk of interspecies transmission with community members could increase self-vigilance and community pressure against members who illegally enter and exploit park resources. Ultimately, the latter approach is the only one that will ensure protection of gorilla populations over the long term.

5. Domestic Animal Disease as a Threat to Gorilla Survival

Just as gorillas are at risk of contracting transmissible diseases from neighboring human populations, associated livestock populations represent another potential source of pathogens. The intense utilization of lands bordering the protected areas allows dynamic physical and ecological interactions between livestock and mountain gorillas. Political instability in the region has contributed to inadequate routine testing, vaccination and treatment of livestock, and to cross-border movements of people, livestock, and their pathogens. Groups of mountain gorillas habituated for tourism and research in Rwanda and Uganda spend significantly more time in agricultural lands outside the park than do unhabituated groups. These forays can last from a few hours up to 8 months and result in dietary changes and reduced daily ranges (Madden, 1998; Butynski, 2001; Goldsmith, 2000; Whittier and Nutter, personal observations). Gorillas outside the park are often in proximity with grazing cattle, sheep, goats, and poultry, which puts them at risk for contracting livestock diseases either directly or via contaminated environments.

Pathogens of concern include a range of bacteria, viruses, and parasites. Previously mentioned studies have documented the presence of the gastrointestinal parasites *Giardia*, *Microsporidia*, and *Cryptosporidium* species that appear genetically identical in gorillas, cattle, and people near the border of the Bwindi Impenetrable Forest, Uganda (Graczyk *et al.*, 2002, Nizeyi *et al.*, 2002a,b). This is objective evidence that pathogens flow among these populations, and contaminated water sources may play a role in pathogen spread since all are commonly waterborne parasites. Tuberculosis and brucellosis have been devastating for wildlife populations in numerous countries throughout the world. The high regional prevalence of *Mycobacterium* and *Brucella* in central African livestock raises concern for the interactions among cattle, sheep, goats, and mountain gorillas. In Uganda and Rwanda, surveys have reported seroprevalence of brucellosis as high as 35% in cattle, and 13–35% in goats (Onekalit 1987; Akakpo *et al.*, 1988; Kabagambe *et al.*, 2001;), with no data available for sheep. In Rwanda, reported tuberculosis prevalence in cattle is 11% (Kabagambe *et al.*, 1988), with no data available for sheep or goats in Rwanda, or for any livestock in Uganda. Published data on livestock diseases is completely lacking for the Democratic Republic of Congo (DRC). Both organisms can survive for months to years in cool, humid environments like that of the Virunga Volcanoes region (Nicolletti, 1998; Bengis, 1999; Woodford *et al.*, 2002). Research is in progress to examine the prevalence of tuberculosis and brucellosis in livestock adjacent to Parc National des Volcans (Rwanda) to help evaluate the potential threat to gorillas. The risk of disease transmission between livestock and gorillas can only properly be addressed if disease prevalences in livestock are known, and if livestock and wildlife management practices can be modified if necessary.

6. Habituation of Gorillas for Tourism

Habituation of gorillas is the slow but increasing methodical exposure of trained trackers to unhabituated gorillas until the gorillas become accustomed to the daily visit and appear to pay little attention and show minimal aggression. With the mountain gorillas the process usually takes about a year, during which it is postulated (but not substantiated) that the gorillas are stressed and potentially more susceptible to disease. The process should be well planned and include a) an extensive health screening of the personnel before and during the habituation to minimize the exposure of the gorillas to disease, and b) the collection of biological samples from known gorillas during the process for objectively assessing the health effects of the habituation. These measures are being undertaken as DFGFI and MGVP, Inc. collaborate on new conservation efforts in the relatively undisturbed habitat of the unhabituated eastern lowland gorilla populations in DRC. Once habituated, the gorillas' exposure to humans greatly increases and veterinary activities as described elsewhere in this chapter should be undertaken.

7. Orphaned Ape Health Management

Young apes continue to be orphaned throughout Africa as a by-product of the bushmeat trade, a result of habitat destruction, and, to a far lesser extent, as the occasional target of poaching for private collections. A number of important geographic, economic, political, and cultural factors contribute to the complexity of this problem that are beyond the scope of this chapter and have been discussed elsewhere (Miles *et al.*, 2005).

The majority of these orphans do not survive very long after separation from their dead mothers. Those that do are usually mentally traumatized, and the even fewer that are recovered by authorities are generally physically weak, malnourished, dehydrated, and are commonly suffering from diarrhea and upper respiratory tract infections. It is critical that these animals receive veterinary/medical care for their immediate survival and that infectious disease issues are considered for their long-term placement. Ultimate outcome can be a highly political issue and will depend on individual case circumstance, but should follow the World Conservation Union (IUCN) guidelines for Placement of Confiscated Animals (IUCN, 2000) and the guidelines for reintroduction programs by IUCN/Species Survival Commission (IUCN, 2000). Even in rare cases when orphans can be quickly returned to their own natal groups, serious consideration must be given to the risk that these individuals will introduce a disease that could jeopardize the health of the population.

Regardless of the ultimate disposition, it is important that confiscated animals receive a quick health assessment, any necessary treatments, and that a proper quarantine is established for the time animals will remain in captivity.

In many cases a full examination under anesthesia is best reserved until the orphan has recuperated to a stable condition. In the meantime, treatment of obvious symptoms and provision of nutrition, hydration, and warmth are a minimum standard of care. Opinions vary, but the consensus of most primatologists, zookeepers, and veterinarians is that the physical and mental health of infant orphan apes benefits from human contact which the gorilla mothers would normally be providing the majority of their time.

A proper quarantine requires significant investment of time, energy, and other resources. The intention of a quarantine is 1) to monitor an animal for evidence of any infectious disease and to allow time for latent infections to manifest, and 2) to contain the animal in an environment that minimizes the potential exposure to any further disease agents. A proper quarantine will also provide opportunity and time for full medical evaluation and analysis of diagnostic samples.

The duration of a quarantine must consider a number of factors, including costs, logistics, personnel, space, availability of diagnostic facilities, specific disease risks, and/or diagnoses and the final intended destination of the animal. Though most captive primate facilities utilize at least a 60-day quarantine (some up to 6 months), orphan cases usually merit different considerations. A balance needs to be struck between allowing enough time to properly insure that an animal is healthy and poses minimal risk to other animals if reintroduced or placed with other captive animals, while also recognizing that apes are highly social species and may suffer detrimental effects from the long-term separation from their own species. A minimum of three weeks can be adequate for quarantine of orphans that appear healthy, receive thorough diagnostic testing, and have no risk of tuberculosis exposure. However, because it is believed to take up to 10–12 weeks from exposure to positive tuberculosis testing in young apes, prolonged quarantines may be required to fully ensure negative TB status.

Examination under general anesthetic should include thorough physical examination with particular emphasis on musculoskeletal injuries, skin condition, and upper respiratory tract function. Serial blood samples should be collected for biochemical analysis, complete blood count, and an appropriate panel of viral antigen and antibody titers. Because these orphans have been exposed to humans in areas where tuberculosis prevalence is high, tuberculosis testing is critical to prevent potential introduction of this disease into a destined population. The best TB screening test is intradermal testing with mammalian old tuberculin, and not the purified protein derivative (PPD) used for human testing. Bacterial or dermatophyte cultures, skin scrapings, and other diagnostics should be utilized when indicated, and appropriate treatments for wounds, parasitic or bacterial infections, or other ailments should be given.

An easily and often forgotten aspect of insuring the health of orphan apes is a proper human preventative health program and screening of any parties that have contact with these apes since their capture. Employee health programs are discussed elsewhere in this chapter, and the general approach can be modified for screening specific orphan contact individuals (e.g., arrested poachers, confiscating authorities, or eventual caretakers). There are many legal and humanitarian issues involved with health screening of non-employees. At a minimum, they should receive a cursory physical exam, even if it is only visual, be questioned about childhood vaccinations, and should be screened for tuberculosis disease. Some of these individuals may have spent considerable time in very confined quarters with the captured orphans, enhancing the potential transmission of tuberculosis. The strong association between HIV status and tuberculosis infection cannot be overstated in these situations. Ideally, blood collection for viral serology screening could help eliminate a number of serious risks, though this is best done with serial titers.

Proper health management of orphaned great apes is not an inexpensive or simple matter. It should be kept in mind that any shortcomings in this management may not only impact the affected individual but could have devastating effects on the larger wild or captive populations where these orphans are placed. Proper planning, preparation, and commitment to orphan management can go a long way towards minimizing secondary effects and preserving these fragile destination populations.

8. Biological Resource Center

Preservation of biological specimens from gorilla populations is of great concern and interest to scientists from many disciplines such as epidemiologists, clinicians, geneticists, and conservationists. When specimens are from known individuals, it allows for potential retrospective studies on emergence of new diseases and accurate determination of changes in prevalence and biodiversity of organisms within a defined population. It also allows for the utilization of new technology as it develops to help resolve historical questions. Existing technology has the potential to store live genetic material long after individuals, if not whole populations, have died out.

Since specimens are being collected for future use when they may be tested using new technology and for yet-unspecified organisms, they are stored in a diverse array of preservatives and storage situations. For example, while fecal samples stored in formalin and polyvinyl alcohol are both good for regular floatation for helminthes, formalin is better for most immunofluorescent assay tests and polyvinyl alcohol is better for polymerase chain reaction assays (Graczyk, unpublished data).

The Mountain Gorilla Veterinary Project now has thousands of samples stored in the Biological Resource Center in Baltimore Maryland. These not only include samples from mountain gorillas, but from humans, domestic animals, and other wildlife that impact gorilla habitat. Many of the samples from the Biological Resource Center have already been utilized by researchers throughout the world for a variety of research and advanced degree projects.

9. Standardized Health Monitoring Systems for Gorillas

Although many versions of health monitoring systems are being utilized by gorilla conservation organizations, there is a desire and logical need to have compatibility so the data can be easily shared. Epidemiology-based information systems should be designed to provide answers to fundamental questions about beneficial or harmful effects of any programmatic or medical interventions. The information system should help to identify critical control points of pathogen flow, and computer modeling should be carried out to provide evidence-based information for sound management recommendations and policy decisions regarding great ape health management, including resource allocation.

From 1986 to 1995, MGVP, Inc. field vets routinely recorded quality but poorly standardized observational health data in field diaries. From 1995 to 2001 MGVP, Inc. utilized the American Association of Zoo Veterinarians MedARKS program designed for captive animal management data. In an effort to expand its usefulness, all of this historic qualitative data has been recently coded and incorporated into an Access database. Since then, MGVP, Inc.'s expansion of scope of work and species studied has resulted in an increase of collected and stored clinical data. To accommodate this increase of data, MGVP, Inc. developed a program called "IMPACT" (Internet-supported Management Program for Assisting Conservation Technologies). The system uses unique identifiers to link several parameters of health information. Data collection of gorilla observations (the daily baseline health monitoring of clinical signs) utilizes handheld computers or paper forms with an MGVP, Inc.–developed program. IMPACT allows for fast and easy entry of observed gorillas, with either normal or abnormal health parameters. Data easily downloaded into IMPACT greatly expands the knowledge of the prevalence of clinical signs and aids with the prognosis of present clinical cases.

IMPACT analyzes the severity, prevalence, and incidence of clinical signs to indicate a disease outbreak possibly triggering a newly developed contingency plan. The system objectively places the outbreak into a low-, medium-, or high-risk category that is reviewed and confirmed by MGVP veterinarians. The risk category and numbers of gorillas involved dictate a local, regional, or international response to minimize the negative impact of the outbreak to the gorillas.

The IMPACT system is designed to minimize data cleaning and automatically generate informational reports to share with stakeholders and partners on an ongoing basis. Since IMPACT is a web-based system, it allows field staff and other partners to have real-time access to the data and output reports, and new demographic information, such as births and deaths, can be updated and coordinated more easily. IMPACT's development is a collaborative effort between those involved in data collection and analysis, which includes epidemiologists/human health professionals, wildlife biologists, MGVP's veterinarians, Mississippi State University, and individuals from other great ape organizations.

10. Summary

The role of conservation medicine in gorilla conservation is evolving and expanding, as is the role of the veterinarian. Wildlife veterinarians will continue to provide basic clinical and pathological services, but the collaborative efforts of wildlife, human health, and local veterinary practitioners will need to continue at the wildlife/domestic animal/human interface to reduce the transmission of disease. Continued health monitoring and health-related research programs are needed to fill data gaps so that newly developed information systems can do epidemiological analysis. Critical control points for disease transmission need to be identified and computer models and historical data analyzed to make health recommendations to park managers. Contingency plans need to be documented and roles well defined to maximize their efficiency at reducing the negative impact if a disease outbreak should occur. Veterinarians and conservation medicine will be important tools to complement other disciplines in the long-term sustainability of gorillas in the 21st century.

11. Authorship

Content contribution and editing: MGVP—Michael R. Cranfield, DVM, Lynne Gaffikin, PhD, Christopher A. Whittier, DVM, Felicia Nutter, DVM; Wildlife Conservation Society—William B. Karesh, DVM. Content contribution: MGVP—Linda Lowenstine. Invited author: MGVP—Michael R. Cranfield, DVM.

Acknowledgments. The authors thank Nina Storch for helping with the manuscript and the field people, of the Ugandan Wildlife Authority, Office Rwandais du Tourisme et des Parcs Nationaux and the Institut Congolais pour la Conservation de la Nature for their dedicated efforts on behalf of the gorillas. MGVP, Inc. is a 501 (C) 3 affiliated with the Maryland Zoo in Baltimore.

References

Adams, H.R., and Sleeman, J.C. (1999). Medical survey of tourists visiting Kibale National Park, Uganda, to determine the potential risk for disease transmission to chimpanzees (*Pan troglodytes*) from ecotourism. *Proceedings of the American Association of Zoo Veterinarians*, pp. 270–271.

Akakpo, A.J., Bornarel, P., Akayezu JMV, *et al.* (1988). Epidemiology of bovine brucellosis in tropical Africa. 5, Serological, epidemiological and bacteriological survey in Rwanda. *Revue de Medicine Veterinaire*, 139(8–9):853–859.

Beck, B. B. Reintroduction of African Great Apes, unpublished manuscript in preparation.

Bengis, R.G. (1999). Tuberculosis in free-ranging mammals. In: Fowler, M.E., and Miller, R.E. (eds.) *Zoo and Wild Animal Medicine: Current Therapy 4*. W.B. Saunders Co., Philadelphia, pp.101–114.

Butynski TM. (2001). Africa's great apes. In: Beck, B.B., Stoinski, T.S., Hutchins, M., Maple, T.L., Norton, B. Rowan, A., Stevens, E.F., and Arluke, A. (eds.) *Great Apes and Humans: The Ethics of Coexistence*. Smithsonian Institution Press, Washington, DC, pp. 3–26.

Butynski, T.M., and Kalina, J. (1998). Gorilla tourism: A critical look. In: Milner-Gulland, E.J., and Mace, R. (eds.), *Conservation of Biological Resources*. Blackwell Science Inc., Oxford, pp. 294–366.

Cranfield, M.R., Gaffikin, L., Sleeman, J.M., and Rooney, M. (2002). The mountain gorilla and conservation medicine. In: Aguirre, A.A., Ostfeld, R.S., Tabor, G.M., House, C., and Pearl, M.C. (eds.), *Conservation Medicine*, Oxford University Press, New York, pp. 282–296.

Foster, W.F. (1993). Health plan for the mountain gorillas of Rwanda. In: Fowler, M. (ed.), *Zoo and Wild Animal Medicine: Current Therapy 3*. W.B. Saunders Company, Philadelphia, pp. 331–334.

Garner, K.J., and Ryder, O.A. (1996). Mitochondrial DNA diversity in gorillas. *Molecular Phylogenetic Evolution* 6(1):39.

Goldsmith, M. (2000). Effects of ecotourism on the behavioral ecology of Bwindi gorillas, Uganda: Preliminary results. (Abstract). *American Journal of Physical Anthropology* (Supplement) 30:161.

Graczyk, T.K., Lowenstine, L.J., and Cranfield, M.R. (1999). Capillaria hepatica (Nematoda) infections in human-habituated mountain gorillas (*Gorilla gorilla beringei*) of the Parc National des Volcans, Rwanda. *Journal of Parasitology* 85(6):1168–1170.

Graczyk, T.K., Mudakikwa, A.B., Cranfield, M.R., and Eilenberger, U. (2000). Hyperkeratotic mange caused by *Sarcoptes scabiei* (Acariformes; Sarcoptidae) in juvenile human-habituated mountain gorillas (*Gorilla gorilla beringei*). *Parasitology Research* 86:104–108.

Hastings, B.E., Kenny, D., Lowenstine, L.J., and Foster, J.W. (1991). Mountain gorillas and measles: ontogeny of a wildlife vaccination program. In: Proceedings of the American Association of Zoo Veterinarians Annual Meeting. Calgary. pp. 98–205.

Homsy, J. (1999). Ape tourism and human diseases: how close should we get? Report for the International Gorilla Conservation Programme IGCP Regional Meeting, Rwanda.

(IUCN) World Conservation Union (2000) Guidelines for the Placement of Confiscated Animals approved by the 51st Meeting of the IUCN Council, February 2000.

(IUCN) World Conservation Union (1995) Guidelines for Re-Introductions, approved by the 41st Meeting of the IUCN Council, May 1995.

Kabagambe, E.K., Elzer, P.H., Scholl, D.T., Horohov, D.W., Geaghan, J.P., Opuda-Asibo, J., and Miller, J.E. (2001). Risk factors for *Brucella* infection in goat herds in eastern and western Uganda. *Preventative Veterinary Medicine* 52 (2): 91–108.

Kabagambe, J., Nshimiyimana, A., Muberuka, J., and Nykligira, J. (1988). Les maladies enzootiques des bovins au Bugesera. *Bulletin Agricole du Rwanda* 21(3):159–168.

Kalema, G., Koch, R.A., and Macfie, E.J. (1998). An outbreak of sarcoptic mange in free-ranging mountain gorillas (*Gorilla gorilla beringei*) in Bwindi Impenetrable

Forest, Southwest Uganda. Proceedings Joint Meeting of the American Association of Zoo Veterinarians and the American Association of Wildlife Veterinarians, pp. 438.

Kalema-Zikusoka, G., Kennedy-Stoskopf, S., Adatu, F., and Levine, J.F. (2003). Tuberculosis survey in people sharing a habitat with mountain gorillas in Uganda: implications for conservation and public health. "Brown Bag" talk. 27 May, Washington, D. C.

Karesh, B.W., and Cook, A.R. (1995). Applications of veterinary medicine to *in situ* conservation efforts. *Oryx* (29):244–252.

Karesh, W.B., Wallace, R.B., Painter, R.L.E., Rumiz, D., Braselton, W.E., Dierenfeld, E.S., and Puche, H. (1998). Immobilization and health assessment of free-ranging black spider monkeys (*Ateles paniscus chamek*). *American Journal of Primatology* 44:107–124.

Kilbourn, A.M., Karesh, W.B., Wolfe, N.D., Bosi, E.J., Cook, R.A., and Andau, M. (2003). Health evaluation of free-ranging and semi-captive orangutans (*Pongo pygmaeus*) in Sabah, Malaysia. *Journal of Wildlife Diseases* 39:73–83.

Madden, F. (1998). *The problem gorilla*. Nairobi: International Gorilla Conservation Programme. Unpublished report.

McCallum, H., and Dobson, A. (1995). Detecting disease and parasite threats to endangered species and ecosystems. *Trends in Ecology and Evolution* 10:190–194.

Miles, L., Caldecott, J., and Nellemann, C. (2005). Challenges to great ape survival. In: Caldecott, J., and Miles, L. (2005) World Atlas of Great Apes and their Conservation. Prepared at UNEP – WCMC. University of California Press, Berkeley.

The Mountain Gorilla Veterinary Project 2002 Employee Health Group (2004). Risk of disease transmission between conservation personnel and the mountain gorillas: results from an Employee Health Program in Rwanda. *Ecohealth* 1, pp. 351–361.

Nizeyi, J.B., Cranfield, M.R., and Graczyk, T.K. (2002a). Cattle near the Bwindi Impenetrable National Park, Uganda, as a reservoir of *Cryptosporidium parvum* and *Giardia duodenalis* for local community and free-ranging gorillas. *Parasitology Research* 88(4):380–385.

Nizeyi, J.B., Sebunya, D., Dasilva, A.J., Cranfield, M.R., Pieniazek, N.J., and Graczyk, T.K. (2002b). Cryptosporidiosis in people sharing habitats with free-ranging mountain gorillas (*Gorilla gorilla beringei*), Uganda. *American Journal of Tropical Medicine and Hygiene* 66(4):442–444.

Nutter, F., and Whittier, C.A. (2001). Improving human health care: benefits for nonhuman primates. Proceedings of The Apes: Challenges for the 21st Century, Brookfield Zoo, Chicago. pp. 52.

Onekalit, A.K.E. (1987). Uganda: bovine brucellosis and brucellosis of small ruminants: diagnosis, control, and vaccination. *Technical Series, Office International des Epizooties* 6:71–72.

Ott-Joslin, J.E. (1993). Zoonotic diseases of nonhuman primates. In: Fowler, M.E. (ed), *Zoo and Wild Animal Medicine: Current Therapy 3*. W.B. Saunders Co., Philadelphia. pp. 358–373.

Roberts, J.A. (1993). Quarantine. In: Fowler, M.E. (ed.) *Zoo and Wild Animal Medicine: Current Therapy 3*, W.B. Saunders Co., Philadelphia, pp. 326–331.

Sibley, C.G., and Ahlquist, J.E. (1984). The phylogeny of the hominid primates as indicated by DNA-DNA hybridization. *Journal of Molecular Evolution* 20:2–15.

Silberman, M.S. (1993). Occupational health programs in wildlife facilities. In: Fowler, M.D. (ed.), *Zoo and Wild Animal Medicine: Current Therapy 3*, W.B. Saunders, Philadelphia, pp. 57–61.

Werikhe, S., Macfie, L., Rosen, N., and Miller, P. (1997). Can the mountain gorilla survive? Population and Habitat Viability Assessment for *Gorilla gorilla beringei*. Kampala, Uganda. Conservation Breeding Specialist Group, Apple Valley, MN, pp. 1–7.

Whittier, C.A., Nutter, F.B., and Stoskopf, M.K. (2001). Zoonotic disease concerns in primate field settings. *The Apes: Challenges for the 21st Century*. Conference Proceedings, May 10–13, Brookfield, IL, pp. 232–237.

Wolfe, N.D., Escalante, A.A., Karesh, W.B., Kilbourn, A., Spielman, A., and Lol, A.A. (1998). Wild Primate populations in emerging infectious disease research: the missing link? *Emerging Infectious Diseases* 4(2):149–158.

Woodford, M.H., Butynski, T.M., and Karesh, W.B. (2002). Habituating the great apes: the disease risks. *Oryx* 36(2):153–160.

The World Bank (2003). The Little Green Data Book, From the World Development Indicators, p. 181.

Chapter 3
Sanctuaries and Reintroduction: A Role in Gorilla Conservation?

Kay H. Farmer and Amos Courage

1. Introduction

The current threats to gorilla (*Gorilla gorilla, Gorilla beringei*) populations, and indeed African wildlife in general, are complex and inextricably interlinked, and include poverty, human population growth, loss of habitat (through logging, mining, and land conversion), and hunting (Butynski, 2001; Teleki, 2001; Nellemann and Newton, 2002). Overexploitation of wildlife is not a new phenomenon and was probably responsible for the historical and ecological extinction of many species (Rao and McGowan, 2002). However, increasing urbanization and associated market economies, modern hunting methods and road networks, have commercialized the bushmeat trade (Kemf and Wilson, 1997; Bowen-Jones, 1998; Robinson and Bodmer, 1999; Wilkie and Carpenter, 1999; Fa *et al.*, 2002; Nellemann and Newton, 2002). The general consensus seems to be that this trade is out of control, unsustainable, and accelerating (Ammann and Pearce, 1995; Kemf and Wilson, 1997; Butynski, 2001), and that gorillas are in danger of becoming extinct in the wild if causal factors are not effectively addressed (Butynski, 2001).

While nursing female primates are killed for the bushmeat trade, the surviving infants may either end up as playthings in a local village (where they die from disease or neglect), or they may be sold to wealthier city dwellers as pets, or to hoteliers and businesses to attract customers (Kemf and Wilson, 1997; Goodall, 2001). In some areas, nursing females are specifically targeted to obtain the infants for sale on the illegal market (Goodall, 2001). Although the capture of wild-born apes for biomedical laboratories, zoo collections, and the entertainment sector has decreased in recent years, in some regions (e.g., United Arab Emirates, South America, and Eastern Europe) the demand remains high and the trade continues unabated (Wallis, 1997; Goodall, 2001). The price tag of US$ 1.6 million for four wild-born infant gorillas (known as the Taiping 4), recently shipped from Nigeria to a Malaysian Zoo with false papers proclaiming captive birth, highlights the financial incentive of this trade (IPPL News, 2002). As wild primate populations decrease, the number of orphaned primates, sanctuaries and attempts to reintroduce primates back to the natural environment are increasing. As a direct consequence, an umbrella organization for African primate sanctuaries called the Pan-African Sanctuary Alliance (PASA) was formed in 2000 (Cox *et al.*, 2000), the IUCN Re-introduction Specialist Group developed a set of specific policy guidelines for primates (Baker, 2002), and in 2006 PASA convened a workshop to bring together African primate sanctuaries and multidisciplinary experts to discuss options and provide guidance for those facilities considering reintroduction as a long-term aim.

This chapter examines the role that *in situ* sanctuaries and reintroduction can play in gorilla conservation. Due to the paucity of systematically collected evaluative data from the sanctuary and reintroduction community per se, this chapter will focus primarily on the potential role that these measures may play in conservation, with specific examples and case studies presented where possible. The first half of this chapter will address the role of sanctuaries in ape conservation, and the second part will focus specifically on reintroduction.

1.1. The What and Where of In Situ Gorilla Sanctuaries

Presently, there are 15 African ape sanctuaries (as defined by PASA membership), and four of these facilities accommodate gorillas (Mills *et al.*, 2005; Table 3.1). Six additional non-PASA facilities in Africa are known to hold apes including three gorillas (Mills *et al.*, 2005). The western lowland gorilla subspecies (*Gorilla gorilla gorilla*) forms the core of the sanctuary gorilla population. The Limbe Wildlife Centre declares one Cross River gorilla (*Gorilla gorilla diehli*). One mountain and one eastern lowland gorilla confiscated in Rwanda are in the care of the Mountain Gorilla Veterinary Project, and one eastern lowland gorilla is in the care of Dian Fossey Gorilla Fund International in the Democratic Republic of Congo (Whittier *et al.*, 2005).

TABLE 3.1. *In situ* sanctuaries with gorillas.

Sanctuary name	Founding organization	Country	Number of gorillas (as of April 2006)
Limbe Wildlife Centre (LWC)	Pandrillus Foundation	Cameroon	11
Cameroon Wildlife Aid Fund (CWAF)	Cameroon Wildlife Aid Fund	Cameroon	13
Projet de Protection des Gorilles (PPG)	John Aspinall Foundation	Republic of Congo	22
Projet de Protection des Gorilles (PPG)	John Aspinall Foundation	Gabon	22

1.1.1. The Limbe Wildlife Centre (LWC), Cameroon

In 1993, Peter Jenkins and Liza Gadsby (co-founders of The Pandrillus Foundation), in co-operation with the then Cameroonian Ministry of Environment and Forestry, established a sanctuary on the grounds of the Limbe (Victoria) Zoo. The LWC was established with the following aims: a) to improve the living conditions of the animals already resident, b) to provide a facility for confiscated wildlife (primarily primates) in Cameroon, and c) to develop a centre for conservation education. The sanctuary has a recently refurbished education centre, an education officer, and an extensive on-site and outreach education program. It is home to several primate (and nonprimate) species, and there are plans to develop a second more remote site to accommodate the growing number of animals. As of April 2006 there are 11 gorillas in the care of LWC.

1.1.2. Cameroon Wildlife Aid Fund (CWAF), Cameroon

In 1996, Chris Mitchell created CWAF, and, in collaboration with Bristol Zoological Gardens and the then Cameroonian Ministry of Environment and Forestry, established a sanctuary on the grounds of the Yaounde (Mvogbetsi) Zoo. The aim of the project was to improve the living conditions of animals at the zoo and to provide an education and awareness program. A second site was later developed in 2000 at Mefou National Park. The two-site sanctuary is home to several primate (and nonprimate) species, and as of April 2006 there are 13 gorillas at the Mefou site. The CWAF has a fulltime education officer and is also part (with Bristol Zoological Gardens and Living Earth Foundation) of a community-based engagement and support program being conducted around the Dja Biosphere Reserve (personal communication, N. Maddison, Head of Conservation Programs, Bristol Zoo Gardens, UK, 2006).

1.1.3. Projet de Protection des Gorilles (PPG), Republic of Congo

In 1981, Yvette Leroy started rescuing wild-born orphaned gorillas in Brazzaville (Republic of Congo). Due to an influx of gorillas, combined with logistical and financial constraints, Yvette Leroy contacted the late John Aspinall, owner and director of Howletts and Port Lympne Zoo (England). Howletts Zoo had the largest captive population of western lowland gorilla and had success in the field of captive breeding. In 1987, a contract was signed between Howletts and Port Lympne Foundation and the Ministry of Economics and Forestry (Congolese Government) to establish and manage a gorilla rehabilitation and reintroduction program.

  In 1989, the John Aspinall Foundation established PPG within the grounds of Brazzaville Zoo. In 1993, PPG established a release site at La Reserve Natural des Gorilles de Lesio-Louna (Lesio-Louna Sanctuary). The aims of the project include rehabilitation and reintroduction of western lowland gorillas, management and protection of the flora and fauna at the release site, and raising awareness nationally and internationally about the bushmeat trade. The center at Brazzaville was evacuated during the 1997 civil war and the gorillas spent 18 months at a Jane Goodall Institute chimpanzee sanctuary (Tchimpounga) just outside of Pointe Noire. After the war, all the activities were transferred to the Lesio-Louna Sanctuary, and the Brazzaville orphanage was closed. In 2004, all gorillas were transferred from Lesio-Louna to Lefini Faunal Reserve as the ecological barriers at Lesio-Louna were unable to stop the gorillas entering villages. As of April 2006, there were 22 gorillas in the care of PPG Congo.

1.1.4. Projet de Protection des Gorilles (PPG), Gabon

During the 1997 civil war in the Republic of Congo, the John Aspinall Foundation researched the possibility of transferring the gorillas to Gabon. Peace returned to Congo in 1998, and the gorillas remained there. However, the foundation decided to continue with a second project in Gabon with the same aims as its sister project in Congo and established a release site in Mpassa Reserve. As of April 2006, there were 22 gorillas in the care of PPG Gabon.

The emphases of the four projects accommodating gorillas differ, but all activities fall within the definition of a PASA sanctuary: "A PASA sanctuary provides a safe and secure home for African apes and other primates in need. The welfare of the individual and the preservation of the species are of prime importance and are considered equally. The sanctuary operates in the context of an integrated approach to conservation, which can include rehabilitation and reintroduction" (Rosen et al., 2001, p. 13).

The LWC and CWAF accommodate a variety of primate species and have a strong emphasis on primate rehabilitation, long-term captive care, and conservation education. Although reintroduction maybe a long-term aim, neither actually practices it. In contrast, the main emphasis of PPG (in both countries) is reintroduction.

2. The Role of Sanctuaries in Gorilla Conservation

The problem of captive primates is frequently seen as tangential to the real conservation issues of habitat loss and commercial hunting, and some wildlife biologists argue that sanctuaries are a waste of money (Mackinnon, 1977; Soave, 1982; Oates, 1999). It has been also suggested that sanctuaries may even exacerbate the trade in live animals if local populations misinterpret project goals by hunting apes with the aim of selling them to the sanctuaries, or if the sanctuaries are viewed as private collections (Karesh, 1995).

However, this debate has continued largely in the absence of solid information from the sanctuaries themselves (Teleki, 2001). Until the recent development of PASA, there was little or no communication between the sanctuaries due to logistical constraints (e.g., distance, irregular and expensive means of communications, and lack of time) and an unwillingness to exchange information (personal observation[1]). This has meant that trial and error processes have predominated, with success defined primarily in terms of individual opinion with no reference to comparative data (Teleki, 2001). However, as the pressures on African apes reach a crisis point, and PASA has emerged as a means to unify sanctuary goals and standards, there is an emerging interest in the potential role that sanctuaries can play in conservation. These roles include law enforcement, conservation by education, captive breeding, and reintroduction.

2.1. Law Enforcement

Undeniably, *in situ* sanctuaries can provide an immediate solution for confiscated animals (Andre, 2002; Farmer, 2002a). Confiscations in response to trafficking and illegal ownership are vital to international law enforcement, and where there are no sanctuaries, there is little or no confiscation (Teleki, 2001). A recent survey of African ape sanctuaries revealed that 61% of gorillas (54% of all apes) were confiscated (Farmer, 2002a). The same survey revealed that 65% of gorillas had been found awaiting sale, and that a quarter had been previously kept as pets. Poor law enforcement lends confusion to the legal status of primate pets and allows the trade to flourish regardless of wildlife laws.

2.2. Conservation by Education

Sanctuaries have been criticized for focusing on the welfare of individual animals rather than the survival of the species (Kleiman *et al.*, 1986). Images of individual primates clinging to human caretakers, circulated by sanctuaries to generate funds, have propagated this perception. Although promoting the welfare of individual animals should not require justification, survival of the species is the ultimate goal, and is one justification given by modern zoos for keeping animals in captivity (Wallis, 1997; Stoinski *et al.*, 2001). Modern zoos claim four inter-related roles: conservation, education, research, and entertainment (Cherfas, 1984; Tudge, 1991; Reade and Waran, 1996), with conservation and environmental education promoted as the primary roles (Reade and Waran, 1996; Stoinski *et al.*, 2001). Brend (2001) argues that *in situ* sanctuaries play a primary role in conservation, and that they can be actively involved in research and education. Although the continued existence of zoos is controversial (Jordan and Ormrod, 1978; Campbell, 1980), zoos provide the only experience that most children (and indeed adults) have with live, exotic animals (Kidd and Kidd, 1996). While the authors are not advocating the opening of zoological collections in Africa, this kind of experience may be particularly important in modern urban Africa where traditional taboos about consuming apes and other species are disregarded, and where the demand for bushmeat is high and its consumption viewed as a luxury item (Wilkie, 2001).

The educational impact that captive animals may have on the viewing public has been poorly studied, but clearly zoos could do more to encourage conservation-related behavioral change in the general public (Reade and Waran, 1996; Wagner et al., 2000; Stoinski et al., 2001). Certain species can more readily act as focal points or flagship species for conservation. Studies in the United States and Brazil have documented preferences for animal species based on their attractiveness (Dietz and Nagangata, 1995). Although cultural preferences exist, factors that were found to positively influence ratings of attractiveness were large size, advanced intelligence, phylogenetic relatedness to human beings, and complex social organization (Dietz and Nagangata, 1995). It is therefore not surprising that apes are among the most popular zoo species and offer powerful opportunities for educating the general public (Gold and Benveniste, 1995; Stoinski et al., 2001). The Mountain Gorilla Program in Rwanda has highlighted how this charismatic species can positively influence conservation. Small groups of students, teachers, and local leaders were taken to habituated groups of gorillas. These visits were found to be highly effective in expanding local interest and support (Weber, 1995). Interest generated by the mountain gorilla, at both national and international levels, has resulted in the gorilla becoming a springboard to promote awareness of wider environmental issues. While the stress of habituation and possibility of disease transmission are potentially life-threatening issues to wild populations (Homsy, 1999; Butynski, 2001), easily habituated sanctuary animals may offer a more practical solution for the local population to learn about their natural heritage. As with wild populations, safeguards to prevent disease transmission and minimize stress in captive populations should be of paramount importance.

Beyond the casual viewing public, American and European zoos claim to offer educational opportunities through on-site formal zoo education programs for children, research training for college/university students, and research opportunities for scientists (Stoinski et al., 2001). Twelve African ape sanctuaries claim to be involved in scientific data collection, behavioral and biological (Farmer, 2002a). Such research is of obvious scientific and conservation importance for those species that have been poorly studied in the wild (e.g., the western lowland gorilla, the bonobo *Pan paniscus*, and the drill *Mandrillus leucophaeus*). Thirteen sanctuaries claim active participation in habitat protection. The location of the sanctuary influences how this involvement is expressed; it ranges from on-site and outreach conservation programs to presence in protected areas, employment of eco-guards, and assisting with snare removal (Farmer, 2002a). Sixteen African ape sanctuaries (including all four with gorillas) claim to actively participate in conservation education through a variety of activities. These activities include on-site visits by the general public and school parties, displays and keeper talks, employment of education officers, nature clubs, activity packs, seminars, workshops, outreach programs, and national and international media attention (Farmer, 2002a).

The importance of education per se in shaping perceptions of nature and biodiversity is widely accepted (Boulton and Knight, 1996; Kellert, 1996; Kidd and Kidd, 1996; Wallis, 1997; Jacobson and McDuff, 1998; McDuff, 2000; Thompson and Mintzes, 2002). Boulton and Knight (1996) argue that formal environmental education and extracurricular activities for children and adults alike in both developed and underdeveloped countries are essential for the future of biodiversity. In developing countries, where few formal environmental education programs exist, informal conservation education may be a valuable alternative (Boulton and Knight, 1996; Jacobson and McDuff, 1998; McDuff, 2000). Wildlife clubs, for example, have successfully represented an entry point into environmental conservation in Kenya and many other parts of Africa (Boulton and Knight, 1996; McDuff, 2000). Clearly, *in situ* sanctuaries, the poor relation of the zoo world, are potentially able to shape perceptions of nature and biodiversity. The following case study from the Limbe Wildlife Centre illustrates the point.

2.2.1. Case Study—The Limbe Wildlife Centre

The Limbe Wildlife Centre in Cameroon is based on the grounds of the Limbe (Victoria) Zoo in a small coastal town, opposite the Limbe Botanic Garden, close to beach resorts, and within easy reach of the large industrial city, Douala. Hence, it is ideally situated to receive the many national and foreign visitors that flock annually to the area. Since the projects' inception in 1993, attendance by the general public has risen steadily and now totals over 30,000 annually. As mentioned earlier, one goal of the LWC is to act as an educational facility to inform Cameroonians about the importance of wildlife conservation. It purports to do this through on-site education and outreach programs (Table 3.2, § Boesch *et al.*, Chapter 5, this volume).

2.2.1.1. On-site education

- LWC is open to the general public seven days a week. A nominal entrance fee is charged (nationals: adults 300 CFA/children 100 CFA, non-nationals 3000 CFA), but donations are encouraged. Each visitor group/family receives a color information leaflet outlining the mission of the LWC, the primary conservation issues, and contact information detailing how people can become involved. A self-guide brochure is being developed. Numerous bilingual signs and displays provide additional information, and a library exists with more than 100 titles.
- A once-weekly nature club is offered to local children and youths where they can learn about wildlife and environmental conservation through

TABLE 3.2. Education program at the Limbe Wildlife Centre, Cameroon.

Sphere of influence	Target group	Activities
On-site	All visitors	- Information displays and species signboards - Guided tours - Free information brochure
	School children and youth groups	- Weekly Nature Club with free membership - Annual workshop - Library
Outreach	School children	- Outreach programs to schools - Assisting schools to develop their own environmental club - Assisting teachers with suitable information and materials
	Adults	- Outreach programs to communities - Conducting lectures to youth groups
	General public	- Local and international publicity

lectures, games, art, poems, singing, drama, and tours of the sanctuary and adjacent botanic garden. Registered members reach approximately 50 per annum.

- LWC holds an annual nine-day workshop conducted for different educational levels (primary, secondary, high school, and university level). Themes have included wildlife conservation; the bushmeat hunting crisis and sustainable hunting; rainforest and sustainable management; man and nature; and aquatic ecosystems and man's influence on them. Various activities focus in and around the centre and field trips. In 2004, more than 200 school children visited fresh and salt water ecosystems as part of the aquatic ecosystems–themed workshop. As part of the 2005 program, LWC implemented "conservation across cultures" with a bushmeat crisis–focused workshop. More than 160 Cameroonian children communicated one-to-one with children in Busch Gardens, Florida, live via the Internet, to debate conservation issues.

2.2.1.2. Outreach programs

- In 2000, the LWC education officer visited more than 100 schools and reached 11,000 school children in and around the Limbe area. Children heard the story of "Pitchou," a gorilla at the centre, how she came to be there, and why species like hers should be protected. Apparently, having heard the story, many children visited the centre specifically to see Pitchou. Furthermore, even a school that had not received a visit from the education unit used Pitchou's story in its end-of-term exam. In 2002, the total number of children reached increased to 20,000.

- The LWC was fundamental in the formation of environmental clubs, with more than 60 members in Limbe schools. A leaflet, "How to Form a Nature Club" was produced and distributed to schools during outreach programs.
- In 2005, a weekly sensitization program that ran for six months across eight schools within Limbe reached more than 800 children. The program included lectures, crossword puzzles, debates, video and drama presentations, and guided tours of the LWC.
- Recognizing that LWC does not have the logistical capacity to regularly visit every school in the area, it developed an education packet to assist teachers delivering conservation messages. Return visits were made to ensure the packet was being used efficiently and materials such as posters and leaflets provided.
- A play devised in collaboration with the local Reformation Theatre Group, aimed to sensitize people to the negative impacts of uncontrolled hunting, was taken to villages throughout Cameroon. After the play, a discussion was held about the consequences of the villages' hunting policies and a film about the work of the LWC was shown. Posters and t-shirts are taken to the villages to propagate the message. In 2002, more than 5,600 people watched the play in villages located in southwest, northwest, and eastern provinces.
- In 2005, a play, "Fruitless Seeds," was taken to eight schools within Limbe and reached more than 800 children. It was a futuristic play about a day when no apes exist in Africa but only in European Zoos. It featured a son asking his father, "What is a chimpanzee?" The father reminisces about the days when chimpanzees existed in the forests of Cameroon.
- The LWC education unit assisted with the development of the PASA education pack that was distributed to all African primate sanctuaries.
- More than 5,000 posters with four themes—bushmeat, orphans of the bushmeat trade, LWC, and the great ape family tree—were produced and distributed to government, schools, and hotels.
- Newsletters, reports, local (e.g., Ocean City and Eden radio) and international (e.g., Radio France International) radio broadcasts, and filming with BBC Panorama, Carte Blanche, Animal Planet, etc. are promoting the work of LWC to a wider audience.

The list of activities presented by the LWC represents an impressive effort to participate actively in Cameroonian conservation issues. Likewise CWAF has developed an education and awareness program with several components that includes:

- Working with school inspectors and teachers in the central province, and supporting environmental education teaching in the national curriculum;
- Informal and formal education sessions for visitors and school parties to both sites;
- Awareness raising, trust building and community support programs for people living in and around protected areas and ape habitat.

The question remains, are the conservation messages delivered being perceived appropriately? Are they successfully filtering down to the local communities? Are they impacting behavior in the short- and long-term? Increased awareness does not guarantee meaningful behavior change, and debates rage among educators concerning the relationship between knowledge, attitudes, and behavior (Jacobson and McDuff, 1998; McKenzie-Mohr and Smith, 1999). The LWC distributed pre- and post-questionnaires to 569 students participating in their 2005 outreach program to evaluate its effectiveness. They found a 44% increase in ability to identify and list three endangered primate species; a 37% increase in the knowledge that having an endangered species as a pet is forbidden by Cameroonian law; 29% increase in ability to identify two human activities that have a negative impact on the environment; and a 44% decrease in the belief that primates make good pets (Costo et al., 2005). The LWC concluded that, while there was impressive improvement in student knowledge, many children living in the rural areas surrounding the LWC are still not aware of the issues, and that the education program should be expanded. Unfortunately, evaluation is not a common component in the design of conservation education programs (or indeed conservation directives generally) due to a lack of time, money, and staff (Jacobson and McDuff, 1998; Ammann, 2001; Fien et al., 2001). It is not possible to say whether the activities of the LWC and CWAF have had any direct impact on the conservation of gorillas in Cameroon, however, the success of the Mountain Gorilla Project education program in helping to change Rwandan attitudes toward conservation (together with tourist-related income and employment) highlights the potential importance of such initiatives (Weber, 1995). The use of attitudinal surveys not only facilitated the development of effective gorilla conservation in Rwanda but also the design of education programs (Weber, 1995).

Even though conservation goals may be focused on biological problems, it is likely that future conservation strategies will focus more on communication and/or education programs designed to affect people's awareness, attitudes, and behaviors toward the natural environment (Jacobson and McDuff, 1998). However, only systematic evaluation of education programs to highlight what is and is not working will enable conservation measures to reach their true potential.

2.3. Captive Breeding of Endangered Species

The potential of incorporating rare and endangered species held in sanctuaries into captive breeding programs as a measure to conserve threatened species may offer further support for the role of sanctuaries in conservation (Kleiman et al., 1986; Brend, 2001). Views on captive breeding as a tool in primate conservation have been mixed (Synder et al., 1996; Oates, 1999; Cowlishaw and Dunbar, 2000). The IUCN Action Plan did not recommend captive breeding for any African primate taxon and only did so in four Asian taxa (Oates, 1986, 1996, Eudey, 1987). In contrast, the IUCN/SSC Global Captive Action Plan for Primates (GCAP) recommends almost half of all primate taxa for captive breeding programs, and 15% as a matter of urgency (Stevenson *et al.*, 1992). However, if captive breeding is considered a viable option, and, assuming that the standard of *in situ* captive care is sufficiently high, it is clearly preferable for captive breeding to occur in the country of origin. *In situ* captive breeding programs can facilitate animal management (e.g., climatic and nutritional variables) and have positive spin-offs such as environmental education, training, local employment etc. (Kleiman *et al.*, 1986; Beck *et al.*, 1994). Although foreign zoos may take primary responsibility for captive breeding due to a lack of investment in *in situ* facilities, sanctuaries within range countries may also play a role by extending numbers and genetic diversity. A prime example is the Drill Breeding and Rehabilitation Centre in Nigeria that holds the largest captive population of the endangered drill (*n* = 120), half of which are captive-bred (Cowlishaw and Dunbar, 2000; Gadsby, 2002).

Concerning gorillas, the small number of western lowland gorillas housed in African sanctuaries (69 gorillas in four facilities; Mills *et al.*, 2005) does not lend this argument a great deal of support. However, this may change in the future if the present rate of habitat destruction and wild population decline continues. PPG is paving the way by developing reintroduction protocols in attempts to release gorillas back to the natural environment. Furthermore, if the current trend in mountain gorilla (*Gorilla beringei beringei*) killings continues, and reintroduction is not considered a viable option for any surviving orphans, these individuals should ideally be placed into an *in situ* sanctuary (as opposed to overseas zoos) for the reasons already described earlier. No mountain gorillas have survived in captivity to date (Whittier *et al.*, 2005), although the death rate of western lowland gorillas in sanctuaries is also high (Farmer, 2002b). A study at PPG found a link between the age of arrival and mortality rate during and after the first two months at the sanctuary. During the first two months, mortality rates are highest for gorillas that arrive when they are more than two years of age, but significantly less for those who are less than six months of age. However, the converse is true for mortality rates following the first two months (King *et al.*, 2005a). It is not known whether mortality in mountain gorillas is due to specific vulnerabilities of the species, to the captive situation/reintroduction process employed, or simply a case of statistics; there are significantly more lowland gorillas in captivity.

2.4. Reintroduction

Protected area management is the primary method employed by most conservation agencies (Stuart, 1991; Oates, 1999; Cowlishaw and Dunbar, 2000). However, loss of habitat and wildlife species, and improvements in captive breeding techniques, served to increase attention on reintroduction as a conservation tool in the 1980s (see Kleiman, 1989; Stanley-Price, 1989;

Gipps, 1991; Beck, *et al.*, 1994; Fisher and Lindenmayer, 2000). This surge of interest was related to the prediction that some species may not survive in the wild without reintroduction programs, and to the many additional spin-offs that can follow, such as increased national and international awareness of conservation issues (Stuart, 1991). The increasing interest in this approach to wildlife management was the main reason for the creation of an IUCN (The World Conservation Union) Species Survival Commission (SSC) Reintroduction Specialist Group in 1988. The group was established to collect and disseminate information on all reintroduction programs (animal and plant) and to provide a set of guidelines to assist in the process (IUCN, 1995, 1998). Recently, the IUCN Reintroduction Specialist Group developed a set of specific policy guidelines for primates, "Guidelines for Nonhuman Primate Re-introductions" (Baker, 2002) and produced a special edition newsletter devoted to primate reintroduction case studies (Sooare and Baker, 2002).

Attempts to reintroduce gorillas have ranged from the addition of individuals to existing populations of conspecifics (known as reinforcement/supplementation; Baker, 2002), to the reintroduction of gorilla groups into areas from which they have been extirpated. There have been five attempts to reintroduce individual infant gorillas to wild groups. Four cases involved mountain gorilla (*Gorilla beringei beringeri*) and one eastern lowland gorilla (*Gorilla beringei graueri*). Mountain gorillas are so threatened that even individual animals are considered important for the continuation and genetic fitness of the subspecies. However, no attempt succeeded and all five gorillas died, disappeared, or had to be returned to captivity. Recommendations for an infant mountain gorilla confiscated in Rwanda, in 2004, included keeping her in captivity with two confiscated eastern lowland gorillas for socialization purposes and then, when the appropriate age was reached (when wild females would naturally migrate), attempting to introduce her to a wild group (Whittier *et al.*, 2005). This recommendation was based on the assumption that other reintroductions may have failed as a consequence of immaturity at release. However, severing group relations that would undoubtedly form between the mountain and eastern lowland gorillas may result in additional complications, given the fragile psyche of gorillas (King *et al.*, 2005a). As of April 2007, the female is doing well in captivity with a growing group of now six eastern lowland gorillas. There is also a second mountain gorilla, still in quarantine, and his arrival will likely change the release strategy, but the female is still too young for release (Dr. C. Whittier, Mountain Gorilla Veterinary Group, *pers comm.*, April 2007).

PPG Congo and Gabon are the only projects currently reintroducing groups of lowland gorillas. Reintroduction is a complex process and involves a great deal of planning and preparation. Program aims and objectives need to be defined, economic and political constraints addressed, suitability of a species and of individuals for reintroduction reviewed, methodology explored (veterinary protocol, quarantine, capture, transfer and release) and established, potential release sites surveyed, and definition of success defined.

This list is not exhaustive, as every aspect and eventuality should be addressed, because inadequate planning can cause a reintroduction to fail. Gorillas are probably one of the most challenging of primates to consider reintroducing to the wild due to their complex social structure and dietary requirements (King *et al.*, 2005a), supported by high mortality rates in sanctuaries (Attwater, 1999; Farmer, 2002b). The following section will describe the reintroduction process employed by PPG, present a summary of gorillas released, and evaluate PPG's contribution to gorilla conservation.

2.4.1. Selecting the Release Site

A reintroduction can only be contemplated if a suitable release site is available. The site must be able to provide sufficient resources for the released individuals without negatively impacting species already present by depletion of key resources. Releasing into areas with wild conspecifics raises questions about disease transmission. Additionally, the site must offer adequate protection from threats, such as logging and hunting, and not expose the released animals to situations of conflict with humans by being located too close to villages and plantations (Tutin *et al.*, 2001; King *et al.*, 2005b).

PPG Congo initially selected the grounds of the Brazzaville Zoo as the base for a gorilla rehabilitation center. The center benefited from the existing zoo infrastructure, acted as an initial quarantine facility away from the release site (prior to release each gorilla underwent a vaccination program and disease surveillance profile including TB testing), and targeted the local population visiting the zoo with conservation education messages. Furthermore, the zoo was surrounded by 25 hectares of secondary forest known as the "Forêt de Patte d'Oie," which served to prepare the gorillas to forest life. The plan was that the gorillas would be rehabilitated at the Brazzaville orphanage and later moved to a suitable site for reintroduction. The search for a suitable reintroduction site led the project to La Reserve Natural des Gorilles de Lesio-Louna (Lesio-Louna Sanctuary). The reserve had a surface area of 45,000 ha, 30% of which was forest and 70% savannah, bordering the Lefini Faunal Reserve on the Batéké Plateau, 120 km north of Brazzaville. It was surrounded by natural barriers (canyons, cliffs, and rivers) that facilitated the management of the gorillas (isolated from human habituation) and also human access to the site. The nearest village was located 12 km away, separated from the project site by an expanse of savannah. Surveys indicated no wild gorilla activity. According to local tradition, gorillas had been extirpated from the plateau in the 1950s, probably because of hunting pressure (King *et al.*, 2005b). However, the former gorilla habitat showed an abundant supply of potential gorilla foods, and seasonally inundated areas of forest rich in herbaceous vegetation. Two brief surveys conducted (Fay, Harris, Moutsambote, and Thomas, cited in Attwater *et al.*, 1991, and Cousins, 2002) identified 81 plant species from 31 families eaten by wild gorillas. Finally, the close proximity to Brazzaville (150 km by paved highway) facilitated the move of the gorillas from the rehabilitation centre and tourism possibilities.

Due to the destruction of the Brazzaville base during the civil war in 1997–1998 and the eruption of violence occurring primarily in Brazzaville, PPG closed the rehabilitation center at Brazzaville. As mentioned earlier, during the war, the gorillas spent 18 months at a Jane Goodall Institute chimpanzee sanctuary (Tchimpounga) just outside of Pointe Noire, and after the war all the activities were transferred to the Lesio-Louna Sanctuary. This not only removed the source of any potential stress due to relocation but also reduced human contact. Frequently, people were found moving through or farming in the "Forêt de Patte d'Oie" that surrounded the zoo, and it was thought that this contact may have been the source of some parasitic infections and gorilla deaths at the orphanage (Cousins, 2002; King et al., 2005b). To accommodate all gorillas at the Lesio-Louna Sanctuary, a second release site was needed. Biological surveys were conducted, and a site within the Lefini Faunal Reserve (still on the Batéké Plateau) was judged suitable, as it had no existing wild gorilla populations, but was former gorilla habitat with abundant potential gorilla foods. Although the first site at Lesio-Louna had some natural boundaries, they were found to be inadequate, and some released gorillas repeatedly returned to base camp and entered local villages (Watkin, 2002; King et al., 2005b). However, the site in Lefini was completely surrounded by natural boundaries; by rivers on the north, east (nearest village 20 km from the release area to the east), and west extremes, and a large expanse of savannah to the south, all impassable by gorillas. As a consequence, the site facilitated staff patrols, as access could only be gained by boat, and allowed more control on people entering the area. In September 2004, all gorillas were transferred to the Lefini Faunal Reserve.

When PPG was seeking a release site in Gabon, similar surveys and rationale were employed. The Mpassa Reserve (Mpassa Sanctuary), an area of 171,800 ha, 110 miles from Franceville, was judged suitable. This area was also on the Batéké Plateau, separated from the Congo site by the national boundary, and therefore also contained no wild gorilla activity but was known to have been former gorilla habitat with sufficient and suitable gorilla foods. The release site was isolated from a neighboring, larger forest mass, where existing wild populations were identified but therefore posed them no threat. Finally, the site was completely isolated from human habitation, as there were no villages in the reserve. PPG Gabon combined the rehabilitation and release site from the outset to avoid any problem of transfer and to facilitate entry into suitable habitat as soon as possible.

2.4.2. Prerelease Training

Often the gorillas arrived at the Brazzaville center in poor physical and psychological shape. However, as soon they were able, the gorillas were placed into age-based groups and taken into the forest surrounding the zoo during the morning (0830–1230 h) and afternoon (1530–1730 h) by human caretakers. This provided the gorillas with the opportunity to locomote on natural substrates, to sample natural vegetation, and develop social relationships with one another. The human caretakers acted as maternal substitutes, providing much needed support to the orphaned gorillas. At midday, all groups were returned to cages for a feed and rest. A botanical survey in 1991 documented 57 plant species, representing 31 families, being consumed by the gorillas in the "Forêt de Patte d'Oie" (M'Passi, cited in Attwater et al., 1991, and Cousins, 2002). However, because the plant species were insufficient to provide nutritionally self-sufficiency, the gorillas also received a combination of fruit, vegetables, and milk approximately three times a day (before morning walk, midday in cages, and following afternoon walk). The gorillas slept in cages at night.

A three-phase, "soft" release strategy was developed for the release sites so that the gorillas could become habituated to the forest prior to supplemental food and shelter being withdrawn. A cage was built at the release site to acclimatize the gorillas to the new environment and provide initial shelter at night. The gorillas moved to the site would initially exist under "controlled liberty." The gorillas continued under the same routine as at the orphanage and remained under constant supervision of human caretakers during the day while having the opportunity to return to a safe place at night. Furthermore, the gorillas would continue to receive supplementary food but be encouraged to feed on natural vegetation. During the second phase of "supervised liberty," the human caretakers would continue to monitor the group's progress but adopt a policy of noninterference. It was hoped that, during this phase, after approximately 18 months, the gorillas would be sleeping in the forest and all supplementary foods withdrawn (Attwater et al., 1991). It was proposed that the third phase, "noninterference liberty," should be reached after a period of three years, by which time the group should be independent and encouraged to move away from the immediate project area whilst remaining in the protected reserve (Attwater et al., 1991). Initially, monitoring consisted of locating the gorillas on a daily basis and noting group dynamics and general health. More recently, PPG has started to collect systematic data on activity patterns, diet and environmental variables (Cousins, 2002; King et al., 2003). However, there is a limited window of opportunity for direct data collection; as the gorillas get older they become increasingly aggressive toward human followers.

Over the intervening years PPG has modified its prerelease training strategies in an attempt to prevent the gorilla-human dependency from developing rather than to have to sever it latterly; the gorillas appear to retain the dependent behavior as exhibited by their attraction to human habitation. While the strategy remains "soft," contact with human caretakers, as far as possible, is one of noninterference from the beginning while still providing necessary support. PPG has found that gorillas benefit behaviorally and emotionally if encouraged to live in the forest full-time as soon as possible; they become less interested in people and more interested in foraging and playing among themselves. Consequently, the gorillas now spend all day in the forest and are encouraged from an early age to build nests and sleep in the forest as soon as possible. Gorillas start folding leaves in nest-building attempts from when they start to walk, and some as young as six months have been observed making functional nests. The youngest age that a gorilla at PPG started to sleep in their forest nest was at approximately two years old. The stage at which the gorillas have supplementary food completely withdrawn is dependent upon age and individual progress in the reintroduction process, but the youngest was approximately three years of age.

2.4.3. Numbers Released and Survivorship

2.4.3.1. PPG Congo

The first gorillas were transferred to the Lesio-Louna Sanctuary in 1994, and since then, 22 individuals have been subject to a three-phase release strategy in three separate stages (stages shown as R1–R3 in Table 3.3). In September 2004, all gorillas were transferred to the Lefini site once it became clear that the ecological barriers at the Lesio-Louna were inadequate. Table 3.3 provides details on the number and sex of gorillas released at each stage, their age on arrival at PPG, time spent in the rehabilitation phase before becoming independent, age at full independence post-release (sleeping in the forest and receiving no supplementary foods), and summary of present status. There were also three females reintroduced in 2006 (not shown in Table 3.3).

The three separate stages were dictated by the formation of similar age-based groups; the average age of gorillas arriving at PPG Congo was 14 months (range of 4 to 36 months [this does not include one infant born on-site]). The average time that the gorillas spent in the rehabilitation phase was 5 years (range 22 months to 8.5 years), and average age at independence was 5 years (range 2.5 to 11 years).

Due to the varying results for each release stage, it is worth describing the background for each one. The first release stage (R1) involved six gorillas that had spent on average over six years in the rehabilitation phase, the majority of which were spent at the Brazzaville center. Four released males were returned to captivity when they exited the natural boundaries of the reserve, strayed close to villages, and displayed aggressive behavior towards villagers and staff (age range at the time was 10–13 years). One male was returned approximately one year post-independence, while the others remained at liberty for three, four, and five years, respectively. During this time, group fissions occurred, and some males roamed solitary. The attraction to people probably reflects over-habituation during the rehabilitation stage at the Brazzaville center. Furthermore, these gorillas remained at the Lesio-Louna Sanctuary throughout the 1997–1998 civil war, and their protection, not reintroduction, became a priority. They were encouraged (with food) to stay around the cage area at the reintroduction site where they could be monitored. This over-habituation probably resulted in both psychological and nutritional dependence. However, the absence of male bachelor groups in western lowland gorillas (Parnell, 2002) (this group did contain one female

TABLE 3.3. Mean age (±SD) of gorillas at arrival, time spent in rehabilitation, and age at independence for each release phase.

Release stage	Sex (n) ♂	♀	Age at arrival Mean	±SD	Time in rehabilitation (pre-release training phase) Mean	±SD	Age at full independence post-release Mean	±SD	Summary of present status (as of April 2006)
Congo									
R1	5	1	1y, 7m	10m	6y, 5m	12m	8y, 2m	18m	4♂ returned to cage after straying close to villages 1 death (♂) (PI) 1♀ successfully released since May 1996, gave birth April 2004 - now with R2
R2	4	3	11m	7m	5y, 6m	18m	6y, 7m	14m	4 fully independent since December 1998. 2 deaths (1♂ 1♀) & 1 disappearance (♀) (all PI)
R3	4	5	14m	7m	4y, 8m	15m	5y, 9m	16m	All 9 fully independent since May 2001
Gabon									
R1	9	9	1y, 1m	13m	3y, 2m	10m	4y, 3m	17m	13 fully independent since 2004. 3 deaths (♂) & 2 disappearances (♀) (1 disappearance PI)
R2	4	5	2y, 9m	20m	2y, 6m (5y, 1m)	7m (21m)	5y, 5m	18m	All 9 fully independent since December 2005. Non-bracket figures for time in rehabilitation phase refer to Gabon site only. Bracket figures include time spent in zoo as part of rehabilitation phase for individuals transferred from UK

Key: SD, Standard deviation; Y, years; M, months; PI, post-independence (no nutritional support or provision of sleeping site).

but the ratio was still significantly skewed), and problems posed by surplus males in captivity (Stoinski *et al.*, 2004a, 2004b), urges caution when forming groups for reintroduction. Had the roving males encountered opportunities for female gorilla-related social openings (male emigration in western lowland populations is the most common male dispersal strategy; Parnell, 2002), perhaps they would not have strayed so close to human habitation. The remaining male of this group survived over five years living as an independent gorilla but died from unknown causes (aged approximately 11 years). The female continues to live independently and has done so for approximately 10 years. She subsequently joined the gorillas released at stage two (R2) and successfully gave birth in April 2004 (King, 2004). Therefore, although the overall success of R1 is disappointing, with only one gorilla remaining at liberty, she has produced the first baby born to a reintroduced gorilla. Furthermore, all the gorillas were able to live independently, in some cases for several years.

Although the R2 group also experienced some pre-release training at Brazzaville, they spent more than one year at the release site, with whole days in suitable gorilla habitat from an earlier age than R1 gorillas. Consequently, R2 gorillas spent less time in the rehabilitation phase and reached full independence at an earlier age (Table 3.3). Four gorillas from this group (two males and two females) have been living independently for more than seven years. The remaining three gorillas also managed to live independently for 1, 1.5, and 2.5 years, respectively; however, one female disappeared and is presumed dead (aged approximately seven years). The remaining male and female died (aged approximately 9 and 10 years, respectively) as a result of injuries inflicted by male gorillas from R1 (read King *et al.*, 2005a, for a discussion concerning psychological versus physical impact of injuries). The scenarios surrounding the attacks are not known, but in the wild, although intergroup fighting between adult silverback gorillas can be fatal, gorillas of blackback age (8–12 years; Parnell, 2002) would normally still be with the natal group, and the resident silverback could intervene in any serious fights. The reintroduced male gorillas are essentially premature solitary silverbacks. While transferring a released male that is threatening females to semi-captivity may be a consideration, as with wild populations, males may take over a group and kill the resident group babies. Clearly, decisions have to be taken on when to intervene or not if behaviors, although unsavoury, represent wild gorilla behavior.

Gorillas released at stage three (R3) have benefited from being located at the release site from the very beginning. All nine R3 gorillas have been successfully living independently for five years since they were approximately six years old. Three gorillas in the pre-release phase have also been located at the release site from the beginning where they were supported with supplementary food and slept in cages at night. They were integrated into R3 in 2006 and are fully independent (these gorillas are not included in Table 3.3). In Table 3.3 it is clear that the average time spent in rehabilitation and age at independence decreased progressively from R1 to R3. This reflects the change in pre-release training protocols—the gorillas spending whole days in the forest versus half days, and encouraging independence as early as possible. As of April 2006, 14 gorillas (64%), 8 females (75%) and 6 males (57%), have been successfully reintroduced in that they have been living independently for an average of 7 years (range of 5–10 years).

2.4.3.2. PPG Gabon

PPG Gabon is much younger than its sister project in the Congo, however, as of April 2006, 22 (82%) gorillas are living independently, 12 females (86%) and 10 males (77%). Three gorillas died; two were captive-born males transferred from Howletts and Port Lympne Zoo in England. One died quite soon after arrival, from an appendicitis (aged approximately three years), whereas the second lived for more two years and was still receiving pre-release support when he died of peritonitis caused by oesophagostomum (aged approximately five years). The third wild-born male drowned (aged approximately 2.5 years); he was also receiving pre-release support when he died. A wild-born female went missing approximately one-year post-independence (aged 9.5 years). Howletts and Port Lympne Zoo has transferred nine gorillas in total from the U.K. to Gabon; two males in 1999 and five males plus two females in 2003. Both males transferred in 1999 died. However, the seven transferred in 2003 are all released and living independently. The rehabilitation process was adapted for the transferred gorillas. For the first few months, only famil-iar people were allowed around the gorillas to minimize stress. While the transferred captive-born gorillas saw the wild-born gorillas in adjacent sleep-ing cages at night, it took several months to gradually introduce them to each other during daytime forest excursions. The rationale for introducing to the two groups was to facilitate the adjustment of the captive-born gorillas to forest life through exposure to more experienced wild-born counterparts.

Table 3.3 shows that, from the very beginning of this project, the time that the gorillas spent in the rehabilitation phase and age at independence is, on average, similar to gorillas from R3 at PPG Congo (if pre-release training phase includes both time in the U.K. zoo and at the Gabon site for trans-ferred captive-born gorillas, then time in rehabilitation increases significantly, see bracketed figures in Table 3.3). This clearly illustrates that PPG Gabon has benefited from lessons learned at PPG Congo and that the new approach of getting the gorillas into the forest as young and as soon as possible is reaping rewards. However, time will tell as the gorillas mature whether this approach has successfully prevented gorilla–human overdependence and subsequent aggression towar humans as witnessed at PPG Congo.

2.4.4. Reintroduction and Its Contribution to Gorilla Conservation

The question remains, has the reintroduction of gorillas contributed to their conservation? The guidelines define the principal aim of any primate reintroduction as the establishment of a viable, self-sustaining population in the wild to an area from which it has become extinct (Baker, 2002). The overall objectives of a reintroduction should include enhancing the long-term survival of a species, maintaining and/or restoring natural biodiversity, providing long-term economic benefits to the local and/or national economy, and promoting conservation awareness. While supplementing individual gorillas into existing groups does meet the principal aims and objectives of reintroduction, these attempts nevertheless highlighted the plight of these endangered species and provided an opportunity to raise conservation awareness. However, very little information has been published on any of these cases and therefore the extent of awareness raised is likely minimal. If and how the IUCN aims and objectives have been met by PPG are discussed in more detail below.

2.4.4.1. Establishment of a Viable, Self-sustaining Population and Enhancing the Long-Term Survival of a Species?

PPG Congo and Gabon have successfully reintroduced 14 and 22 gorillas, respectively, that have been living independently (for a few months ranging to 10 years) in the Lefini and Mpassa Reserves. Preliminary ad libitum observations indicate the gorillas are living normally in terms of behavior and diet (viable and self-sustaining) (Cousins, 2002; King et al., 2003; King, 2004). However, the reintroduced groups are presently of artificial composition with relatively similar-aged, random sex-ratios, and unrelated individuals, which, at least upon the time of full independence post-release (phase three in the reintroduction process), form juvenile or sub-adult groups. In contrast, naturally occurring lowland gorilla groups are generally composed of a single "silver-back" adult male, plus several adult females, sub-adult "black-back" males, juveniles, and infants (Parnell, 2002).

Central to long-term viability of species survival is successful reproduction, however, only a small number of gorillas at PPG Congo are reaching sexual maturity, and none at PPG Gabon are yet mature. Revisiting this question in a few years will provide more answers on reproduction rate and ability to rear young. However, the first gorilla birth at PPG Congo over two years ago has so far demonstrated successful ability to rear young. The mother was relatively old when she was orphaned and this may have afforded her some of the necessary skills to competently care for the baby. If she is successful, this group will become the first typical mixed-age family group within the project (King, 2004).

2.4.4.2. Maintaining and/or Restoring Natural Biodiversity?

PPG has successfully reintroduced a species to an area from which wild populations have been extirpated for over 50 years (Courage et al., 2001; Cousins, 2002; King, 2004). Unfortunately, there has been little systematic census of mammal populations undertaken in the Lesio-Louna, Lefini, or Mpassa Reserves to provide detailed pre- and post-reintroduction comparisons. Early studies indicate that large mammal densities were low at Lesio-Louna due to heavy hunting pressure (Attwater *et al.*, 1991). However, anecdotal evidence suggests that, since the project occupied the core area of the Reserve in 1994, numbers of various mammal species have increased (Watkin, 2001; King, 2005). Preliminary studies of the avifauna identified nearly 300 species in Lesio-Louna and 210 species at Mpassa. Importantly, the project presence has served to focus attention on regional conservation issues. This resulted in the Lesio-Louna Sanctuary being decreed a Reserve in 1999, and Mpassa Sanctuary in Gabon being incorporated into the Batéké Plateau National Park in 2002, lending protection to all wildlife in the area. A detailed survey examining the specific impact of reintroduced gorillas on flora and fauna in these areas is needed. However, overall, it does seem as if the reintroduction is helping to restore natural biodiversity (King, 2005). Importantly, part of a unique central African landscape is now being protected because of the presence of PPG.

2.4.4.3. Long-Term Economic Benefits to the Local and/or National Economy?

Benefits of PPG to the economy presently extend only to the local level through employment. PPG Congo and Gabon employ about 40 workers (recruited from the villages surrounding the Reserve/Park), contributing to the local economy in salaries and subsequent spending.

One objective of PPG is to develop the necessary infrastructure to promote gorilla tourism at both the local and international level, which, if successful, may generate significant income to both the local communities and the government. Reintroduction projects and sanctuaries are affected by the same problems that afflict all conservation projects in Africa with regard to long-term stable and cost-effective tourism. Political instability in Congo impacts PPG as much as any other protected area in the country, except that the site is near enough to the capital Brazzaville to be visited in one day (150 km on paved highway). PPG Gabon has more potential for tourism because Gabon is politically stable and the authorities have the political will and power to enforce conservation laws. In 2002, 13% of Gabonese territory was classified as protected, including the Batéké Plateau National Park.

PPG predicts tourism on two levels: individuals who wish to visit the Reserve in general and gorilla tourism. It is important that these sites are attractive tourist destinations in their own right, quite apart from the presence of the reintroduced gorillas. Both sites have been tourist destinations for expatriates living near the plateau since the 1950s. While a tourism component is part of the agreement with the Congolese authorities, in Gabon PPG is not allowed to develop tourism at the site. In theory, sanctuary-reared gorillas are less likely to be disturbed by tourist viewing and also have some degree of immunity to viruses, such as the common cold. Prior to reintroduction, all gorillas are vaccinated against several infectious diseases transmitted by humans. Conversely, although the majority of the reintro-duced gorillas have shown nutritional independence, there have been problems with some individuals (particularly solitary males) entering villages, stealing food, and even being aggressive toward villagers and project staff (Watkins, 2002; King, 2005; King *et al.*, 2005b). Unfortunately, the reintro-duced gorillas do not respond with fear to noise as with wild gorillas. In Rwanda, villagers have learnt what not to plant to avoid human-wildlife conflict; for example, gorillas do not like potatoes but are attracted to bananas. However, it is not known whether it is food that is attracting the released gorillas. As mentioned earlier, when PPG Congo was in its infancy, the reintroduction procedure probably practiced over-habituation. The suc-cess of gorilla tourism at PPG Congo will ultimately depend upon the success of preventing rather than severing the psychological tie of gorillas to humans.

2.4.4.4. Promotion of Conservation Awareness?

The proximity of PPG to Brazzaville has allowed the project to take advan-tage of the Congolese media. Since 1998, when the displaced gorillas returned from Pointe Noire, a long-running television reality-documentary on the project was shown on television, also abridged for broadcast by Congo Radio. Exposés about the gorillas at Lesio-Louna and Lefini and the plight of their wild counterparts have been presented at the French Cultural Centre in Brazzaville. These have helped to raise PPG's profile in Brazzaville and fur-ther afield. The long-term presence of the projects in both countries has contributed significantly to raising public awareness about the gorilla and the sanctuaries, but this has been largely due to daily contact with PPG agents rather than through a systematic and sustained education program. Structured outreach education programs are missing from the PPG strategy.

3. Conclusion

Our inability to critically evaluate the role of sanctuaries in conservation emphasizes the need for evaluative techniques to be incorporated into program design. However, sanctuaries clearly can play a role in conservation, particularly with regard to reintroduction, law enforcement, and conservation education.

Formerly, PPG produced little published material except for a small number of articles in the popular press. However, with the instigation of systematic data collection, this is changing. Valuable data are becoming available to allow behavioral comparisons to wild conspecifics and reveal the extent of post-release adaptation. The future of reintroduction as a tool to manage wildlife depends upon careful planning, generalizing the results from successful projects to reduce costs, and then documenting results and experiences (Stanley-Price, 1989). Published protocols will contribute toward a better understanding of all the issues involved by broadening our limited knowledge about gorilla reintroduction (e.g., the selection of the release site, medical protocols, and rehabilitation techniques). The history of the gorillas at PPG (except those transferred from the U.K.) are probably representative of many gorillas presently being cared for in African sanctuaries, and revised methodologies can help those considering reintroduction as an option. This project has shown that gorillas can become nutritionally self-sufficient in the natural environment but can easily become over-habituated if captivity and human contact is prolonged. Results indicate that pre-release training at the release site and provision of forest experience as soon as possible with minimal human contact can facilitate transition back to the wild.

The illegal trade in wildlife presents a serious threat to the survival or conservation of many endangered species, but, despite this, law enforcement has traditionally been a neglected directive in the conservation community. Sanctuaries can fill this void by providing a facility in which to place confiscated wildlife. The much publicized case of the "Taiping 4" and the recent repatriation of two gorillas from Nigeria to the LWC in Cameroon served to highlight that the illegal trade is still a problem and the central role that sanctuaries can play in facilitating law enforcement (Dewar, 2003). Only now, more than a decade after a law was passed in Cameroon prohibiting the trade in endangered species, has the first national received a jail sentence for trying to sell a chimpanzee (Ngwa Niba, 2003). The chimpanzee is now living at the LWC, yet without this facility, not only would there have been nowhere to place her, but also it is highly unlikely that sufficient interest would have been generated to attempt such a conviction. The placement of orphaned confiscated wildlife within *in situ* sanctuaries, and the national and international media attention generated, illustrates the potential role that individual animals can play in promoting conservation messages.

An essential component, and key to successful conservation efforts, is education (Tutin and Vedder, 2001). Properly managed *in situ* sanctuaries, with an understanding of local values and attitudes, can play an important role in environmental education and in nurturing respect for animals and their environment (Karesh, 1995). This form of localized education theoretically has far more practical conservation potential than zoos in countries that do not have indigenous primate populations. Educational messages must be relevant to the visiting public, and clearly American and European audiences may feel distant from complex African issues (Stoinski *et al.*, 2001). However, as it is the people of Africa who will ultimately decide the fate of their natural heritage, surely they should also be the targets of conservation messages. The discrepancy between the promoted and supported role of American and European zoos in conservation and the neglected role of *in situ* sanctuaries is clearly prejudicial and outdated.

To conclude, this chapter has presented evidence to suggest that sanctuaries can play an important role in gorilla conservation. This book has highlighted the many different approaches adopted in efforts to conserve wild gorilla populations, and clearly most conservation problems are too complex to yield to a single solution. Perhaps then we should follow the example of the modern-day medical practitioner and accept differing methodologies as complementary rather than as competing alternatives. The survival of gorillas in the wild may depend upon it.

Acknowledgments. The authors thank Dan Bucknell, Tony Chasar, David Lucas, Felix Lankester, and Ateh Wilson of the LWC; Neil Maddison of Bristol Zoo Gardens/CWAF; and Liz Pearson and Tony King of PPG, for their invaluable contributions to this chapter. We also thank Hannah Buchanan-Smith, James Anderson, Liz Williamson, Anita Rennie, and an anonymous reviewer for commenting on the manuscript. Finally, thanks to Tara Stoinski for inviting this hotly debated topic alongside more traditional approaches to conservation.

References

Ape Alliance. (1998). The African bushmeat trade—a recipe for extinction. London: Ape Alliance.

Attwater, H. (1999). *My Gorilla Journey: Living with the Orphans of the Rainforest.* Sidgwick & Jackson, London.

Attwater, M., Blake, S., Hudson, H., and Kopf, H.O. (1991). *Gorilla Protection Program—Proposal Document.* Unpublished report.

Ammann, A. (2001). Bushmeat hunting and the great apes. In: Beck, B.B., Stoinski, T.S., Hutchins, M., Maple, T.L., Norton, B., Rowan, A., Stevens, E.F., and Arluke, A. (eds.), *Great Apes and Humans, the Ethics of Coexistence.* Smithsonian Press, Washington, pp. 71–85.

Ammann, K., and Pearce, J. (1995). *Slaughter of the Apes: How the Tropical Timber Industry Is Devouring Africa's Great Apes.* World Society for the Protection of Animals, London.

Andre, C. (2002). *The Bonobo Sanctuary of Congo: Why Sanctuaries?* Paper presented at the Support for African/Asian Great Apes symposium. 14–17 November 2002. Inuyama, Japan.

Baker, L.R. (2002). *IUCN/SSC Re-Introduction Specialist Group: Guidelines for Nonhuman Primate Re-introductions.* IUCN, Gland, Switzerland.

Beck, B.B., Rapaport, L.G., Stanley-Price, M.R., and Wilson, A.C. (1994). Reintroduction of captive-born animals. In: Olney, P.J.S., Mace, G.M., and Feistner, A.T.C. (eds.), *Creative Conservation: Interactive Management of Wild and Captive Animals.* Chapman & Hall, London, pp. 265–286.

Bowen-Jones, E. (1998). *The African Bushmeat Trade—A Recipe for Extinction.* Ape Alliance, London.

Brend, S. (2001). Do sanctuaries have a role in primate conservation? *Primate Eye* 75:44–45.

Boulton, M.N., and Knight, D. (1996). Conservation education. In: Spellerberg, I.F. (ed.), *Conservation Biology,* Longman, Harlow, Essex, pp. 69–79.

Butynski, T.M. (2001). Africa's great apes. In: Beck, B.B., Stoinski, T.S., Hutchins, M., Maple, T. L., Norton, B., Rowan, A., Stevens, E.F., and Arluke, A. (eds.), *Great Apes and Humans, the Ethics of Coexistence*. Smithsonian Press, Washington, pp. 3–56.

Campbell, S. (1980). *Lifeboats to Ararat*. McGraw-Hill/New York Times Book Company, New York.

Cherfas, J. (1984). *Zoo 2000—A Look Beyond the Bars*. British Broadcasting Corporation, London.

Costo, S., Bern, A.W., Atemnkeng, W., and Lankester, F. (2005). Report on the nature club and outreach program 2005: addressing the bushmeat crisis in Cameroon. The Government of Cameroon and The Pandrillus Foundation. Unpublished.

Courage, A., Henderson, I., and Watkin, J. (2001). Orphan gorilla reintroduction: Lesio-Louna and Mpassa. *Gorilla Journal* 22:33–35.

Cousins, D. (2002). Natural plant foods utilized by gorillas in the former Brazzaville Orphanage and the Lesio-Louna Reserve. *International Zoo News* 49(4):210–218.

Cowlishaw, G., and Dunbar, R. (2000). *Primate Conservation Biology*. The University of Chicago Press, Chicago.

Cox, D., Rosen, N., Montgomery, C., and Seal, U. (2000). *Chimpanzee Sanctuaries: Guidelines and Management Workshop Report*. Conservation Breeding Specialist Group (SSC/IUCN), Apple Valley, MN.

Dewar, J. (2003). Gorillas in the news: The Taiping 4/The Kano 2. *Gorilla Gazette* 16(1):54–55.

Dietz, L.A., and Nagangata, E. (1995). Golden lion tamarin conservation program: a community educational effort for forest conservation in Rio de Janeiro State, Brazil. In: Jacobson, S.K. (Ed.), *Conserving Wildlife—International Education And Communication Approaches*, Columbia University Press, NY, pp. 64–86.

Eudey, A.A. (1987). *Action Plan for Asian Primate Conservation*. IUCN/SSC Primate Specialist Group, Gland, Switzerland.

Fa, J., Peres, C., and Meeuwig, J. (2002). Bushmeat exploitation in tropical forests: an intercontinental comparison. *Conservation Biology* 16:232.

Farmer, K.H. (2002a). Pan-African Sanctuary Alliance: status and range of activities for great ape conservation. *American Journal of Primatology* 58:117–132.

Farmer, K.H. (2002b). *The Behaviour and Adaptation of Reintroduced Chimpanzees (Pan troglodytes troglodytes) in the Republic of Congo*. Unpublished doctoral dissertation. University of Stirling, Scotland.

Fien, J., Scott, W., and Tilbury, D. (2001). Education and conservation: lessons from an evaluation. *Environmental Education Research* 7(4):379–395.

Fischer, J., and Lindenmayer, D.B. (2000). An assessment of the published results of animal relocations. *Biological Conservation* 96:1–11.

Gadsby, E.L. (2002). Preparing for re-introduction: 10 years of planning for drills in Nigeria. In: Soorae, P.S., and Baker, L.R. (eds.), *Reintroduction News*. Newsletter of the Reintroduction Specialist Group (21), IUCN/SSC Reintroduction Specialist Group, Abu Dhabi, UAE, pp. 20–23.

Gipps, J.H.W. (1991). *Beyond Captive Breeding Reintroducing Endangered Mammals to the Wild*. Clarendon Press, Oxford.

Gold, K., and Benveniste, P. (1995). *Visitor Behaviour and Attitudes Towards Great Apes at Lincoln Park Zoo*. Annual Proceedings of the American Zoo and Aquarium Association Conference: 152–182.

Goodall, J. (2001). Problems faced by wild and captive chimpanzees: finding solutions. In: Beck, B.B., Stoinski, T.S., Hutchins, M., Maple, T.L., Norton, B., Rowan, A., Stevens, E.F., and Arluke, A. (eds.), *Great Apes and Humans, the Ethics of Coexistence*. Smithsonian Press, Washington, DC, pp. xiii–xxiv.

Homsy, J. (1999). *Ape Tourism and Human Diseases: How Close Should We Get?* International Gorilla Conservation Programme, Nairobi. Unpublished.

IUCN/SSC Re-Introduction Specialist Group. (1995). IUCN Guidelines for Re-introductions. IUCN, Gland, Switzerland.

IUCN/SSC Re-Introduction Specialist Group. (1998). IUCN Guidelines for Re-introductions. IUCN, Gland, Switzerland.

IPPL News (2002). Animal dealer offers gorilla babies. *IPPL News* 29(1):3–4.

Jacobson, S.K., and McDuff, M.D. (1998). Conservation education. In: Sunderland, W.J. (ed.), *Conservation Science and Action*. Blackwell Science, Oxford, pp. 237–255.

Jordan, B., and Ormrod, S. (1978). *The Last Great Wild Beast Show*. Ebenezer Baylis and Son, London.

Karesh, W.B. (1995). Wildlife rehabilitation: additional considerations for developing countries. *Journal of Zoo and Wildlife Medicine* 26(1):2–9.

Kellert, S. (1996). *The Value of Life*. Island Life, Washington.

Kemf, E., and Wilson A. (1997). *Great Apes in the Wild*. World Wide Fund for Nature, Gland, Switzerland.

Kidd, A.H., and Kidd, R.M. (1996). Developmental factors leading to positive attitudes toward wildlife and conservation. *Applied Animal Behaviour Science* 47:199–125.

King, T. (2004). Reintroduced western gorillas reproduce for the first time. *Oryx* 38(3):251–252.

King, T. (2005). Gorilla reintroduction program, Republic of Congo. *Gorilla Gazette* 18:28–31.

King, T., Boyen, E., and Muilerman, S. (2003). Variation in reliability of measuring behaviours of reintroduced orphan gorillas. *International Zoo News* 50(5):288–297.

King, T., Chamberlain, C., and Courage, A. (2005a). Rehabilitation of orphan gorillas and bonobos in the Congo. *International Zoo News* 52(4):198–209.

King, T., Chamberlain, C., and Courage, A. (2005b). Reintroduction in gorillas: reproduction, ranging and unresolved issues. *Gorilla Journal* 30:30–32.

Kleiman, D.G. (1989). Reintroduction of captive mammals for conservation. *Bioscience* 39:152–161.

Kleiman, D.G., Beck, B.B., Dietz, J.M., Dietz, L.A., Ballou, J.D., and Coimbra-Filho, A.F. (1986). Conservation program for the golden lion tamarin: captive research and management, ecological studies, educational strategies, and reintroduction. In: Benirschke, K. (ed.), *Primates—The Road to Self-sustaining Populations*. Springer-Verlag, New York, pp. 959–979.

Mackinnon, J. (1977). The future of orang-utans. *New Scientist* 74:697–699.

McDuff, M. (2000). Thirty years of environmental education in Africa: the role of the wildlife clubs of Kenya. *Environmental Education Research* 6(4):383–396.

McKenzie-Mohr, D., and Smith, W. (1999). *Fostering Sustainable Behavior*. New Society Publishers, Gabriola Island, Canada.

Mills, W., Cress, D., and Rosen, N. (2005). Pan African Sanctuary Alliance (PASA) 2005 Workshop Report. Conservation Breeding Specialist Group (SSC/IUCN), Apple Valley, MN.

Nellemann, C., and Newton, A. (2002). *The Great Apes—The Road Ahead. A Global Perspective on the Impacts of Infrastructural Development on the Great Apes.* United Nations Environment Programme, Nairobi, Kenya.

Ngwa Niba, F. (2003). Jail for Cameroon chimp trafficker. BBC World Edition Online 24.07.03. http://news.bbc.co.uk/2/hi/uk_news/3091315.stm

Oates, J. (1986). *Action Plan for African Primate Conservation.* IUCN/SSC Primate Specialist Group, Gland, Switzerland.

Oates, J. (1996). *African Primates Status Survey and Conservation Action Plan.* IUCN/SSC Primate Specialist Group, Gland, Switzerland.

Oates, J. (1999). *Myth and Reality in the Rain Forest. How Conservation Strategies Are Failing in West Africa.* University of California Press, Berkeley. 310 pages.

Parnell, R.J. (2002). Group size and structure in western lowland gorillas (*Gorilla gorilla gorilla*) at Mbeli Bai, Republic of Congo. *American Journal of Primatology* 56:193–206.

Rao, M., and McGowan, P.J.K. (2002). Wild meat use, food security, livelihoods and conservation. *Conservation Biology* 16(3):580–583.

Reade, L.S., and Waran, N.K. (1996). The modern zoo: how do people perceive zoo animals? *Applied Animal Behaviour Science* 47:109–118.

Robinson, J., and Bodmer, R. (1999). Towards wildlife management in tropical forests. *Journal of Wildlife Management* 63(1):1–13.

Rosen, N., Seal, U.S., Montgomery, C., and Boardman, W. (2001). *Pan-African Sanctuaries Alliance (PASA) Workshop Report.* Conservation Breeding Specialist Group (SSC/IUCN), Apple Valley, MN.

Soave, O. (1982). The rehabilitation of chimpanzees and other apes. *Laboratory Primate Newsletter* 21(4):3–8.

Soorae, P.S., and Baker, L.R. (2002). *Re-introduction News*: Special Primate Issue, Newsletter of the IUCN/SSC Re-introduction Specialist Group, Abu Dhabi, UAE. No. 21: 60 pp.

Stanley-Price, M.R. (1989). *Animal Reintroductions: The Arabian Oryx in Oman.* Cambridge University Press, Cambridge.

Stuart, S.N. (1991). Reintroductions: to what extent are they needed? In: Gipps, J.H.W. (ed.), *Beyond Captive Breeding Reintroducing Endangered Mammals to the Wild.* Clarendon Press, Oxford, pp. 27–37.

Stevenson, M., Baker, A., and Foose, T.J. (1992). *Global Captive Action Plan for Primates.* IUCN/SSC Captive Breeding Specialist Group report.

Stoinski, T.S., Kuhar, C.W., Lukas, K.E., and Maple, T.L. (2004a). Social dynamics among captive groups of bachelor gorillas. *Behaviour* 141:169–195.

Stoinski, T.S., Lukas, K.E., Kuhar, C.W., and Maple, T.L. (2004b). Factors influencing the formation and maintenance of all-male gorilla groups in captivity. *Zoo Biology* 23:189–204.

Stoinski, T.S., Ogden, J.J., Gold, K.C., and Maple, T.L. (2001). Captive apes and zoo education. In: Beck, B.B., Stoinski, T.S., Hutchins, M., Maple, T.L., Norton, B., Rowan, A., Stevens, E.F., and Arluke, A. (eds.), *Great Apes and Humans, the Ethics of Coexistence.* Smithsonian Press, Washington, DC, pp. 113–132.

Synder, N.F.R.S.R., Derrickson, S.R., Beissinger, J.W., Wiley, T.B., Smith, W.D., and Miller, B. (1996). Limitations of captive breeding in endangered species recovery. *Conservation Biology* 10:338–348.

Teleki, G. (2001). Sanctuaries for ape refugees. In: Beck, B.B., Stoinski, T.S., Hutchins, M., Maple, T.L., Norton, B., Rowan, A., Stevens, E.F., and Arluke, A. (eds.), *Great*

Apes and Humans, the Ethics of Coexistence. Smithsonian Press, Washington, DC, pp. 133–149.

Thompson, T.L., and Mintzes, J.J. (2002). Cognitive structure and the affective domain: on knowing and feeling in biology. *International Journal of Science Education* 24(6):645–660.

Tudge, C. (1991). Last animals at the zoo: how mass extinction can be stopped. Hutchinson Radius, London.

Tutin, C.E.G., Ancrenaz, M., Paredes, J., Vacher-Vallas, M., Vidal, C., Goossens, B., Bruford, M., and Jamart, A. (2001). Conservation biology framework for the release of wild-born orphaned chimpanzee into the Conkouati Reserve, Congo. *Conservation Biology* 15(5):1247–1257.

Tutin, C.E.G., and Vedder A.(2001). Gorilla conservation and research in Central Africa: a diversity of approaches and problems. In: W. Weber, L.J.T. White, A. Vedder, and L. Naughton-Treves (eds.), *African Rain Forest Ecology and Conservation*. New Haven, CT, Yale University Press, pp. 429–448.

Wagner, K., Ogden, J., and Vernon, C. (2000). Zoos and aquariums—do we make a difference? *Communiqué* (Sept):10.

Wallis, J. (1997). From ancient expeditions to modern exhibitions: The evolution of primate conservation in the zoo community. In: Wallis, J. (ed.), *Primate Conservation: The Role of Zoological Parks*. American Society of Primatologists, San Antonio, TX, pp. 1–27.

Watkin, J.R. (2001). *A Guide to the Lésio-Louna Reserve and Projet Protection des Gorilles République du Congo*. Unpublished.

Watkin, J. (2002). Going Ape. *Swara* (East African Wildlife Society) 25(3):24–28.

Weber, W. (1995). Monitoring awareness and attitude in conservation education: the mountain gorilla project in Rwanda. In: Jacobson, S.K. (ed.), *Conserving Wildlife — International Education and Communication Approaches*, Columbia University Press, New York, pp. 28–48.

Wilkie, D.S., and Carpenter, J.E. (1999). Bushmeat hunting in the Congo basin: an assessment of impacts and options for mitigation. *Biodiversity and Conservation* 8:927–955.

Wilkie, D.S. (2001). Bushmeat trade in the Congo Basin. In: Beck, B.B., Stoinski, T.S., Hutchins, M., Maple, T.L., Norton, B., Rowan, A., Stevens, E.F., and Arluke, A. (eds.), *Great Apes and Humans, the Ethics of Coexistence*. Smithsonian Press, Washington, DC, pp. 86–112.

Whittier, C., Mbula, D., Mudakikwa, T., Fawcett, K., and Gray, M. (2005). Confiscated Mountain Gorilla Maisha: Report on care and disposition recommendations. Prepared by the Scientific Technical Steering Committee (Mountain Gorilla Veterinary Project, Office Rwandais du Tourisme et Parcs Nationaux, Institute Congolais pour la Conservation du Nature, Dian Fossey Gorilla Fund International, International Gorilla Conservation Program). Unpublished.

Chapter 4
Responsible Tourism: A Conservation Tool or Conservation Threat?

Carla A. Litchfield

1. Introduction

The year 2003 marked the 30th anniversary of organized gorilla tourism, with tens of thousands of international visitors catching a precious glimpse of the gorillas' fragile equatorial ecosystem (Weber, 1993). For three decades, gorilla trekkers have stepped into a breathtakingly beautiful African landscape, steeped in human and gorilla blood. Gorilla lives and deaths have been played out against a backdrop of human war, genocide, poverty, and disease, seemingly unnoticed by the international community at large (Stanford, 2001; Weber and Vedder, 2001). During these 30 years, global international tourist arrivals per year have increased by about 500 million (World Tourism Organization, 2000), and more than 30 new diseases have emerged (World Health Organization, 2002). Ebola hemorrhagic fever decimates gorilla populations in western equatorial Africa (Walsh *et al.*, 2003), and the rapid global spread of SARS coronavirus, shows how easily new diseases may be spread by international (air) travelers (World Health Organization, 2003a).

While tourism can help fund conservation, the 100th anniversary of the discovery of mountain gorillas in 2002 (Schaller, 1963), coinciding with the *International Year of Ecotourism*, was marred by terrorist attacks in Djerba, Bali, and Mombasa. Unpredictable global tourism trends, as a result of the war in Iraq and continued terrorist attacks, mean that conservation managers in Africa cannot afford to rely on gorilla tourism funds alone (Blom, 2001a).

Most people associate gorilla tourism with the mountain gorillas (*Gorilla beringei beringei*) of Rwanda/Uganda/Democratic Republic of Congo (DRC), but the first organized gorilla tourism project was established six years earlier at Kahuzi-Biega National Park in eastern DRC with Grauer's (*Gorilla beringei graueri*) or eastern lowland gorillas (Weber, 1993; Meder and Groves, 2005). The intervening years have seen these early sites devastated by regional conflicts, with two fledgling Ugandan sites emerging as the premiere gorilla tourism sites (Litchfield, 2001a). New western lowland gorilla (*Gorilla gorilla gorilla*; Meder and Groves, 2005) tourism sites have sprung up in Central African Republic, Gabon, Congo-Brazzaville, Cameroon, and equatorial

Guinea (Aveling, 1999; Blom, 2001a; Djoh and van der Wal, 2001). Only the rare Cross River gorillas (*Gorilla gorilla diehli*; Meder and Groves, 2005), restricted to small populations at the Nigeria–Cameroon border, remain to be visited by tourists, although plans to habituate the remaining gorillas (<30) for tourism at Afi Mountain Wildlife Sanctuary have been proposed by the Cross River State Forestry Commission (Dunn, 2005).

Can *responsible tourism* save gorillas, as war, deforestation, mining, disease, and the bushmeat trade rapidly push them toward extinction? Tourism has been recognized as an important conservation management tool to protect gorillas, yet itself may pose a threat to their survival (Homsy, 1999; Butynski, 2001). For the purposes of this report, the term *responsible tourism* (which cares for the earth, and means, simply, not exploitation, but sharing), rather than *ecotourism* will be used (Litchfield, 1997, 2001b). The term *ecotourism* has been attached to ventures that have resulted in "deleterious impacts," and the nature of the *ecotourism* industry as a whole has been "ill-defined" (Rabinor, 2002).

2. Tourism Trends and Sustainability Issues

More than three million international tourists per year travel to West and Central Africa, whilst a further six million visit East Africa. By 2020, 77 million international visitors are expected to visit Africa (World Tourism Organization, 2000). "Post-September 11 syndrome" did not impact on international tourist arrivals to sub-Saharan Africa (World Tourism Organization, 2003a), but continuing international tensions may have a considerable, if short-term, impact. Although the World Tourism Organization figures reflect mass tourism numbers rather than select groups likely to visit gorillas, overall trends could help managers anticipate increasing numbers of tourists, or conversely, low numbers.

Three main goals of ecotourism (in its ideal form) apply to *responsible* gorilla tourism. First, local communities must benefit "without overwhelming their social and economic systems" (Dawson, 2001, p.41). Second, all aspects of the "resource base" (gorillas and their habitat, and cultures of local communities) must be protected. Third, ethical behavior of tourists and tourism operators is required. Only if the ecotourism venture is "limited in scale" and minimizes environmental and social impacts can it be considered a form of sustainable development (Dawson, 2001).

Researchers are still unable to agree on an adequate measure of sustainability (Rao, 2000), but warn that global environmental issues will impact upon all forms of tourism within the next two decades (Mann, 2000). Planes and other forms of transport contribute to greenhouse gas emission, and strategies, such as reforestation programs, must be put in place now (Gössling, 2000). Local community involvement (with leadership provided by local authorities), and incorporation of principles and guidelines of Local

Agenda 21 in all aspects of planning and management is more likely to result in sustainable tourism (United Nations Environment Programme, 2002, 2003).

Embraced as a nonconsumptive and low-impact means of poverty alleviation for developing countries, tourism does not appear to be increasing minimum standards of living, despite increases in tourist numbers (Rao, 2000). The World Tourism Organization is attempting to address this problem with its recently launched Sustainable Tourism-Eliminating Poverty (ST-EP) program.

"Even the most environmentally conscientious tourist will have some degree of impact, however small" (Cater, 1995, p.77). Despite holding pro-environmental *attitudes*, people in developed nations engage in environmentally destructive *behavior* (Tenbrunsel *et al.*, 1997). North America, Europe, and Japan (15% of the world's population) are responsible for up to 80% of the consumption of world resources, and contribution to toxic pollution (Gladwin *et al.*, 1997). About 80% of all international travelers (Mann, 2000) come from these developed nations, and, as tourists, they consume wood, water, energy, and food at unsustainable levels (Rao, 2000).

Wildlife tourism in protected areas is beset with a number of problems and negative impacts (direct and indirect) that are well documented in the literature (Roe *et al.*, 1997; Weaver, 2000; United Nations Environment Programme, 2002). However, there is one overriding problem unique to gorilla (and other great ape) tourism, namely the gorillas' susceptibility to human diseases, as a result of our genetic closeness. International tourists, en route to gorillas, have passed through other countries or even continents. "This represents, from an epidemiological point of view, a very effective means of transport for an increased number of exotic germs due to the speed and diversity of modern transport systems" (Homsy, 1999, p.v). The global outbreak of SARS in 2003 reflects this only too well.

The adoption of strict guidelines and rules at gorilla tourism sites, or other regulatory action to avoid environmental disaster, even in the absence of clear scientific evidence, is compatible with the *Precautionary Principle* (O'Riordan and Cameron, 1994; O'Riordan *et al.*, 2001). Higginbottom (2002) stresses that the *Precautionary Principle* should be adopted if: "(i) a population decline may be difficult or impossible to reverse by the time it is reliably detected; (ii) a population is small and geographically restricted, (iii) a species is of particular conservation and/or public concern" (p.4). This is clearly the case for gorillas. Taking extra precautions is not only prudent but well advised.

3. Gorilla Tourism during Times of Crisis

All tourist sites should develop integrated crisis management plans for dealing with natural or human-induced disasters, based on *worst-case scenarios*. In the event of a disaster, if coupled with honesty and transparency in communication, good crisis management techniques can speed up the process of tourists returning to the destination (World Tourism Organization, 2003b). Worst-case scenarios already experienced at gorilla tourism sites include outbreak of war (e.g., Rwanda and DRC), death of gorillas from Ebola (e.g., Lossi), and kidnap and killing of tourists and park personnel (e.g., Bwindi). If a crisis arises, pre-existing funding and strategies should be activated to protect the gorillas, their habitat, and key personnel on the ground.

Excessively high visitor numbers can result in a number of problems, as witnessed at the two Ugandan mountain gorilla sites in 1997 and 1998, when these were the only sites in the region officially open to tourists (Macfie, 1997). At Mgahinga Gorilla National Park and Bwindi Impenetrable National Park, pushy tourists and tour leaders attempted to bribe park staff to increase visitor numbers to look for gorillas, or allow double visits (Macfie, 1997). In order to manage the day-to-day problems associated with the large number of tourists, rangers may have neglected duties such as antipoaching and boundary patrols, as well as preventing wood cutting in the park (Karlowski and Weiche, 1997). Uganda alone was unable to meet the demands of the tourism industry.

During times of conflict (when tourism is not possible), gorilla conservation organizations can provide training to relief/humanitarian agencies on environmental management during refugee operations (Lanjouw, 2000). This may minimize loss of natural habitat (35 km^2 of Parc National des Virunga was deforested in <2 years). It may also reduce poaching of gorillas (18 mountain gorillas were killed between 1995 and 1998), and transmission of human diseases to wildlife from improper disposal of human and medical waste (used syringes, human waste, and bloodstained materials were dumped within Parc National des Virunga). "Ranger Based Monitoring," developed in 1996 as part of a rehabilitation program for Parc National des Virunga, employs field staff to collect basic information about gorillas, elephants, and humans in the park, thereby informing park management decisions. This simple and cost-effective program now serves as a strategic protected area management tool (Lanjouw, 2000; Gray and Kalpers, 2005). It must be stressed that the success of this tool in the past has rested on the bravery of field staff, who risked and in some cases gave their lives to protect gorillas (e.g., at Karisoke).

Lanjouw's (2000) examples of the strategies implemented by the International Gorilla Conservation Programme provide alternative scenarios that can be incorporated into an integrated crisis management plan for all gorilla tourism sites. Such an integrated crisis management plan should also incorporate contingency plans for excessively high or low visitor numbers in order to prevent overwhelming problems from developing. These might include access to a database for travel agents or tourists wanting to visit gorillas with updates on sites within a region that are best suited to cater for tourists at any given time.

An integrated crisis management plan should also develop strategies for how to protect gorillas and local communities if the tourism project fails.

Several successful gorilla tourism programs have suddenly collapsed, or been disrupted for extended periods of time. Local communities cannot afford to become dependent upon income generated by tourism alone. In the event of a crisis, the question arises, should funding also be made available to local communities involved in the tourism project? The establishment of trust funds (or other stable and sustainable sources of revenue) by international donors and conservation organizations may ensure that gorilla conservation does not rely solely on tourism funds or other "alternative economic ventures" (Blom, 2001b, p.41). Ideally, *all* gorilla tourism sites should receive sufficient funding, irrespective of tourist numbers. A specific fund for the long-term care of gorillas habituated for tourism could provide money to care for gorillas during times when tourism is not possible. Reserve funds set up from tourist income could be used to support site(s) financially during times of tourist scarcity. U.K. tourists seem "prepared to pay a premium of perhaps 5% for guarantees of responsible and sustainable travel" (World Tourism Organization, 2001, p.10). Thus, *responsible tourism* might provide a potential source of funding for conservation, but conservation managers must determine how much money is needed to support each gorilla tourism site, and whether a 5% contribution by each tourist to a reserve fund is a realistic figure.

4. The Economics of Gorilla Tourism

In the 1970s, gorilla tourism was considered to be the "only immediate option capable of galvanising sufficient and immediate support to save mountain gorillas from poaching, habitat encroachment and possible extinction" (Homsy, 1999, p.1). It is still viewed by many as the most "lucrative" and effective gorilla conservation tool (Homsy, 1999). Without mountain gorilla tourism in Uganda, it is unlikely that the tiny Mgahinga Gorilla National Park (<40 km^2) would even exist today.

Some studies have estimated the value of a specific animal to the economy of a country, or the financial value (from tourism revenue) of a park per hectare in its protected state (Ceballos-Lascurain, 1996). However, viewing gorilla conservation or tourism projects merely in terms of monetary gains may not communicate the right messages. It may not make clear that gorillas and their ecosystems (particularly forests) play a vital role in preventing soil erosion, protecting water catchment areas, stabilizing local climates and compensating for greenhouse gas emissions (Werikhe *et al.*, 1997).

4.1. Potential Economic Benefits of Gorilla Tourism

In 1990, prior to the war in Rwanda and the DRC, tourists paid nearly US$2 million in entry fees to visit gorillas. Rwanda's gorilla revenues alone represent more than half of the money earned during the same period at Amboseli National Park in Kenya, but with 430% fewer tourists than at Amboseli

(Weber, 1993). Moyini and Uwimbabazi (2000) provide a detailed economic analysis of gorilla tourism in Uganda during the 1990s, in order to "assess the economic value of the mountain gorilla as a tourism resource" (p.13). More than 17,500 people (predominantly from outside Uganda) visited gorillas at Mgahinga and Bwindi between 1994 and 1999. The average expenditure per visitor was approximately US$768.

Moyini and Uwimbabazi's (2000) estimates are based on full capacity utilization of the three habituated groups of gorillas. At full capacity (8,760 visitors per year—6,570 at Bwindi and 2,190 at Mgahinga), the annual benefit to the Uganda Wildlife Authority is US$2.1 million (gorilla permits and park fees), with benefits for local communities estimated at US$678,000 (20% of entrance fees plus direct tourist expenditures). Moyini and Uwimbabazi (2000) stress that the combination of an over-representation of lower income spectrum of tourists in their study sample, and the likelihood of tourists engaging in other tourism activities in Uganda, suggest that their estimates are rather conservative.

Since 1994, income generated from gorilla tourism has been shared (sporadically) with local communities in Uganda, who have been able to build schools and clinics (Archabald and Naughton-Treves, 2001; Adams and Infield, 2003). During the pilot phase (April 1993 to June 1994) at Bwindi Impenetrable National Park, local communities received approximately US$15,000 (Meder, 1996). For the first few years, at Mgahinga Gorilla National Park, 20% of entrance fees supported local community projects, and park rangers' salaries were covered by proceeds from tourism, resulting in considerable financial independence, security, and confidence of local communities (Karlowski and Weiche, 1997). Money from tourism projects, as well as funding from the *Mgahinga and Bwindi Impenetrable Forest Conservation Trust* (MBIFCT), help to compensate local communities for loss of access to resources (Meder, 1999). Archabald and Naughton-Treves (2001) suggest that tourism revenue sharing appears to have improved attitudes towards protected areas in Western Uganda, but they also point out that changes in legislation resulted in no sharing of revenue with local communities between 1998 and 2001.

A number of local, national and international groups have been vying for a greater share (or some share) of funds generated by gorilla tourism in Uganda (Adams and Infield, 2003). Originally intended to benefit local communities, gorilla-tourism revenues have helped fund national parks throughout Uganda. The perception of local communities at Mgahinga continues to be dominated by a sense of economic loss (Archabald and Naughton-Treves, 2001; Adams and Infield, 2003), suggesting that their share of gorilla tourism revenues is inadequate (Brown *et al.*, Chapter 10, this volume).

4.2. Economic Costs of the Habituation Process and Infrastructure

The potential economic benefits of gorilla tourism, based on income generated by mountain gorilla tourism during times of stability (and at maximum capacity), are likely to feature prominently as the prime incentive for establishment of new sites (Djoh and van der Wal, 2001). Yet, the costs of the habituation process and infrastructure may be prohibitive, unless well funded by large international organizations over a prolonged period (Blom, 2001a, 2001b). The time needed to habituate a gorilla group can vary between six months and 14 years (Goldsmith, 2005). Blom (2001b) estimates that a two-year habituation period for one group of gorillas, based on the process used at Dzanga-Sangha (Central African Republic), would cost at least US$250,000. This figure excludes the budget for the health-monitoring program that should be conducted prior, during, and after habituation.

Blom (2001b) points out that the potential revenues that gorilla tourism can generate for protected area management, as a self-financing strategy, in the central African region is limited at best. The mountain gorilla tourist sites have set a high standard of service to tourists, resulting in expectations of well-organized tracking and good gorilla viewing. These sites have also generated large and well-publicized revenues during stable periods, which may set unrealistic financial expectations in other countries. Western lowland gorillas living in forests that are logged, and where they are hunted for bushmeat, may be difficult to habituate. Djoh and van der Wal (2001) emphasize that "the gorillas have many bad memories of their encounters with men...If they have too many bad memories, it could very well be impossible to regain their trust" (p.34). Western lowland gorillas may also live in closed and flat lowland forests with dense undergrowth, which makes tracking more difficult and risky, since trackers unexpectedly may stumble into gorillas (Djoh and van der Wal, 2001).

5. "New" Western Lowland Gorilla Tourism Sites

Despite the constraints imposed by environment and human activities (e.g., hunting and logging), western lowland gorillas have been habituated for tourism at a number of sites. These include Dzanga-Sangha in Central African Republic, Lopé Wildlife Reserve in Gabon, Maya north saline (Odzala National Park), Lossi Gorilla Sanctuary (about 15 km from Odzala National Park) in Congo-Brazzaville, and Monte Alen National Park in Equatorial Guinea. Habituation trials have also taken place at Lomié (near Dja Wildlife Reserve) in Cameroon (Aveling, 1999; Blom, 2001a; Djoh and van der Wal, 2001). These sites provide opportunities to observe gorillas in a variety of ways. At Maya north saline, gorillas and other large mammals (e.g., forest elephants, buffaloes, bongos, and sitatungas) feeding in forest clearings can be viewed from observation hides. At Lossi, tourists could once view gorillas feeding on fruit above them in the trees (prior to the gorillas' decimation by Ebola). Unlike mountain gorilla and eastern lowland gorilla tourism sites, not all western lowland gorilla tourism sites are located within traditional protected areas (e.g., national parks and reserves). The Lossi site still belongs to its traditional owners, the villagers of Lengui-Lengui, who developed gorilla tourism as part of a community conservation initiative (Aveling, 1999). The Lomié site is located partly within a production forest, and partly within a forest requested as a community forest by the neighboring villages of Karagoua and Koungoulou (Djoh and van der Wal, 2001).

The Lossi Gorilla Sanctuary now serves as a stark reminder of the vulnerability of gorillas and gorilla tourism. For almost a decade, Magdalena Bermejo and Germain Ilera have been there, studying and monitoring eight families of gorillas (139 individuals). Two of these groups, the first lowland gorillas to be habituated in central Africa, were habituated for tourism. At the end of 2002, Ebola virus was confirmed in four gorilla and two chimpanzee carcasses. Since then, all but a handful of gorillas have disappeared from the study area—victims of an Ebola epidemic (Aveling, 2003). Ebola appears to pose an even greater threat to western lowland gorillas than the bushmeat trade. The combined impact of Ebola and hunting has resulted in an estimated 56% decline in the gorilla and chimpanzee population numbers in Gabon and Congo Republic, between 1983 and 2000 (Walsh *et al.*, 2003).

6. Evidence for Potential Threats to Gorillas from Tourism

This section examines some of the potential threats that tourism may pose to gorillas. These potential threats fall into three main categories—the process of habituating gorillas for tourism, increased risk of disease transmission, and inappropriate tourist purchases. Before embarking on an analysis of potential threats of tourism, it should be noted that the high profile of "Virunga" mountain gorillas (which excludes "Bwindi") and protection afforded them by regular tracking and monitoring for tourism and research (260 of the total 359 + individuals) have resulted in an increase in mountain gorillas over the last three decades (annual growth rate of 1.0–1.3% per year) despite prolonged armed conflict in the area (Kalpers *et al.*, 2003). "Many of the warring factions have actually shown commitment and invested resources to ensure that the gorillas were not harmed," as a direct result of recognition of these gorillas as an "important resource (through tourism)" (Kalpers *et al.*, 2003, p.335). Nonetheless, not all mountain gorillas habituated for tourism and research have fared equally well. Whereas four of the gorilla groups in Rwanda increased in size by 76% over a decade (1989–2000), seven habituated groups in DRC declined in size by almost 20% in four years (1996–2000; Kalpers *et al.*, 2003).

6.1. The Habituation Process as a Potential Threat to Gorillas

The Nkuringo group of gorillas at Bwindi Impenetrable National Park (Uganda) has been habituated for tourism since 1997, but has not yet been included in the tourism program. At the end of 1998, Michele Goldsmith and colleagues (2006) conducted a three-month study of the behavioral ecology of this group of 16 gorillas (2 adult silverback males, several adult females, juveniles and infants), monitoring daily habitat use, diet, daily path length, and group cohesion. The researchers were particularly interested in how much time the gorillas spent outside the National Park boundaries, and whether their behavioral ecology differed once outside the park.

Over a 36-day follow, the Nkuringo group of gorillas nested within the park on only one occasion and, on two other days, nested almost 1 km outside the park, within meters of the main road. Spending most of their time outside the park, the gorillas consumed large amounts of nonforest food (domestic banana pith, eucalyptus bark, and sweet potato leaves). Almost half the trails outside the park were in open agricultural areas. Unlike gorillas inside the park, this group traveled shorter daily distances, and demonstrated a type of "home-base" nesting strategy (nesting cohesively and often reusing sleeping sites over consecutive nights).

These preliminary findings provide clear evidence that the Nkuringo Group explores and exploits human-inhabited areas. Increased contact with humans in agricultural areas around National Parks may increase the gorillas' risk of contracting diseases (e.g., outbreak of scabies in the Nkuringo group, Nkurunungi, 2001). Increases in crop raiding may lead to further conflict with local communities, already a concern to communities around Bwindi (Nkurunungi, 2001; Biryahwaho, 2002).

While the habituation process may pose a threat to gorillas, gorillas that are overly habituated to humans (e.g., rehabilitated orphaned gorillas) may pose a danger to tourists by responding unpredictably. For example, a gorilla at Lefini Reserve (Republic of Congo) charged, "attacked," and jumped onto a pirogue (small boat) containing people (King et al., 2005). Larger "buffer" distances between rehabilitated gorillas and tourists may be necessary. Some researchers suggest that "de-habituation" of rehabilitated gorillas should take place, with all human contact minimized or eliminated (Carlsen et al., 2006, p.33).

6.2. Disease Transmission as a Potential Threat to Gorillas

Regular checks on the website of the World Health Organization (WHO), with its information about the latest disease outbreaks, can make even the hardiest of individuals feel uneasy. As a result of international travel and trade, emerging infectious diseases have the potential to spread globally, as witnessed by the new coronavirus (SARS), which traveled globally within its incubation period. For example, one infected man traveled by plane from Singapore to New York to Germany before he was hospitalized (World Health Organization, 2003a, 2003b). The WHO took the unprecedented step of advising travelers to avoid Hong Kong, southern China, and Toronto. Disease risks vary depending on the type of travel undertaken (e.g., package tours or independent), and the type of tourist (e.g., businessperson, soldier, backpacker). Nevertheless, all tourists "may be susceptible to diseases transmitted during travel, and these may be more common than is presently recognized . . . all such infections may be transported around the world within their incubation period" (Green and Roberts, 2000, p.560).

Genetically similar to humans, gorillas are extremely vulnerable to human diseases, with a common cold potentially life threatening to wild populations, which may have no natural immunity. Similarly, although the risk may be small, the potential exists for humans to be exposed to potentially deadly new viruses (Homsy, 1999). Sick tourists and staff are prohibited from tracking gorillas in Uganda and Rwanda. However, this "sickness rule" cannot be monitored effectively, since most people are only capable of recognizing obvious symptoms of illness (e.g., coughs, sneezes, rashes or stomach ailments), and could be shedding viruses or bacteria before or after symptoms have appeared (Homsy, 1999). The self-report rule depends on the honesty of the individual tourist and staff member. The provision and use of facemasks (and safe collection and disposal of them afterwards) when viewing gorillas may bring home these issues more emphatically.

The minimum distance of five meters or 15 feet (the "buffer distance" rule) previously in place at the tourist sites in Uganda was considered inadequate to protect gorillas from the risk of disease transmission, and has been increased to seven meters (Homsy, 1999; Lanjouw et al., 2001). In the absence of wind, sneeze particles can travel 6 m (20 feet), influenza can be transmitted up to 20 m, and other airborne organisms may travel even further in favorable wind and ultraviolet light conditions (Homsy, 1999). To protect primates from human diseases in zoos, Plexiglas structures are often built as a barrier. The only protection that is afforded to wild gorillas is the strict enforcement of an adequate minimum distance. Unfortunately, it is the one rule that the guides and park staff report that they have the most difficulty enforcing (Homsy, 1999).

Homsy (1999) points out that researchers have tended to focus on the risk of tourists passing on respiratory infections to gorillas (e.g., measles, tuberculosis, pneumonia, influenza, and respiratory syncitial virus). Disturbingly, measles microbes can travel great distances in the open (especially if it is windy), and polio microbes can survive in the soil for several months (Homsy, 1999). There are, however, many other diseases that can be contracted by gorillas if they come in contact with human faeces or fomites (inanimate objects). Hepatitis A and B viruses, shigella, trichuris, herpes simplex, scabies, polio, and intestinal worms may pose an even greater threat to the ultimate survival of wild populations of gorillas (Homsy, 1999). The habituation process may cause stress, which in turn may exacerbate diseases, such as scabies or sarcoptic mange (McNeilage, 1996; Woodford et al., 2002).

Research following the baseline studies of intestinal parasite fauna of mountain gorillas prior to tourism (e.g., Ashford et al., 1990, 1996) suggests that exposure to tourists and other humans in the parks (thereby making it impossible to determine whether the origin of the parasites is from tourists or other groups of people) has introduced new parasites or altered the natural parasite fauna of the mountain gorillas. New parasites found include *Entamoeba, Trichuris, Chilomatix,* and *Endolimax nana* (Homsy, 1999). More recent baseline studies measure differences in prevalence of infection in different primate species within the same area (e.g., baseline study of intestinal parasites of western lowland gorillas, chimpanzees, agile mangabeys, and humans working at the Mondika Research site at Dzanga-Ndoki National Park, Central African Republic, Lilly *et al.*, 2002).

Evidence for disease symptoms, for lack of current vaccinations, and for ongoing infectious diseases in both tourists and a local community was found in a study conducted at the Kanyanchu chimpanzee tourism site (Kibale National Park) in Uganda (Adams *et al.*, 2001). This study is of relevance to gorilla tourism, since more than two-thirds (67%) of the total tourists surveyed had either already visited, or were planning to visit, gorillas or other chimpanzees. In Uganda, it is possible for tourists to visit orphan chimpanzees at Ngamba Island Chimpanzee Sanctuary, as well as wild chimpanzees and gorillas at several tourist sites all within one to two weeks (Litchfield, 2001a).

Based on the self-reported medical histories of 62 tourists (predominantly European), Adams *et al.* (2001) found that few were currently vaccinated against influenza (3%), mumps (21%), or measles (37%). Almost half were not vaccinated against tuberculosis, and about a third were not vaccinated against viral hepatitis A or polio. Symptoms of illness experienced during their visit to Africa included diarrhea (>50%), coughing (>10%), fever, vomiting, and general illness (all >5%). Disturbingly, five cases of herpesvirus, six of influenza, and one of chickenpox were considered infectious at the time of visit. With respect to tuberculosis, less than 50% of the tourists had ever been tested. Three people (10%) reported having had positive intradermal skin test results, indicating that they may have been infectious at the time of their visit. This study provides evidence that the potential exists for tourists to spread infection to more than one wild group of gorillas and/or chimpanzees.

Wallis and Lee (1999) point out that researchers (and visitors) who work with laboratory apes in the United States undergo stringent testing procedures for tuberculosis (at least annually) and usually wear gloves and masks if they come in contact with the apes. Yet, ironically, these same people can visit gorillas in the wild, without having to take similar precautionary measures.

Tourists are not the only humans encountered by wild populations of gorillas (Butynski and Kalina, 1998). A number of gorilla groups are also exposed to gorilla conservation workers, local communities, and illegal extant communities (MGVP/WCS, Chapter 2, this volume).

6.3. *Tourist Purchases as a Potential Threat to Gorillas*

Poaching may be encouraged inadvertently if tourists purchase inappropriate souvenirs, such as souvenir drums made of supple duiker (antelope) skin rather than cow hide in Uganda. Duikers are trapped illegally in wire snares set in forests and National Parks, and gorillas (and chimpanzees) are maimed, crippled, or killed by these snares (Weber and Vedder, 2001). Many African souvenirs are made of animal products (bones, skulls, and skins), and tourists must make a concerted effort to find out what they are buying (Friends of Conservation, 2002).

Well-meaning tourists may buy malnourished and suffering orphaned gorillas or chimpanzees, inadvertently supporting trade in great ape infants. Money made this way encourages dealers and poachers to obtain other infants illegally. Currently, African primate sanctuaries (of the Pan African Sanctuary Alliance or PASA) care for 80 orphaned gorilla infants, 700 + chimpanzees, and 45 bonobos—"by-products" of the bushmeat trade, surviving the slaughter of mothers or other members of their community (Carlsen *et al.*, 2006).

Curiosity of some tourists to taste "exotic" meat (even that of endangered species), potentially increases demand for bushmeat (Barlow, 2001). Tourist consumption of domestic animal meat in Africa may contribute to destruction of habitat for cattle grazing (Goodall and Bekoff, 2002). Early index cases of human disease epidemics may be traced to close contact with infected animals that are butchered and eaten. Cases of human plague have been attributed to consumption of raw infected camel liver in Saudi Arabia (Bin Saeed *et al.*, 2005), and humans have contracted SARS-CoV from a restaurant serving palm civets in China (Wang *et al.*, 2005). In gorilla habitat countries, Ebola kills people who butcher and eat infected or dead chimpanzees and gorillas, and eating nonhuman primates or keeping them as pets has allowed HIV and Simian Foamy Virus to emerge in humans (Wolfe *et al.*, 2005).

Sanctuaries play an important role in conservation education and tourism, with orphaned apes serving as powerfully emotive messages for the plight of African apes (Farmer and Courage, Chapter 3, this volume). Open and informed discussion and dissemination of information about disease transmission, and unsustainable and unsafe consumption of meat, is vital for gorilla and human health and survival (Wolfe *et al.*, 2005). Recommending that tourists avoid eating meat whilst in Africa may be a good message to promote. As with information about disease transmission, visitors should be provided with details regarding appropriate tourist purchases prior to their trip.

7. Strategies to Minimize Potential Threats to Gorillas

A number of strategies have been implemented to attempt to minimize the potential threat to gorillas that tourism may pose, as mentioned in the previous section. This section will examine these, as well as other general approaches that have been employed to help protect gorillas.

7.1. Minimizing the Potential for Human–Gorilla Conflict as a Result of the Habituation Process

The habituation of gorillas for tourism has the potential to increase human–gorilla conflict, since it may lead to gorillas that crop raid as a result of spending increased amounts of time outside protected areas. The International Gorilla Conservation Programme's "Human Gorilla Conflict Force" (HUGO) was developed to deal with the problem of crop raiding by gorillas. Special ranger groups (local community members and park rangers) have been trained to patrol boundary areas, and use loud noises (e.g., bells or drums) to "herd" gorillas back into the national park (Lanjouw *et al.*, 2001). Despite such attempts to alleviate conflict between gorillas and farmers, gorillas may still be killed while crop raiding. Recently, a three-year old gorilla (Bahati) was killed in a corn field near the border of Parc National des Virunga in DRC, after stones and wood were hurled at his family who had consumed 235 corn stalks (Kiyengo and Binyeri, 2003). If a site is to be developed for gorilla tourism, opportunities for creating buffer zones (e.g., nonpalatable crops), or physical barriers (Lanjouw *et al.*, 2001) should be explored and implemented before the habituation process begins.

7.2. Minimizing the Potential for Disease Transmission and Inappropriate Tourist Purchases

As long as humans enter the habitat of the gorilla, the potential for disease transmission exists. Relatively long incubation periods and rapid travel between countries make the adoption of standardized guidelines and rules at *all* gorilla sites vital. How do researchers themselves fit into the current picture? Are they a special category of tourist? Unlike tourists who are restricted to a one-hour visit with gorillas, researchers may sometimes conduct "nest-to-nest" follows, which involve observing gorilla behavior from morning (when they leave their nests), until evening (when they build and retire to their nests). Strict monitoring of health and inoculations of all visitors and workers should form part of regular routines at all gorilla sites (Adams *et al.*, 2001). The Dian Fossey Gorilla Fund International (DFGFI) has a two-week quarantine period for any researcher working with habituated gorillas (Tara Stoinski, personal communication).

Standardized guidelines and rules (health and tourist behavior and purchases), particularly the reasons behind them, must be provided prior to the tourist or researcher's arrival at gorilla sites (Macfie, 1997; Litchfield, 2001a). As Homsy (1999) maintains, "the best hope for a least damaging tourism programme resides in the widespread sensitisation, awareness and understanding of the catastrophic consequences of unconscious gorilla tourism" (p.57). Educational interpretation, which moves beyond provision of basic information to a more challenging, engaging, and explicit discussion of latest research, current threats (including those posed by tourism), concrete suggestions for activism (on behalf of gorillas) may help to alleviate problems associated with gorilla tourism (Russell, 2001).

7.3. *Other Strategies for Protecting Gorillas*

Gorilla tourism has the potential to serve as a model for *responsible* tourism with endangered species. As a result of the serious threat that human diseases pose to gorillas, perhaps the most stringent guidelines for any form of wildlife tourism exist. Vets and other researchers linked with the Mountain Gorilla Veterinary Project and the Wildlife Conservation Society (MGVP/WCS, Chapter 2, this volume) have been able to pioneer the application of conservation medicine principles to ecosystem and human health and gorilla health. This holistic ecosystem health approach recognizes the importance of health monitoring of all humans, domestic animals, and wildlife. A *Population and Habitat Viability Assessment* (PHVA) serves as a first and most vital step in formulating a practical conservation management program for the survival and recovery (e.g., at Lossi) of gorillas in a particular country or region (Miller *et al.*, Chapter 8, this volume). The mountain gorilla PHVA held in Uganda stressed the importance of tourism as part of the overall conservation management strategy (Werikhe *et al.*, 1997; Litchfield, 2001a).

Responsible tourists and the tourism sector might serve to highlight the threat that the illegal bushmeat trade poses to gorilla conservation and international animal and human health (Peeters *et al.*, 2002). In 2001, spot checks at London's Heathrow airport netted 5.5 tons of bushmeat, including "bits of gorilla" (Lawrance, 2002, p.2). As much as 17,484 tons of illegal meat enters the United Kingdom per year from outside the European Union (a conservative estimate), with West Africa and South Africa as the top five contributors to the total flow (Department of the Environment, Food and Rural Affairs, 2003). In Australia, it is illegal to bring in meat, and 48 teams of detector dogs at Australian international airports sniff out such items (Department of Agriculture, Fisheries and Forestry, 2006). Following a pilot scheme at Heathrow airport of two detector dogs, there are still only six teams of such dogs in the UK (Duggan and Jarvis, 2003; National Audit Office, 2005). By engaging, educating, and encouraging proactive behavior in *responsible* tourists, they may be enlisted to help protect gorillas.

8. Conclusion

"Yes, we assured him, there were thousands of crazy white people out there who would pay a lot of money to hike through the cold rain and steep terrain to sit with wild gorillas. The director laughed at the notion. *Beaucoup d'abazungus fous?* Yes, that much we could vouch for: the world was full of crazy white people" (Weber and Vedder, 2001, p.157).

Despite a history spanning 30 years, it is too soon to determine whether responsible tourism with gorillas is a sustainable option—at any site, or in any country or region. In the short-term, mountain gorilla tourism has been adversely affected in Rwanda and the Democratic Republic of Congo (and to a lesser extent in Uganda following, for example, the massacre of tourists at Bwindi). Within its shorter history, lowland gorilla tourism has been destroyed at the Lossi Gorilla Sanctuary (Republic of Congo), and the same threat of Ebola (and the bushmeat trade) hovers over other western lowland gorilla tourism sites. Whether tourism with Grauer's gorillas at Kahuzi-Biega National Park can be resurrected is unclear at the present time. Nevertheless, the local chiefs and thousands of children from the town of Bukavu appear keen to protect the remaining gorillas and this World Heritage Site as a future "main pillar of tourism" (Iyatshi and Schuler, 2003, p.3; Kyalangalilwa *et al.*, 2003).

If governments are supportive, standardized guidelines are followed, comprehensive health monitoring is implemented, local communities are involved, rebel activity is contained, and public interest (both local and global) in conserving gorillas is aroused, tourism may become a sustainable option at some sites. Since the ultimate survival of mountain gorillas depends on the political and social situations prevalent in three countries (Uganda, Rwanda, and DRC), a regional approach to conservation management and tourism is most likely to succeed (Lanjouw, Chapter 13, this volume). It could be argued that survival of *all* gorillas may depend on a more collaborative regional or transboundary approach (ECOFAC lowland gorilla tourism sites already collaborate), which allows investment costs to be shared (e.g., training, marketing, and policy development), potentially leading to increases in tourism revenue and avoidance of competition between sites (Lanjouw *et al.*, 2001). All gorilla tourism sites could then use the stringent and carefully developed standardized guidelines for mountain gorilla tourism, similar interpretative or educational material, and successful training procedures for staff and tourism-linked enterprises for local communities, and health monitoring programs.

For economic and ecological sustainability to be achieved, an optimal number of visitors must visit gorilla tourism sites—a steady stream of low numbers. Every human visitor can make a difference to the ultimate survival or demise of gorillas. The interest and support of *responsible* tourists at *responsible* sites may help some populations of gorillas survive, and can help promote the concept of Heritage Species status for gorillas and other apes (Wrangham, 2000; Wrangham *et al.*, Chapter 14, this volume). Tourism itself can promote world peace and support global Peace Parks or transfrontier protected areas (Lanjouw *et al.*, 2001; International Institute for Peace Through Tourism, 2006). Responsible tourists can serve as ambassadors for gorillas, raising awareness of their plight.

Responsible tourism with gorillas: conservation tool or conservation threat? Many conservationists do not believe that gorillas should have to be the focus of tourism—paying for their own conservation (Werikhe *et al.*, 1997).

Ethical considerations (e.g., ape rights affording gorillas ethical and legal protection) may one day call a halt to gorilla tourism and some field research (not directly of benefit to welfare and conservation of gorillas), which might be considered too intrusive and exploitative (Butynski, 2001; Goldsmith, 2005). Funding mechanisms independent of tourism may provide the gorillas' greatest hope of survival, but until these become a reality, and since "eco-tourism is unlikely to go away" (Russell, 2001, p.41), well-informed *responsible* tourists and researchers may serve as an increasingly effective international voice for gorilla conservation.

Acknowledgments. Special thanks to my mother, Anita O'Hair, for her help in editing this chapter, and to Professor Richard Wrangham for his invaluable comments and support over the years. Thanks also to Dr. Tara Stoinski for her patience and hospitality, to Dr. Michele Goldsmith for copies of her most recent publications, and to Dr. Liz Williamson for sharing her views. Special thanks to Hannah Lawson for her friendship and inspiration (as one of the original team of scientists to monitor parasite loads in gorillas at Bwindi). Finally, I dedicate this chapter to my daughter, Kaitie Afrika, who has experienced the beauty and heartbreak of life on the African continent, and who serves as a daily reminder of all that is wonderful in the world!

References

Adams, H.R., Sleeman, J., Rwego, I., and New, J.C. (2001). Self-reported medical history survey of humans as a measure of health risk to chimpanzees (*Pan troglodytes schweinfurthii*) of Kibale National Park, Uganda. *Oryx* 35:308–312.

Adams, W.M., and Infield, M. (2003). Who is on the gorilla's payroll? Claims on tourist revenue from a Ugandan National Park. *World Development* 31(1):177–190.

Archabald, K., and Naughton-Treves, L. (2001). Tourism revenue-sharing around national parks in Western Uganda: early efforts to identify and reward local communities. *Environmental Conservation* 28(2):135–149.

Ashford, R.W., Reid, G.D., and Butynski, T.M. (1990). The intestinal faunas of man and mountain gorillas in a shared habitat. *Annals of Tropical Medicine and Parasitology* 84:337–340.

Ashford, R.W., Lawson, H., Butynski, T.M., and Reid, G.D.F. (1996). Patterns of intestinal parasitism in the mountain gorilla *Gorilla gorilla* in the Bwindi-Impenetrable Forest, Uganda. *Journal of the Zoological Society of London* 239:507–514.

Aveling, C. (1999). Lowland gorilla tourism in central Africa. *Gorilla Journal* 18:18–20.

Aveling, C. (2003). An ebola epidemic on the borders of the Odzala National Park (Congo Brazzaville). News release, ECOFAC, Gabon (March 2003); (http://www.ecofac.org/Divers/EbolaEN.htm, accessed 1 October 2006).

Barlow, Z. (2001). In Vietnam, a battle to save endangered monkeys. *The Boston Globe*, 24 July.

Bin Saeed, A.A., Al-Hamdan, N.A., and Fontaine, R.E. (2005). Plague from eating raw camel liver. *Emerging Infectious Diseases* 11:1456–1457.

Biryahwaho, B. (2002). Community perspectives towards management of crop raiding animals: experiences of CARE-DTC with communities living adjacent to Bwindi Impenetrable and Mgahinga Gorilla National Parks, Southwest Uganda. In: Hill, C., Osborn, F., and Plumptre, A.J. (eds.), *Human-Wildlife Conflict: Identifying the Problem and Possible Solutions*. Albertine Rift Technical Report Series Vol. 1. Wildlife Conservation Society, New York, pp. 46–57.

Blom, A. (2001a). The monetary impact of tourism on protected area management and the local economy in Dzanga-Sangha (Central African Republic). *Journal of Sustainable Tourism* 8:175–189.

Blom, A. (2001b). Potentials and pitfalls of tourism in Dzanga-Sangha. *Gorilla Journal* 22:40–41.

Butynski, T.M. (2001). Africa's great apes. In: Beck, B.B., Stoinski, T.S., Hutchins, M., Maple, T.L., Norton, B., Rowan, A., Stevens, E.F., and Arluke, A. (eds.), *Great Apes and Humans: The Ethics of Coexistence*. Smithsonian Institution Press, Washington D.C., pp. 3–56.

Butynski, T.M., and Kalina, J. (1998). Gorilla tourism: a critical review. In: Milner-Gulland, E.J. and Mace, R. (eds.), *Conservation of Biological Resources*. Blackwell, Oxford, pp. 280–300.

Carlsen, F., Cress, D., Rosen, N., and Byers, O. (2006). *African Primate Reintroduction Workshop Final Report*. IUCN/SSC Conservation Breeding Specialist Group, Apple Valley, Minnesota.

Cater, E. (1995). Ecotourism in the Third World—Problems and Prospects for Sustainability. In: Cater, E. and Lowman, G. (eds.), *Ecotourism: A Sustainable Option?* John Wiley and Sons, Chichester, pp. 69–86.

Ceballos-Lascurain, H. (1996). *Tourism, Ecotourism, and Protected Areas: The State of Nature-Based Tourism Around the World and Guidelines for its Development*. IUCN (The World Conservation Union), Gland, Switzerland.

Dawson, C.P. (2001). Ecotourism and nature-based tourism: one end of the tourism opportunity spectrum? In: McCool, S.F., and Moisey, R.N. (eds.), *Tourism, Recreation and Sustainability*. CAB International, Wallingford, Oxon, UK.

Department of Agriculture, Fisheries and Forestry (2006). *Detector Dogs*. Department of Agriculture, Fisheries and Forestry, Canberra (http:// www.daff. gov.au/content/output.cfm?ObjectID=68ED30F2-40A7-4071-B7ACB496 21780C0A&contType=outputs; accessed 1 October 2006).

Department of the Environment, Food and Rural Affairs (2003). *Risk Assessment for the Illegal Importation of Meat and Meat Products*. Department of the Environment, Food and Rural Affairs, London.

Djoh, E., and van der Wal, M. (2001). Gorilla-based tourism: a realistic source of community income in Cameroon? Case study of the villages of Goungoulou and Karagoua. *Network Paper 25e*, July 2001. Rural Development Forestry Network, Overseas Development Institute, London, pp. 31–37.

Duggan, J., and Jarvis, S. (2003). Illegal imports of meat and other foodstuffs. Minutes of the Heathrow Airport Consultative Committee meeting held on 26 March, pp. 9–10 (http://www.lhr-acc.org/documents/23Min26Mar.pdf; accessed 1 October 2006).

Dunn, A. (2005). Update on Nigeria: Recent work by the Wildlife Conservation Society. *Gorilla Journal* 30:16–17.

Friends of Conservation. (2002). *Conservation Code*. Friends of Conservation, London.

Gladwin, T.N., Newburry, W.E., and Reiskin, E.D. (1997). Why is the Northern elite mind biased against community, the environment, and a sustainable future? In:

Bazerman, M.H., Messick, D.M., Tenbrunsel, A.E. and Wade-Benzoni, K.A. (eds.), *Environment, Ethics, and Behavior: The Psychology of Environmental Valuation and Degradation*. The New Lexington Press, San Francisco, pp. 234–274.

Goldsmith, M., Glick, J., and Ngabirano, E. (2006). Gorillas living on the edge: literally and figuratively. In: Newton-Fisher, N.E., Notman, H., Paterson, J.D., and Reynolds, V. (eds.), *Primates of Western Uganda*. Springer, New York.

Goldsmith, M.L. (2005). Habituating primates for field study: Ethical considerations for African Great Apes. In: Turner, T.R. (ed.), *Biological Anthropology and Ethics*. State University of New York Press, New York.

Goodall, J., and Bekoff, M. (2002). *The Ten Trusts: What We Must Do to Care for the Animals We Love*. HarperCollins, New York.

Gössling, S. (2000). Sustainable tourism development in developing countries: some aspects of energy use. *Journal of Sustainable Tourism* 8(5):410–425.

Gray, M., and Kalpers, J. (2005). Ranger Based Monitoring in the Virunga-Bwindi region of east-Central Africa: A simple data collection tool for park management. *Biodiversity and Conservation* 14:2723–2741.

Green, A.D., and Roberts, K.I. (2000). Recent trends in infectious diseases for travellers. *Occupational Medicine (London)* 50(8):560–565.

Higginbottom, K. (2002). Principles for Sustainable Wildlife Tourism, with particular reference to dolphin-based boat tours in Port Phillip Bay. Report to the Victorian Department of Natural Resources and Environment, July 2002. CRC for Sustainable Tourism, Gold Coast, Queensland, Australia.

Homsy, J. (1999). *Ape Tourism and Human Diseases: How Close Should We Get?* International Gorilla Conservation Programme, Kampala, Uganda.

International Institute for Peace Through Tourism (2006). *Credo of the Peaceful Traveler*. International Institute for Peace Through Tourism, Stowe, Vermont.

Iyatshi, I., and Schuler, C. (2003). Good news from Kahuzi-Biega. *Gorilla Journal* 26:3.

Kalpers, J., Williamson, E.A., Robbins, M.M., McNeilage, A., Nzamurambaho, A., Lola, N., and Mugiri, G. (2003). Gorillas in the crossfire: Population dynamics of the Virunga mountain gorillas over the past three decades. *Oryx* 37:326–337.

Karlowski, U. and Weiche, I. (1997). Mgahinga Gorilla National Park. *Gorilla Journal* 15:15–16.

King, T., Chamberlan, C., and Courage, A. (2005). Reintroduced gorillas: Reproduction, ranging and unresolved issues. *Gorilla Journal* 30:30–32.

Kiyengo, C.S., and Binyeri, D.K. (2003). Primates at the edge of the abyss. *Gorilla Journal* 26:6–7.

Kyalangalilwa, J.F., Ntabarusha, I., and Barhigenga, B. (2003). Children fight for gorilla conservation. *Gorilla Journal* 26:3–4.

Lanjouw, A. (2000). *Building Partnerships in the Face of Political and Armed Crisis*. International Gorilla Conservation Programme, Nairobi, Kenya.

Lanjouw, A., Kayitare, A., Rainer, H., Rutagarama, E., Sivha, M., Asuma, S., and Kalpers, J. (2001). *Beyond Boundaries: Transboundary Natural Resource Management for Mountain Gorillas in the Virunga-Bwindi Region*. Biodiversity Support Program, Washington, D.C.

Lawrance, C. (2002). Illegal meat importation. Minutes of the Heathrow Airport Consultative Committee meeting held on 30 January, pp. 2–3 (http://www.lhr-acc.org/documents/22Min30Jan.pdf; accessed 1 October 2006).

Lilly, A.A., Mehlman, P.T., and Doran, D. (2002). Intestinal parasites in gorillas, chimpanzees and humans at Mondika Research Site, Dzanga-Ndoki National Park, Central African Republic. *International Journal of Primatology* 23:555–573.

Litchfield, C.A. (1997). *Treading Lightly: Responsible Tourism with the African Great Apes*. Travellers' Medical and Vaccination Centre, Adelaide, Australia.

Litchfield, C. (2001a). Responsible tourism with great apes in Uganda. In: McCool, S.F., and Moisey, R.N. (eds.), *Tourism, Recreation and Sustainability*. CAB International, Wallingford, Oxon, UK.

Litchfield, C. (2001b). Competing for responsible tourists in the right numbers. *Gorilla Journal* 22:38–40.

Macfie, E.J. (1997). Gorilla tourism in Uganda. *Gorilla Journal* 15:16–17.

Mann, M. (2000). *The Community Tourism Guide: Exciting Holidays for Responsible Travellers*. Earthscan Publications, London.

McNeilage, A. (1996). Ecotourism and mountain gorillas in the Virungas. In: Taylor, V.J., and Dunstone, N. (eds.), *The Exploitation of Mammal Populations*. Chapman and Hall, London, pp. 334–344.

Meder, A. (1996). Report from Uganda. *Gorilla Journal* 12:8–9.

Meder, A. (1999). A different conservation concept. *Gorilla Journal* 19:10.

Meder, A., and Groves, C.P. (2005). Where are the gorillas? *Gorilla Journal* 30:21–28.

Moyini, Y., and Uwimbabazi, B. (2000). *Analysis of the Economic Significance of Gorilla Tourism in Uganda*. International Gorilla Conservation Programme, Kampala, Uganda.

National Audit Office (2005). *H.M. Customs and Excise: Stopping Illegal Imports of Animal Products into Great Britain*. The Stationery Office, House of Commons, London.

Nkurunungi, J.B. (2001). Habituation of Bwindi Mountain Gorillas. *Gorilla Journal* 22:37–38.

O'Riordan, T., and Cameron, J. (1994). *Interpreting the Precautionary Principle*. Cameron and May, Earthscan, London.

O'Riordan, T., Cameron, J., and Jordan, A. (2001). *Reinterpreting the Precautionary Principle*. Cameron May, London.

Peeters, M., Courgnaud, V., Abela, B., Auzel, P., Pourrut, X., Bibollet-Ruche, F., Loul, S., Liegeois, F., Butel, C., Koulagna, D., Mpoudi-Ngole, E., Shaw, G.M., Hahn, B.H., and Delaporte, E. (2002). Risk to human health from a plethora of Simian immunodeficiency viruses in primate bushmeat. *Emerging Infectious Diseases* 8(5):451–457.

Rabinor, Z.D. (2002). Sustainable development and management of ecotourism in the Americas: preparatory conference for the International Year of Ecotourism, 2002. *Journal of Environment and Development* 11:103–109.

Rao, N. (2000). Sustainable tourism: A southern perspective. In: Dodds, F. (ed.), *Earth Summit 2002: A New Deal*. Earthscan Publications, London.

Roe, D., Leader-Williams, N., and Dalal-Clayton, B. (1997). *Take Only Photographs, Leave Only Footprints: The Environmental Impacts of Wildlife Tourism*. Wildlife and Development Series No.10, International Institute for Environment and Development, London.

Russell, C. (2001). Can eco-tourism help tourists understand conservation? *Gorilla Journal* 22:41.

Schaller, G.B. (1963). *The Mountain Gorilla: Ecology and Behavior*. University of Chicago Press, Chicago.

Stanford, C. (2001). *Significant Others: The Ape-Human Continuum and the Quest for Human Nature*. Basic Books, New York.

Tenbrunsel, A.E., Wade-Benzoni, K.A., Messick, D.M. and Bazerman, M.H. (1997). Introduction. In: Bazerman, M.H., Messick, D.M., Tenbrunsel, A.E. and Wade-Benzoni,

K.A. (eds.), *Environment, Ethics, and Behavior: The Psychology of Environmental Valuation and Degradation*. The New Lexington Press, San Francisco, pp. 1–9.

United Nations Environment Programme. (2002). *Sustainable Tourism in Protected Areas: Guidelines for Planning and Management*. UNEP/IUCN/WTO, Paris.

United Nations Environment Programme. (2003). *Tourism and Local Agenda 21: The Role of Local Authorities in Sustainable Tourism (Case Studies and First Lessons)*. UNEPTIE, Paris.

Wallis, J., and Lee, D.R. (1999) Primate conservation: The prevention of disease transmission. *International Journal of Primatology* 20:803–826.

Walsh, P.D., Abernethy, K.A., Bermejo, M., Beyers, R., De Wachter, P., Akou, M.E., Huijbregts, B., Mambounga, D.I., Toham, A.K., Kilbourn, A.M., Lahm, S.A., Latour, S., Maisels, F., Mbina, C., Mihindou, Y., Obiang, S.N., Effa, E.N., Starkey, M.P., Telfer, P., Thibault, M., Tutin, C.E.G., White, L.J.T., and Wilkie, D.S. (2003). Catastrophic ape decline in western equatorial Africa. *Nature* 422:611–614.

Wang, M., Yan, M., Xu, H., Liang, W., Kan, B., Zheng, B., Chen, H., Zheng, H., Xu, Y., Zhang, E., Wang, H., Ye, J., Li, G., Li, M., Cui, Z., Liu, Y.-F., Guo, R.-T., Liu, X.-N., Zhan, L.-H., Zhou, D.-H., Zhao, A., Hai, R., Yu, D., Guan, Y., and Xu, J. (2005). SARS-CoV infection in a restaurant from palm civet. *Emerging Infectious Diseases* 11:1860–1865.

Weaver, D. (2000). Tourism and national parks in ecologically vulnerable areas. In: Butler, R.W., and Boyd, S.W. (eds.), *Tourism and National Parks: Issues and Implications*. John Wiley and Sons, Chichester, pp. 107–124.

Weber, W. (1993). Primate conservation and ecotourism in Africa. In: Potter, C.S., Cohen, J.I., and Janczewski, D. (eds.), *Perspectives on Biodiversity: Case Studies of Genetic Resource Conservation and Development*. AAAS Press (American Association for the Advancement of Science), Washington, D.C., pp. 129–150.

Weber, B., and Vedder, A. (2001). *In the Kingdom of Gorillas: Fragile Species in a Dangerous Land*. Simon and Schuster, New York.

Werikhe, S., Macfie, L., Rosen, N., and Miller, P. (1997). *Can the Mountain Gorilla Survive? Population and Habitat Viability Assessment for Gorilla gorilla beringei*. SSC/IUCN Conservation Breeding Specialist Group, Apple Valley, Minnesota.

Wolfe, N.D., Daszak, P., Kilpatrick, A.M., and Burke, D.S. (2005). Bushmeat hunting, deforestation, and prediction of zoonoses emergence. *Emerging Infectious Diseases* 11:1822–1827.

Woodford, M.H., Butynski, T.M., and Karesh, W.B. (2002). Habituating the great apes: the disease risks. *Oryx* 36:153–160.

Wrangham, R.W. (2000). The Other Apes: Time for Action. A View on the Science: Physical Anthropology at the Millennium. *American Journal of Physical Anthropology* 111:445–449.

World Health Organization (2002). *Understanding the BSE Threat*. World Health Organization, Geneva.

World Health Organization (2003a). *World Health Organization issues emergency travel advisory: Severe Acute Respiratory Syndrome (SARS) spreads worldwide*. Press release, Geneva (March 15, 2003); http://www.who.int

World Health Organization (2003b). Severe acute respiratory syndrome (SARS): status of the outbreak and lessons for the immediate future. World Health Organization, Geneva (20 May 2003).

World Tourism Organization (2000). *Tourism Market Trends 2000: Long-term Prospects*. World Tourism Organization, Madrid.

World Tourism Organization (2001). *The British Ecotourism Market.* World Tourism Organization, Madrid.

World Tourism Organization (2003a). *Tourism Highlights 2002.* World Tourism Organization, Madrid.

World Tourism Organization (2003b). *Crisis Guidelines for the Tourism Industry.* World Tourism Organization, Madrid.

Copies of my booklet for tourists, *Treading Lightly: Responsible Tourism with the African Great Apes,* can be obtained from: Travellers' Medical and Vaccination Centre Group, 29 Gilbert Place, Adelaide 5000, Australia. Phone: 61 – 8 – 8212 7522; Fax: 61 – 8 – 8212 7550; http://www.tmvc.com.au. E-book versions (1997 and 2007) will be available at http://www.primatesplus.com

Chapter 5
Chimpanzee Conservation and Theatre: A Case Study of an Awareness Project Around the Taï National Park, Côte d'Ivoire

Christophe Boesch, Claude Gnakouri, Luis Marques, Grégoire Nohon, Ilka Herbinger, Francis Lauginie, Hedwige Boesch, Séverin Kouamé, Moustapha Traoré, and Francis Akindes

1. Introduction

Do educational activities designed to increase awareness of wildlife and conservation issues actually lead to behavioral changes that promote conservation and protect wild populations? For the most part, this question remains unanswered for a variety of reasons. Evaluations of educational programs are often not conducted and, when they are, results addressing behavioral change are rarely included or are unclear. For example, the relationship between changes in knowledge and attitudes that often accompanies educational programs and changes in conservation-related behavior is not well understood (Stoinski *et al.*, 2001). Additionally, in many field situations, conservation activities are often multifaceted, and thus it is difficult to quantify the effectiveness of individual components on changes in human practices (Oates, 1999). Specifically, gaining information on the effectiveness of educational activities aimed to influence the local population is important, as local support is mandatory for the long-term success of conservation programs.

We present here an assessment of an educational program that occurred in villages around the Taï National Park, Côte d'Ivoire. The focus of the program was a play performed by the company "Ymako Teatri" and organized by the Wild Chimpanzee Foundation, a nonprofit organization created to address issues related to the decline of chimpanzee populations. The play focused on the problems related to the co-existence of chimpanzees and humans and was designed to confront the people with a serious conservation concern, namely, the threat to chimpanzee survival, in an original, attractive, and lively approach. Because theater had not been used before in this region of Africa to communicate conservation messages to the local population, the efficiency of the program was assessed through an independent evaluation done before and after the performances.

The chimpanzee is our closest living relative on earth and is strikingly similar to humans with respect to its natural biology, behavior, and cognitive capacities (Jones *et al.*, 1992; Tomasello and Call, 1997; Boesch and Boesch-Achermann, 2000). Recent results reveal chimpanzees and humans share many sophisticated abilities, such as tool use and tool making, cooperative hunting, empathy, and the understanding of others' knowledge and beliefs (Goodall, 1986; Boesch and Boesch-Achermann, 2000; Hare *et al.*, 2000, 2001). Despite such similarities, chimpanzee populations are threatened throughout the African tropical belt from Tanzania to Senegal. Chimpanzees are listed as endangered on the IUCN Red List (IUCN, 2002), and have already gone extinct within the last 40 years in 5 of the 22 countries of their original distribution (Teleki, 1989; Kormos and Boesch, 2003).

Christophe and Hedwige Boesch have been studying the chimpanzees of the Taï National Park, Côte d'Ivoire, since 1979, documenting the unique abilities and social complexities of this population (for a complete list of references see Boesch and Boesch-Achermann, 2000). Through contact with the local people, it became obvious that their knowledge about chimpanzees was very limited, but at the same time they were fascinated by the exceptional abilities of this species. Thus, the Wild Chimpanzee Foundation developed an awareness program aimed at informing local people about the plight of the chimpanzee. Because of the attraction of humans to theatric representations of life situations, particularly in societies with oral traditions, we decided to use a play as one of our first methods of communication with the local community (Farmer and Courage, Chapter 3, this volume). Theater seemed a promising way to improve the perception of the chimpanzee and to address the issue of its coexistence with the local human population.

The company Ymako Teatri, located in Abidjan, seemed well suited for this project. They had a history of presenting plays to villagers in several West African countries focused on real-life themes, such as the consumption of drugs, AIDS, religious sects, and the position of women in society. We decided the message of the play would be "chimpanzees are our cousins in the forest, do not kill them." This theme was based on the fact that, in forest regions extending from Liberia to the Democratic Republic of Congo, there are some families who believe they are related to chimpanzees because of dramatic events that occurred in the past and, thus, do not kill them. This tradition of keeping a "totem" is still very active in most villages and was a key concept of the play.

To develop a theme that would be both convincing and attractive to the local populations, Claude Gnakouri and Luis Marques, the directors of Ymako Teatri, visited the chimpanzees of Taï National Park to develop an understanding of chimpanzee behavior and motion. Additionally, they visited nearby villages to obtain detailed information from the local people on the problems related to the killing of chimpanzees and the destruction of the forest.

The structure of the play was a mixture of theatric actions, mimes, dances, music, and songs. Repeated consultations with people, dancers, and musicians living in the target region were done to select locally appealing music and songs, which were then adapted to the purpose of the play. The play consisted of different scenes lasting approximately 45 minutes and included lyrics and dances that reiterated the overall message of chimpanzees and humans as cousins. The play was presented in French, but phrases in local dialects were embedded throughout.

Ymako Teatri used the "forum" format whereby the play was built to reach an intense conflict between the protagonists. At that moment, the play was interrupted and the audience was asked to give its opinion on how the situation should be resolved. The response of the audience decided the outcome of the play. If their response conflicted with the main message, the play was finished accordingly but with the original message being reinforced. For example, in the single village that said that eating chimpanzees was normal, the actors reinforced the message about sparing the chimpanzee's life because they are our cousins. Once the play was over, staff of the Wild Chimpanzee Foundation led a discussion with the public, answering the questions raised by the play.

2. Content of the Play

2.1. Protagonists

- *The hunter family:* They are having a funeral in the village, where all the families will be present, but there is no meat in the kitchen. The wife threatens to leave her husband, Zaipodo, if he does not provide her with meat immediately. She is not going to face the shame of having guests and no meat to offer. An uncle supports her, and Zaipodo, who wishes to respect the law, has no choice but to go on a hunt illegally in the nearby national park.
- *The Oussé family:* This family lives in the same village as the hunter family but considers itself to be related to the chimpanzees because in the past one of the daughters went into labor alone in the forest. She and the baby survived through the assistance of a big male chimpanzee. To show its gratitude, the family decided to take the chimpanzee as a totem and to respect as a family member. The yearly ceremony to commemorate this event is performed on stage.
- *The chimpanzee family:* A group of chimpanzees is shown using branches and stones as hammers to pound nuts, and the mothers share the kernels with their youngsters. Two females drink water out of a tree trunk with the help of sponges made of fresh leaves while a male eats ants he fishes with sticks. After socializing for a while, they hear a red colobus monkey call. They form a well-organized team and successfully hunt the monkey. Meat is shared with all group members.

2.2. The Drama

As the chimpanzees, unaware of the imminent danger, share and eat the meat, Zaipodo approaches silently and shoots a mother with a youngster. The chimpanzees run away but after a while stop when they notice that the mortally wounded female is unable to follow. Well hidden, Zaipodo observes with disbelief and fascination the emotions and behavior displayed by the chimpanzee family over the suffering and the death of the wounded female. All the chimpanzees surround the dying female, groom her, and lick her wound. Once she has died, they continue to try to provoke reactions from her body, chase the flies away, and remain quietly nearby. Eventually, they cover her body with branches. Meanwhile, the newly orphaned youngster screams desperately beside the body of his mother and is eventually adopted by his mother's closest friend. Zaipodo is highly moved by these observations, and he cannot imagine ever shooting or eating a chimpanzee again—"it is like murder." He is unable to bring the dead female back to the village, but his younger brother that followed him comes across the dead chimpanzee and carries it back to the village.

Shortly before Zaipodo leaves the chimpanzee, a daughter of the Oussé family who is looking for snails in the forest, sees Zaipodo with the dead chimpanzee. Shocked by what she perceives as murder, in tears she beats the hunter and runs back to the village. A violent discussion then unfolds between the two families: the Oussés want to give the dead chimpanzee a traditional funeral, claiming that she is a member of the family that died within the land belonging to the Oussé. The Zaipodo's family views the chimpanzee simply as food and wants to begin preparing the meat for the big funeral. Unable to agree, a fight between the elders starts.

At that moment the play is interrupted and the actors ask the audience how the conflict between the two families could be solved. What should be done? Should the chimpanzee be buried or cooked? Should the hunter go to jail? The public is given time to discuss the issue, and after reaching an agreement, the chief of the village gives the answers to the actors. The play is then finished accordingly.

3. Effectiveness of the Play

The play was performed for a total of 8,000 individuals in 16 villages located around the park during May 2002. Villages were selected to include different ethnic groups and socioeconomic situations of the region. Only one (6%) of the 16 villages suggested that Zaipodo's family was correct and that the chimpanzee should be eaten. In the remaining villages (94%), the public said that Zaipodo's family should give the dead chimpanzee back to the Oussé family so that a traditional funeral could take place. In four of these villages (25%) it was added that the park authorities should be informed of the chimpanzee's killing.

To assess the effectiveness of the play, a group of sociologists led by Professor Francis Akindes, from the University of Bouaké, interviewed 75 people in 5 of the 17 villages that observed the play. The evaluation was conducted in two phases. First, people were asked about their perceptions of chimpanzees a month prior to seeing the play (Akindes et al., 2002a). Four months after the play, a second evaluation was done to look for any attitudinal changes (Akindes *et al.*, 2002b). The sociologists were completely independent of the awareness team and were not involved in the play's creation or performance. They did, however, watch a performance of the play in Abidjan and developed the second evaluation in consultation with Christophe Boesch.

During the second evaluation, a new sample of 75 individuals from five villages were interviewed; 57% of the participants were local to the area whereas the remaining 43% originated from other regions of Côte d'Ivoire and West Africa (mainly Burkina Faso and Mali). Of the local people, 74% had a tradition of hunting for meat, but only 25% of those from outside the region hunted. Similarly, 67% of the locals reported sharing a totemic relationship with the chimpanzees, whereas only 25% of the second group had such a relationship (Akindes *et al.*, 2002). In general, 96% of the people thought the play was a good way of presenting a problem because it reaches a large audience, is easy to understand, faithful to reality, and facilitates empathy with the actors. Overall, people perceived the play as representing a real situation; 80% of the interviewed participants agreed with the reality of the situation proposed to them.

Table 5.1 presents the main messages remembered by audience members four months after seeing the play. The close similarity between chimpanzees and humans was sometimes expressed by people even before the play, but was accompanied by a negative impression of "incompleteness" in the chimpanzee's humanity when compared to human. Chimpanzees were described as being too 'savage', 'ugly' and 'uncontrolled' as compared with humans (Akindes *et al.*, 2002a). After the play, however, there was recognition of similarities between the two species based on shared traits of intelligence, an organized social life, the expression of feelings, attention to wounded and dead individuals, and the adoption of orphans. Interviewees expressed these sentiments through statements like "a chimpanzee is like a human" and 'the killing of chimpanzees by poachers is murder'. It is interesting that the former expression is used by elders in the population when referring to chimpanzees. Thus, the play appears to support traditional perceptions of chimpanzees and to increase awareness of this perception within the younger generation.

In terms of knowledge gained, 80% of the interviewees said they learned something new about chimpanzees from the play. Additionally, 57% of hunters and 65% of people without totemic relationships with chimpanzees

TABLE 5.1. Message remembered by interviewees four months after the play's presentation (after Akindes *et al.*, 2002b).

Message	Percentage of interviewees
The chimpanzee is like a human	62%
Chimpanzees have rich, daily lives	18%
Chimpanzees need to be protected	11%
Threats to chimpanzees	5%

TABLE 5.2. Behavioral/attitudinal changes after seeing the play in the Taï region (Akindes *et al.*, 2002b).

Behavioral/Attitudinal changes	Percentage of interviewees
New conception of chimpanzees	41%
Stopped consuming chimpanzees	27%
More affinity to the chimpanzee	15%
Need to protect the chimpanzee	10%
Killing chimpanzees is a crime	5%
Less fear of the chimpanzee	3%

reported gaining knowledge, suggesting that the play was effective in reaching a diversity of audience members.

The play was also effective in promoting attitudinal and behavioral changes towards chimpanzees, as shown in Table 5.2. Most striking is the fact that four months after the play, 27% of the interviewees reported they had stopped consuming chimpanzee meat and 10% said chimpanzees should be better protected. Seventy- nine percent of interviewees said people in the village changed their behavior following the play. Changes reported, in order of importance, included decreases in poaching of chimpanzees, less chimpanzee meat sold in the villages, and criticism by children of their parents for consuming chimpanzee meat. In all the visited villages, children were said to refer to their parents as "man-eaters" when they ate chimpanzee meat.

4. Conclusion

We believe the success of our play convincingly shows that theater can be a very effective medium for promoting positive cognitive, attitudinal, and behavioral change toward wildlife among local people. Allowing people to visualize the complex lives of chimpanzees created a greater understanding of their unique abilities and facilitated both an increased awareness of the need to protect the species and behavioral changes towards such a goal. An important aspect of this success is presenting a clear, culturally relevant message that is understood by the audience. We felt this was achieved in our case through developing the concept with an African company whose members were from the same region where the play was performed, consulting local people about music and songs, and testing the credibility of the situation with local people. The popularity of the play produced requests among the audience for additional activities, and thus the Wild Chimpanzee Foundation is continuing its educational efforts in these villages through a variety of programs, including a newsletter, discussion group, and video presentations on the behavior of the Tai chimpanzees. Additionally, the positive response of the audience to the play as revealed by the evaluation, indicates that this could be an efficient tool for conservation programs in many target regions. Therefore, the Wild Chimpanzee Foundation plans to export the play to other regions with protected forests and remaining chimpanzee populations. Additionally, we are planning to have Ymako Teatri members train secondary school pupils from the targeted regions to perform the play. In this way, the message will hopefully spread in regions where chimpanzees urgently need effective protection.

Since September 2002, political unrest in Côte d'Ivoire has limited the enforcement of wildlife laws. Such a situation increases the need for regular visits from conservationists to reinforce messages concerning wildlife, a point that has been emphasized to us by the local people in the last year. Fundraising agencies should be aware that in times of political instability such actions need particular support to ensure the active involvement of local people in conservation and research projects.

5. Postscript

In May 2003, one year after the play's presentation and eight months into a period of extreme civil unrest, we visited one of the test villages. Despite having to evacuate the village twice because of approaching looters, children were still rehearsing and singing parts of the play. Discussions with village members, including the chief, women, children, and a poacher, made it clear that the message of the play was still a vivid part of the village's memory.

Acknowledgments. We thank the Ivorian authorities for supporting the projects of the Wild Chimpanzees Foundation from the beginning, especially the "Ministère de l'Environnement," the "Direction de la protection de la Nature," the PACNPT (Projet Autonome pour la Conservation du Parc National de Taï), and the WWF. We like to thank the following people for support: Camille Troh Dji, Touré Zoumana, Honora Néné Kpzahi, and Paul Zouhou. Particular thanks go to Daniel Pauselius for his skill in filming the performances. We thank the following organization for logistic and financial support of this project: Swiss Center for Scientific Research (CSRS) in Côte d'Ivoire, Worldwide Wildlife Fund–Germany, Leipzig Zoo, Tierschutz Zürich, USFW-Great Apes Conservation Fund, and UNEP-Great Ape Survival Project.

References

Akindes, F., Kouamé, S., and Touré, M. (2002a). Évaluation de la perception des villageois de la périphérie du PNT des chimpanzés et de la présence du Parc. Laboratoire d'Economie et de Sociologie Rurales, Université de Bouaké, June 2002, p. 25.
Akindes, F., Kouamé, S., and Touré, M. (2002b). Évaluation des premiers impacts d'une représentation théâtrale sur les perceptions sociales du chimpanzé et sur les

comportements à son égard dans les villages de la périphérie du PNT. Laboratoire d'Economie et de Sociologie Rurales, Université de Bouaké, October 2002, p. 21.

Boesch, C., and Boesch-Achermann, H. (2000). *The Chimpanzees of the Taï Forest: Behavioural Ecology and Evolution*. Oxford: Oxford University Press.

Goodall, J. (1986). The chimpanzees of Gombe: Patterns of behavior. Cambridge: The Belknap Press of Harvard University Press.

Hare, B., Call, J., Agnetta, B., and Tomasello, M. (2000). Chimpanzees know what conspecifics do and do not see. *Animal Behaviour* 59(4):771–785.

Hare, B., Call, J., and Tomasello, M. (2001). Do chimpanzees know what conspecifics know? *Animal Behaviour* 61:139–151.

IUCN (2002). *The IUCN Red list of Threatened Species*. Cambridge: IUCN

Jones, S., Martin, R., and Pilbeam D. (1992). *Human Evolution*. Cambridge: Cambridge University Press.

Kormos, R., and Boesch, C. (2003). Regional Action Plan for the Conservation of Chimpanzees in West Africa. Washington, DC: Conservation International.

Oates, J. (1999). *Myth and Reality in the Rain Forest: How Conservation Strategies Are Failing in West Africa*. Berkeley: University of California Press.

Stoinski, T.S, Ogden, J., Gold, K., and Maple, T.L. (2001). Captive apes and zoo education. In: Beck, B., Stoinski T.S., *et al*. (2001). *Great Apes and Humans: The Ethics of* coexistence. Washington, DC: Smithsonian Institution Press, pp. 113–132

Teleki, G. (1989). Population status of wild chimpanzees (*Pan troglodytes*) and threats to survival. In: Heltne, P., and Marquardt, L. (eds.), *Understanding Chimpanzees*. Cambridge: Harvard University Press, pp. 312–353.

Tomasello, M., and Call, J. (1997). *Primate Cognition*. Oxford: Oxford University Press.

Chapter 6
The Value of Long-Term Research: The Mountain Gorilla as a Case Study

Netzin Gerald Steklis and H. Dieter Steklis

1. Introduction

During the 1930s and 1940s, the early days of primatological research, "prolonged" field studies typically consisted of a single field season of a few months' duration. At best, these early studies were comprised of a series of two to three such field seasons (see Carpenter, 1964). Short-term field studies provided valuable first documentation, or a "snapshot," of a population's or a single social group's ecological setting, its size and structure, and, if habituation permitted, its basic behavioral repertoire. Beginning in the late 1950s, however, there was a trend for primate field studies to increase in duration. In surveying the literature over a five-year period during the late 1980s, Dobson and Lyles (1989) found that the median duration of primate field studies on 53 populations, representing 18 species, was 1.5 years. Though the length of some of these studies reflects the average time required to complete doctoral research, the short duration is probably also a reflection of real constraints (e.g., logistical, political, financial). Longer-term field work requires the sustained motivation and dedication of effort by one or more lead investigators, and an ongoing influx of human and monetary resources alongside a host country's political stability and goodwill. Despite these costs and constraints, there are now several examples of ongoing, long-term primate field studies—spanning decades—on a variety of taxa. Best known among these are the chimpanzees of Gombe National Park, Tanzania (Goodall, 1986), the baboons of Amboseli National Park (Altmann and Altmann, 1970), and the mountain gorillas of Rwanda's Virunga Volcanoes (Fossey, 1983). Such long-term research must provide significant benefits that justify the considerable investment of financial and human resources. And yet, there are remarkably few publications that make explicit these benefits (notable exceptions are Strum, 1986; Wright and Andriamihaja, 2003).

In this chapter, we compare the relative benefits of short-term and long-term field research, recognizing the difference between "long-term, continuous research" and "long-term, serial research" (see below). Using the mountain gorilla as a case study, we illustrate the scientific benefits of continuous long-term research with three examples: 1) documentation of rare events, 2) acquisition of accurate life history information, and 3) facilitation and stimulation of further research. The second example includes results of new analyses on female age at first reproduction that underscore the dependence of such life history analyses on continuous long-term research and, in turn, their significance for modeling population dynamics. Subsequently, we consider additional benefits of long-term field research, including economic and conservation benefits. We conclude with an assessment of the relative scientific benefits as well as some costs of continuous versus serial long-term research, and we compare these to the costs and benefits viewed from economic and conservation perspectives.

The mountain gorilla of the Virunga Volcanoes region of central-east Africa serves as an appropriate case study for exploring the value of long-term research because it is one of the longest running projects on any long-lived mammal. Begun by Dian Fossey at the Karisoke Research Center in 1967, the decades of research conducted at Karisoke by scientists from all over the world have yielded a rich, cumulative database on the biology (i.e., physiology, behavior, ecology, life history, and demography) of this unique and highly threatened gorilla subspecies (*Gorilla beringei beringei*) (see Robbins *et al.*, 2001).

2. Benefits of Short-Term Versus Long-Term Research

The benefit or value of a short- or long-term research project depends on a *valuer*'s interest. The benefits of a research project clearly serve a scientific interest, but may also serve conservation or economic interests. For example, basic research on feeding behavior (serving a purely scientific interest) at the same time can provide a conservation benefit simply through observer presence and increased monitoring. Further, the results of research serving primarily a scientific interest can inform conservation managers or wildlife authorities about the size and composition of a study population, its habitat use, and habitat characteristics, all of which may also serve economic interests (e.g., ecotourism). Interests, of course, may also conflict (e.g., scientific research versus economic development, such as ecotourism or habitat conversion), though such conflicts must necessarily be avoided in order for a research project to become a successful long-term project. In Table 6.1 we compare the benefits of both short- and long-term field research from the standpoint of different stakeholder interests, as well as the degree to which each benefit can be realized in a short- versus long-term research project.

We differentiate between three different types of field research projects. We define a "short-term research project" to include a single field season of a few months' duration or less (e.g., length of a reconnaissance study), a series of two to three such field seasons, or one extended field season of a year or two (e.g., length of a dissertation study). A long-term research project is one in which a population has been followed for a decade or more, and the resulting

TABLE 6.1. Potential benefits of a primate field research project.

Potential benefit	Short-term field research project	Long-term, serial field research project	Long-term, continuous field research project	Interest
Basic description of behavior and ecology and snapshot of demography	√√	√√	√√	s
Applications to conservation and management of primates	√√	√√	√√	c, e
Contributions to scientific theory (general principles)	√	√√	√√	s
International publicity	√	√√	√√	e
Protection of primate population and habitat	–	√	√√	e, c
Habituation of primates	–	√	√√	s, e
Local and national economic benefits	–	√	√√	e
Develop local capacity	√	√	√√	e, c
Heighten local conservation awareness	–	√√	√√	c
Record rare events	–	√	√√	s
Facilitate and stimulate more research and varied research	–	√	√√	s, c
Facilitate and stimulate varied initiatives (e.g., economic development, education, tourism)	–	–	√√	e, c
Accurate individual-based life history data	–	–	√√	s

Contribution toward potential benefit: – = little/none; √ = some; √√ = large.
Motivational perspectives: s = scientific; c = conservation; e = economic.

data collection and archiving is coordinated. Customarily, long-term research provides individual-based information. We further define two subcategories. One, a "long-term, serial research project," is one in which the ongoing research is punctuated by significant periods of absence. While there may be a field station or center at or near the site, there is no continuous researcher/observer presence. Second, a "long-term, continuous research project" is one in which there is a permanent researcher/observer based at the field research station.

3. Scientific Benefits

3.1. Recording of Rare Events

One of the advantages of long-term, continuous field research is the higher likelihood of recording significant rare events. For mountain gorillas, events such as female agonism, intergroup interactions, or twin births can occur so rarely that they are likely to be missed altogether in a short-term study and not observed with sufficient frequency in a long-term, serial study. Also missed may be unusual or novel behaviors such as innovative tool use, or snare trap detection and destruction. The occurrence of male-practiced infanticide is a particularly apt example, in that it took many years to document it sufficiently to understand its centrality for socioecological theories of gorilla (and other primate) social groups.

First described in mountain gorillas by Fossey (Fossey, 1984), it took many more years of documentation in gorillas and other species before infanticide became accepted as a male reproductive strategy and hence as a force in primate social evolution generally. Prior to infanticide being recognized as an important factor, enhanced protection from predation and effective intergroup competition for resources were considered as the two key advantages of social group formation in primates (e.g., see review by Wrangham and Rubenstein, 1986). Because mountain gorillas have no natural predators and groups do not compete for the abundant, evenly distributed food, the question soon arose as to the selective advantages to gorilla females of living in heterosexual groups.

Female protection from male-practiced infanticide became the most likely answer, but because it is an infrequent event and rarely directly observed, it required nearly 20 years of observations to record a sufficient number of infanticide cases to evaluate their functional significance (Watts, 1989). Moreover, the relative contribution of infanticide to social grouping could only be tested once sufficient data on gorilla feeding ecology (i.e., resource distribution, abundance, and competition) were obtained (Watts, 1985; Vedder 1989). There had not been any evidence of natural predation (other than poaching by humans), and, given the gorilla's large body size, predation was generally ruled out as a significant reason for the formation of gorilla groups.

In his careful analysis of 19 cases of observed, potential, or suspected infanticides since 1967, Watts (1989) found strong support both for infanticide as a male reproductive strategy and as the principal force of gorilla gregariousness. The earlier work on gorilla socioecology had established that there were low levels of feeding competition, and hence female gregariousness was unlikely the consequence of between-group feeding competition. Infanticide, however, had a high cost for females, accounting for at least 38% of infant mortality for that time period, a cost that could be reduced or eliminated by the protection provided by large silverback males. In short, the cumulative evidence unequivocally supported the idea that mountain gorilla females live in heterosexual groups to reap the benefits of infanticide protection, and socioecological models generally incorporated infanticide as an additional factor (Sterck *et al.*, 1987). Long-term observations thus became critical in establishing accurate estimates of the occurrence of rare events such as infanticide.

3.2. Accurate Life History Data

A unique benefit of long-term, continuous field research is the compilation of accurate life-history data from known individuals—critical for studies of individual variation in life history traits and for understanding population dynamics (Strier *et al.*, 2006). For life history data, such as a female primate's sequence and timing of births, short-term field research cannot capture an individual's reproductive history, especially for long-lived primates. A limitation of long-term, serial research is that there can be no guarantee that an event in the reproductive history was not missed (i.e., birth and death of an infant in the interval between observations) or the accurate timing of an event was recorded (i.e., birth and death dates). Thus, "accurate" life history data consists of a nearly complete accounting of an individual's life history events and accurate dates for these events. In addition, where event dates are estimated, reliability is improved by providing a measure of error around the date. In the following examples, we will show that careful documentation of life history data is important for understanding evolutionary processes, population dynamics, and, ultimately, socioecological theory.

Focused, continuous study of the details of an individual's life course is critical to understanding the causes of variation in life history traits (Morbeck, 1997). Moreover, accurate data from a large number of known individuals are needed to provide a reliable representation of the full range of individual variation in life history traits. In turn, because heritable individual variation in growth, survival, and reproduction results in variation in fitness (i.e., reproductive success), it serves as a source of natural selection (Stearns, 1992). As reproductive success (number and survival of offspring) is most reliably measured over the course of an individual's reproductive career (Clutton-Brock, 1987), studies examining the source of fitness variation among individuals need to be long term.

In wild chimpanzees, for example, a female's reproductive life can cover several decades, and hence the factors that account for individual differences in reproductive success can only be examined from longitudinal data. Consistencies in female chimpanzee dominance rank relationships, for example, may only become apparent by examining behavioral interactions over the course of many years. At Gombe, Pusey and colleagues (Pusey *et al.*, 1997, 2005) examined agonistic interactions (pant grunts) between pairs of females over several decades, and, in this way, were able to demonstrate the influence of dominance rank on reproductive success: High-ranking females have higher body weights, their infants have higher survival rates, and their daughters reach sexual maturity earlier. These results, alongside those from a two-year analysis of dominance relationships among Ivory Coast chimps (Wittig and Boesch, 2003), contradict previous widely held notions of chimpanzees as relatively egalitarian, in the sense that contest competition is low and hence dominance relationships are unimportant (Williams *et al.*, 2002). Similarly, a longitudinal analysis of agonistic interactions among female mountain gorillas (Robbins *et al.*, 2005) revealed patterns of stable long-term rank relationships not consistently gleaned previously from analyses of shorter time intervals. The longitudinal analysis showed that rank improves with age and, compared with low-ranking females, females of high rank had shorter interbirth intervals and higher offspring survival, which translates into a weak, though positive difference in reproductive success (Robbins *et al.*, 2006; 2007).

Life history traits and their variability are fundamental to understanding or modeling population dynamics (e.g., PVAs, Miller *et al.*, Chapter 8, this volume). The accuracy and predictive power of these models depend on population life history profiles developed from long-term study (Robbins and Robbins, 2004). For mountain gorillas, for example, we require data from several generations to reduce the error around life history variables to a level acceptable for reliable analyses or models of gorilla population structure and growth (Gerald, 1995). Some key variables, such as mortality, require the accurate recording of the age at death from a sufficient number of individuals. While shorter-term data often permit the estimation of life history traits through cross-sectional (or time-specific) analyses, these are not as reliable or powerful as cohort (age-specific) analyses (Gerald, 1995). As we will show below, individual variation in life history traits is linked to variation in population structure and composition (demography), which, in turn, affects a population's extinction risk (Dobson and Lyles, 1989; IUCN, 1996).

Several studies have examined demographic and life history traits of mountain gorillas (Harcourt and Fossey, 1981; Harcourt *et al.*, 1981; Webber and Vedder, 1983; Watts, 1990; Gerald, 1995; Robbins, 1995; Gerald-Steklis and Steklis, 2001; Kalpers *et al.*, 2003). The value of continuous, long-term data collection on this population is apparent in reviewing the results of these studies. Because subject sample sizes were small for studies prior to 1990, many important parameters in these analyses necessarily were estimated rather than empirically determined. For example, population dynamics models are normally based on demographic parameters of the females in the population. For mountain gorillas, however, demographic analyses previous to 1990 were not sex specific due to small sample size. Similarly, important age classes, such as infancy and adulthood, had to be broadly defined because subdivision of the age class would have reduced samples to an unusable level. Later analysis (Gerald, 1995) showed that use of such broad age classes fails to detect important demographic details, such as uneven mortality rates that only become apparent with further age and sex subdivision. Later population models were based on these more accurate age/sex-specific mortality rates (Miller *et al.*, 1998; Robbins and Robbins, 2004).

The importance of long-term continuous data in contributing to our understanding of population dynamics is well illustrated in a study by Robbins and Robbins (2004). They used long-term data to develop an agent-based model of the Virunga mountain gorilla population dynamics. Though population models have been constructed in the past employing the usual factors such as age-specific mortality and fecundity/reproductive schedule (Weber and Vedder, 1983; Miller *et al.*, 1998), Robbins and Robbins' agent-based model included not only the more accurate aforementioned data for these typical factors, but also included factors that are only acquired through long-term observations (i.e., rates of female transfer, male emigration, group fission). As a result, the model not only provided predictions of the standard population model parameters (i.e., population growth rate and age structure) but went further in making predictions about the distribution of gorillas in social groups. Because their agent-based model relied on accurate life-history data from the Karisoke gorilla groups, the resulting "base simulation" closely matched the dynamics of this subpopulation.

In addition to their importance for population modeling, accurate life history data can provide novel insights into the relationship between life history traits and group demographics. In the following example, we show how female age at first reproduction (also a key life history variable for population modeling; Caswell, 2001) varies with group demography. This finding was first discovered serendipitously by Gerald (1995) in her analysis of mountain gorilla life history traits and demography, but, as we show here, our further analyses and understanding of this result are owed to the reliability and detail of the continuous, individually based, long-term data.

First births (especially if the infant does not survive long) can easily be missed or birth date accuracy can be compromised by discontinuous group monitoring. While previous estimates of female age at first reproduction have largely been in agreement on a mean age of 10 years, until recently, there had not been sufficient data to explore the sources of variation around this mean. In analyzing the available demographic records from 1967 to 1994, Gerald (1995) revealed a striking finding concerning variation of female mean age at first reproduction: Females in multi-male groups first conceived and reproduced at a statistically significantly younger age than females in single-male groups (approximately one year difference). For the present purpose, we re-examined this finding by including all available data on female first births from 1967 to 2005. We excluded one female, Shangaza, from a multi-male group, as she

had been shown to have an unusual hormonal profile and irregular reproductive cycles (Czekala and Sicotte, 2000). We also excluded two young adult
females, who were not observed for more than a year due to war conditions,
and thus may have had their first offspring undetected. This analysis
reinforced the earlier finding of a significant difference in age at first
reproduction in single-male (median age = 10.9 years, $n = 10$) versus multi-
male (median age = 9.5 years, $n = 35$) groups (Mann-Whitney $U = 82.00, p = .01$;
Figure 6.1). A result of this difference is that by age 10 years, nearly 70% of
the "primiparous female population" in multi-male groups had given birth to
their first offspring, while only 30% of such females in single-male groups
had launched their reproductive careers (Figure 6.2).

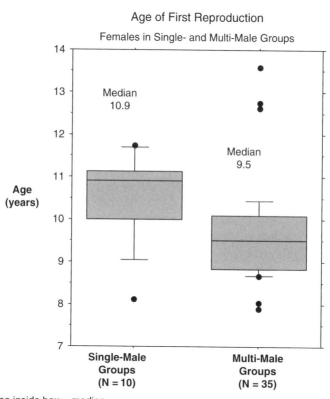

FIGURE 6.1. Age (in years) of first reproduction for females in single- versus multi-
male groups. Boxes contain data points between 25th and 75th percentiles, bars
outside boxes delimit data points between 10th and 90th percentiles, and horizontal
lines indicate median values. Data points with values outside 10th and 90th percentiles
(outliers) are shown as dots.

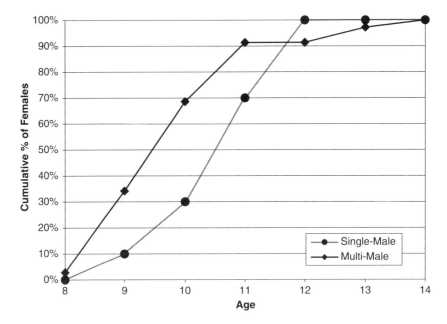

FIGURE 6.2. Cumulative proportion of females who have reproduced by a given age in single–versus multi-male groups.

Because female sample sizes for this comparison remain relatively small and unequal between single- and multi-male groups, some caveats are in order. One concern is that, if the larger number of females (35) in the multi-male group sample is drawn from more groups than is the case for the females (10) from single-male groups, then the latter sample may not be equally representative of the population of single-male groups. In the present analysis, however, this concern would appear negligible, in that the sample of single-male group females represents six gorilla groups compared with seven groups for the multi-male sample.

A second caveat concerns the accuracy of female (mother) age determination. Because of the near-continuous monitoring of the gorilla groups, most of the females in this sample (32 of 45) had accurate birth dates (24 were known to within a few days, 7 known to within a month, 1 known to within 2 months). The remaining 13 females were assigned estimated birthdates based on their narrow age class when first observed as an immature (i.e., infants, juveniles, or sub-adults) with potential errors up to ±1.5 yrs. This raises the possibility that the median difference of 1 year and 5 months in female age at first reproduction in single-male vs. multi-male groups, though statistically robust, is an artifact of the relative proportion of females with accurate versus estimated birth dates in each comparison group. The sample of single-male groups does contain a greater proportion of females (mothers) with estimated birth dates (7 of 10, or 70%) than that for multi-male groups (6 of 35, or 17%).

Accuracy of first offspring birthdates is also a potential source of error, but in this sample 43 of 45 of all first offspring birthdates were known to within one month. Again, this high proportion of accurate life history data is a testament to continuous research. The other two offspring, one each in a multi-male and single-male group, had birth date estimates of ±6 months (with the mother's age accurate) and ±1.5 months (with the mother's age estimate of ±1.5 years), respectively. As a result, the age of first reproduction of both of these mothers was treated as an estimate. In other words, we classified the mother-offspring pairs into two categories: one containing mothers and offspring both with accurate birthdates, the other containing mothers or offspring whose birthdates were estimates. This grouping had little effect on the proportion of female/offspring pairs with estimated birthdates in single-male (7 of 10, 70%) compared to multi-male groups (7 of 35, 20%).

If the difference in female age at first reproduction between single- and multi-male groups is an artifact of an unequal proportion of female/offspring pairs with accurate versus estimated birthdates, then the difference should no longer be significant when only female/offspring pairs with estimated birthdates are compared. This is not the case, however, as the age of first reproduction for this subgroup of females in single-male groups ($n = 7$) remains significantly higher than that for the equivalent subgroup of females in multi-male groups ($n = 7$) (Mann-Whitney $U = 7.00$, $p = .02$). (Note that simply restricting the data to accurate female/offspring pairs reduces the sample size to 3 for single-male groups versus 28 for multi-male groups—an unusable sample size for comparison.) The results of this comparison argue against a birth error confound, and thus support the validity of the overall difference in age at first reproduction.

Finally, it is worth noting that the accuracy of the birthdates for males in the group must also be considered. Males over 12 years old were considered silverbacks and were thus counted in order to classify a group as single- or multi-male. The cases of estimated birthdates for males posed no problems in group classification.

The consistent difference in age at first reproduction between single- and multi-male groups raises a question about the mechanisms responsible for the reproductive differences. One possibility is that the differences between single- and multi-male groups affect the physiology of female sexual maturation. Various socio-ecological factors that affect female sexual maturation (e.g., nutrition, social stress, strange male) have been documented in other mammals, however, there are no studies to our knowledge that have linked group composition to differences in sexual maturation. Detailed endocrinological work with gorillas in these groups is needed to test this possibility. Another possibility is that females in both types of groups mature at the same age, but have different reproductive opportunities. For example, a higher adult sex ratio (adult females to adult males) in single-male groups might provide any given female with fewer opportunities for copulation during estrus compared with an estrous female in a multi-male group. This possibility could be tested by comparing copulation rates of young females in single- versus multi-male groups and by comparing the adult group sex ratio at conception to the age at first conception.

Given the difference in age at first reproduction for females in single-versus multi-male groups, we might expect female lifetime reproductive fitness to vary with group type. Females in multi-male groups enjoy a higher level of protection from infanticide (Watts, 1989), which may explain the finding that females prefer to transfer to multi-male groups (Watts, 2000). More recent analysis of long-term records analyzing interbirth intervals, infant survivorship, and surviving infant birth rates, shows that infant mortality is higher in single-male groups even after the exclusion of infanticide cases. Overall, higher ranking females in larger multi-male groups have higher surviving birthrates (Robbins *et al.,* 2007).

There are also benefits to males living in multi-male groups. Watts (2000) found that there were greater fitness payoffs for males to remain in multi-male groups (as "followers") than to leave them to form their own groups (see also Robbins, 2001; Robbins and Robbins, 2005). The combined fitness benefits for males and females living in multi-male groups is supported by the consistent documentation of a significant proportion of multi-male groups comprising the Virunga population over the years (14–53% based on seven population censuses between 1971 and 2003; see Gray *et al.*, 2005). In addition, recently some multi-male groups have swelled to historic sizes (e.g., Pablo's Group numbers 60) or have contained a record number of silverback males (e.g., Shinda's Group with 7 silverbacks) (Fawcett, personal communication).

Such changes in population structure (demography) have clear implications for projections or models of population growth in light of the significant female reproductive performance differences in single versus multi-male groups (e.g., Robbins and Robbins, 2004). While there is no clear association over time between total population size and number of multi-male groups (see Kalpers *et al.*, 2003), population growth rate will necessarily be affected by the relative percentage of females within multi-male groups at any one time. Variation in life history variables, such as age at first reproduction, will significantly affect gorilla population dynamics models that use 100 years or more time-spans. Hence, population modelers (Miller *et al.*, Chapter 8, this volume) will need to take into account this additional complexity of population structure. Variability in life history traits and comprehensive population models requires accurate life history information that can only be provided by continuous long term research.

3.3. Stimulate and Facilitate Research

A benefit of a long-term research project—continuous or serial—is that it stimulates and facilitates further research in three ways. First, the long-term focus on a study population can provide a stimulus for expanding the scope of research on the same population (e.g., adding genetic or physiological components to a behavioral-ecological focus), thus deepening our under-standing. Second, because of its inherently cumulative nature, the growing documented knowledge on a population serves as a permanent benchmark, inviting comparison with other populations or species or diachronic compar-isons of the population. For example, accumulating observations on a primate population and its ecology often reveals patterns and processes that only become evident over longer time spans (e.g., climate change, demographic change), and thus may reveal adaptive behavioral and social flexibility to long-term environmental change (Altmann and Altmann, 1979; Alberts *et al.*, 2005). Third, a long-term research project provides opportunity for broaden-ing the research program to include other species or wider aspects of the ecosystem generally. In addition, the presence of a permanent field center facilitates research generally in providing logistical support and research resources (e.g., herbarium, library, basic equipment).

The research record over the past four decades at the Karisoke Research Center serves as a good example of the stimulation and facilitation of research. As is evident in Table 6.2, a chronology of researchers and research topics at Karisoke, initial work there focused on basic gorilla natural history and expanded in scope to include studies of, among others, social relationships, com-munication, paternity, and behavioral endocrinology. There is little doubt that the establishment of Karisoke as a permanent research center along with the heavy investment in habituation of gorilla groups facilitated the impressive succession of researchers and expansion of the scope of research.

In serving as a benchmark, the cumulative, long-term research data from a population stimulates and facilitates comparative intra- and interspecific research. For example, research at Karisoke provided a stimulus and basis for comparative investigations of the nearby eastern lowland gorilla (*G. beringei graueri*) (e.g., Yamagiwa *et al.*, 2003), the Bwindi gorilla population (*G. beringei ssp?*) (e.g., Robbins and McNeilage, 2003), as well as the Western gorilla (*G. gorilla gorilla*, see *American Journal of Primatology* special issue on the behavioral ecology of Western gorillas, 2004, vol. 64, issue 2). Several of the researchers who went on to participate in or set up these projects across equatorial Africa had previously worked at Karisoke, which enhanced their field research skills and pointed them toward further research direc-tions. In the west of Africa, several field sites were established to examine, from a comparative standpoint, the socio-ecology of Western lowland goril-las (see Taylor and Goldsmith, 2003). For example, the Mondika Research Center was set up in 1995 by Dian Doran, former Director of and researcher at the Karisoke Research Center, with the explicit goal of testing predictions derived from the socio-ecology of mountain gorillas against the very differ-ent ecology of Western lowland gorillas.

Comparative research is also facilitated by having data organized in a for-mat that is readily accessible for analysis (e.g., a computerized database). An electronic database can be shared amongst internationally based scien-tists with a common research interest. For example, life history data from

TABLE 6.2. Karisoke Research Center: Researchers 1967–2005.

Residency period	Researcher	Nationality	Research topics
1967–1985*	**Dian Fossey**	USA	Mountain gorilla behavior and socioecology, including infant development, ranging, migration/transfers, vocalizations, feeding ecology, reproduction, infanticide, population dynamics including census work and demography.
1971–1974 1980–1983	**Alexander Harcourt**	UK	Mountain gorilla behavior and socioecology, including social relationships, social structure, migration/transfers, feeding ecology, reproductive behavior, vocalizations, population dynamics including census work and demography.
1973–1977* 1981–1983	**Kelly Stewart**	USA	Mountain gorilla behavior and socioecology, including social relationships, social structure, migration/transfers, feeding ecology, reproductive behavior, vocalizations, social development and immature/adult relationships.
1976–1978	Ian Redmond	UK	Mountain gorilla intestinal parasites
1970–1971 1987–1989	**Alan Goodall**	UK	Mountain gorilla foraging and socioecology and comparison to gorillas of Mt. Kahuzi-Biega region
1974	Tim Caro	USA	Mountain gorilla lone silverback ranging behavior.
1978–1979	Amy Vedder	USA	Mountain gorilla feeding ecology, including ranging/habitat selectivity, nutrients of foliage, population dynamics including census work and demography
1978–1979	Bill Weber	USA	Socioeconomics of habitat and wildlife preservation; mountain gorilla population dynamics including census work and demography
1978–1979	**David Watts**	USA	Mountain gorilla behavior and socioecology, including social relationships, social structure, migration/transfers, feeding ecology/foraging strategy, sexual behavior, reproduction, adolescent behavior, twins, infanticide, life history strategies.
1984–1987* 2–3 months in 1989, 1991, 1992			
1979–1980	**Peter Veit**	USA	Mountain gorilla sexual behavior
1983	Robert Seyfarth	USA	Mountain gorilla vocalizations
1983	Dorothy Cheney	USA	Mountain gorilla vocalizations
1981	Augustin Mutamba	Rwanda	*Hagenia abyssinica* distribution and contributing factors.
1984	Joseph Munyaneza	Rwanda	Insect abundance in the *Hagenia* forest; Acrididae ecology.
1986			
1985	**Wayne McGuire**	USA	Mountain gorilla female competition.
1986	Juichi Yamagiwa	Japan	Mountain gorilla behavior in all-male groups, including intra- and inter-group interactions.

Dates	Name	Country	Research
1986	Jorg Hess	Germany	Mountain gorilla families and twins.
1987–1990* 1993–1994 1996–1998*	**Pascale Sicotte**	Canada	Mountain gorilla behavior, including silverback-infant social relationships, infant development, female transfer, neigh vocalization; female reproductive hormonal assessment; traditional perceptions of the forest and gorillas.
1987	Froduald Hakizimana	Rwanda	Buffalo habitat utilization, nutrition and altitudinal distribution.
1987	Evariste Nsanzurwimo	Rwanda	Frugivorous bird feeding strategies in the Virunga and Nyungwe forests.
1988	Stephane Dondeyne	Belgium	Soil chemistry and vegetation on Visoke; natural regeneration of the Afro-Alpine vegetation on Karisimbi.
1988	David Tangishaka	Rwanda	Bushbuck ecology.
1988	Kristine Smets	Belgium	Local community use of forest products; lava-zone characteristics and problems.
1996 1999–2001*	**Andrew Plumptre**	UK	Vegetation-herbivore dynamics; changes in ungulate populations due to war and hunting.
1989	Richard Byrne	UK	Mountain gorilla foraging skills and handedness.
1989	Jennifer Byrne	UK	Mountain gorilla foraging skills and handedness.
1992 1999–2005*	Nancy Czekala	USA	Mountain gorilla reproductive hormones; tourism impact on stress hormones
1989–1991	**Diane Doran**	USA	Mountain gorilla locomotor behavior.
1990–1993 2003	Alastair McNeilage	UK	Mountain gorilla ecology and carrying capacity; mountain gorilla census.
1990–1993 2004–2005*	Alison Fletcher	UK	Mountain gorilla social development; development of gorilla twins
1990	Eric Knox	USA	Botanical reference / herbarium.
1990–1992 1999–2005*	Martha Robbins	USA	Mountain gorilla male social behavior and hormones, male and female reproductive strategies and success; paternity analyses
1991–2005*	**Dieter Steklis**	USA	Mountain gorilla facial expressions and vocalizations; GIS analyses of gorilla ranging, habitat characteristics, and poaching-patterns; Mountain gorilla population dynamics; Tourism impact on mountain gorilla behavior and hormones.
1991–2005*	Netzin Gerald Steklis	USA	Mountain gorilla vocalizations; GIS analyses of gorilla ranging, habitat characteristics, and poaching-patterns; Mountain gorilla population dynamics, demography and life history traits; female reproductive strategies and success; Tourism impact on mountain gorilla demography; Biodiversity inventory.

(Continued)

TABLE 6.2. Karisoke Research Center: Researchers 1967–2005—Cont'd.

Residency period	Researcher	Nationality	Research topics
1992	John Mitani	USA	Mountain gorilla vocalizations.
1992	Samson Kajonjoli	Rwanda	Rodent inventory and altitudinal ranges.
1992	François Karamuka	Rwanda	Mountain gorilla spatio-temporal variation of activities.
1993	Jos Milner	UK	Tree hyrax ecology.
1995–1996	Ymke Warren	UK	Mountain gorilla social dynamics of bi-male groups; Mountain gorilla infant carrying by males and carrying of dead infants.
1996–2002	**Elizabeth Williamson**	UK	Mountain gorilla silverback social dynamics in a large group: infant carrying by males and carrying of dead infants.
1996	Andrew Routh	UK	Mountain gorilla syndactyly.
2000	Chloe Wilson	UK	Mountain gorilla feeding ecology and its implications for conservation.
2002	T. Munyangabe	Rwanda	Community participation in conservation management of protected areas in Rwanda.
2002–2005	**Katie Fawcett**	UK	Tourism impact on mountain gorilla behavior and hormones; Mountain gorilla census; Bird census; Biodiversity inventory; Golden monkey socioecology.
2002–2005	Serge Nsengimana	Rwanda	Biodiversity inventory; local use of medicinal plants; plant parts eaten by large herbivores.
2003–2005*	Tara Stoinski	USA	Male mountain gorilla social dynamics and hormones in multi-male groups
2003–2004	Chloe Hodgkinson	UK	Questionnaire study of tourist evaluation of mountain gorilla tracking experience.
2003–2004	Carla Venturoli	Italy	Mountain gorilla copulations during pregnancy.
2003–2004	Eugene Kayijamahe	Rwanda	GIS analysis of poaching patterns and gorilla ranging.
2003–2004	Theogene Ngaboyamahina	Rwanda	Human activities and park interface.
2003–2004	Aimable Nsanzurwimo	Rwanda	Bamboo ecology.

* Includes some significant periods of absence from Karisoke Research Center.

Notes: Researchers in bold also served as Director of Karisoke Research Center. This table includes researchers who were based at Karisoke and whose independent research and observations resulted in a written publication or report (i.e., contribution to the scientific record). We excluded, for example, research assistants who collected data under the direct supervision of the project's principal investigator. The information in this table is derived from publications of research, filed university dissertations, and/or Karisoke Research Center documents. Several listed research projects were conducted in collaboration with other resident researchers and/or researchers from other institutions and conservation organizations that did not reside at Karisoke.
All studies were conducted in the Virunga Conservation Area (Volcanoes National Park in Rwanda, Virunga National Park in Democratic Republic of Congo, and Mgahinga National Park in Uganda) and immediate surrounding area.

long-term, individual-based field studies, on a diversity of primate taxa, can be used to answer comparative ecological and evolutionary questions (Strier et al., 2006).

A long-term research project, particularly one based at a permanent field station, provides opportunity for a diversity of research projects beyond the focal study population. Research at Karisoke, despite a continuing focus on the gorillas themselves, over the years expanded, through a mutual facilitation process, to encompass all aspects of the fauna and flora of the Virungas (Table 6.2). It covered a striking diversity of projects (plants, birds, insects, rodents, ungulates, geology). Many of the more significant research projects that were not focused on gorillas, nevertheless, depended on the accumulating knowledge of gorilla behavior and ecology. For example, studies of feeding competition among gorillas and other large herbivores (Plumptre, 1996) depended and built upon previous work on the feeding ecology of mountain gorillas (Watts, 1984, 1988; Vedder, 1989). Similarly, this earlier work on gorilla feeding ecology provided a basis for the GIS classification of Virunga habitat types, their relationship to gorilla group ranging patterns, and an estimate of habitat carrying capacity—also relevant to park and wildlife management (McNeilage, 1995; Steklis et al., Chapter 11, this volume). Indeed the diversity of research at Karisoke reflects host country needs, such as habitat and biodiversity inventories and impact of tourism on gorilla behavior and biology. Such diversification to address host country needs is also evident at other long-term primate research sites (e.g., baboon crop raiding and aversion conditioning; Strum, 1986). Overall, the breadth of these research projects is testimony to the potent stimulus provided by a cumulative research process anchored to a permanent location.

4. Additional Benefits: Conservation and Economic

Beyond the scientific value, long-term research often has conservation and economic benefits that may have been unforeseen at the inception of the project. Conservation benefits often stem from applying research techniques and results to population management issues. In this regard, the cumulative record of a study population serves as a "reference book" for both conservation management and the measurement of conservation effectiveness (Wright and Andriamihaja, 2003). As we have already pointed out, research on population dynamics models and habitat use, for example, can help protected area managers monitor population growth and habitat carrying capacity. An indirect conservation benefit of long-term research is the protection of the primate population provided by field staff's presence (Steklis and Gerald-Steklis, 2001). Moreover, as in the case of the Virunga mountain gorilla, antipoaching and veterinary intervention programs were launched by Dian Fossey and became a necessary adjunct to research in order to insure the survival of the study population (Fossey, 1983; MGVP/WCS, Chapter 2, this volume). Long-term research can also foster local conservation awareness and provide a sense of its importance through increased social contact and social relationships between researchers and local people (Wright and Andriamihaja, 2003). At Karisoke and elsewhere, conservation education programs are a vital component of long-term research (Nsengimana *et al.*, 2004; Boesch *et al.*, Chapter 5, this volume). Lastly, the employment and training of local field assistants for the research project builds long-term, in-country conservation interest and expertise.

A long-term research project, with its logistical needs (e.g., staff, supplies, vehicles, permits), also provides economic enhancement at both local and national levels. As an example, Karisoke employs a staff in excess of 60 nationals, infusing nearly $300,000 into the local economy in 2005. In the Parc National de Ranomafana of Madagascar, research brought a greater economic value to the region than ecotourism (Wright and Andriamihaja, 2003). For many primate sites, ecotourism is a major source of revenue. These ecotourism programs depend on habituation techniques developed by researchers, and in some cases involve the same groups habituated for research. Further, research often attracts international media attention which in turn attracts tourists and revenue from filming fees (e.g., the publicizing of Dian Fossey's research by National Geographic photographer Bob Campbell; Campbell, 2000). Finally, the resident research expertise can be harnessed by local communities and government agencies in solving problems of economic interest. In Rwanda, for example, the use of GIS for gorilla habitat studies served to launch a GIS Center at the National University of Rwanda and to provide training and instruction in the general application of this technology for both conservation in other national parks and government development initiatives. Likewise, the research expertise from the Gilgil Baboon Project assisted Kenyan farmers in developing appropriate agricultural techniques and in controlling baboon pests (Strum, 1986).

5. Discussion and Conclusions

Throughout this chapter we have concentrated on the benefits of long-term research from the perspective of science, conservation, and economics. However, before drawing any firm conclusions about the overall benefits of long-term research, we must also examine its costs. Similar to the way we examined the benefits, we frame our discussion with a consideration of the potential costs to science, followed by a discussion of the costs from conservation and economic perspectives.

One potential cost to science is the bias that may be introduced through an intensive focus on a single study population. For a species with subpopulations that occupy different habitats or have had exposure to unique historical events (e.g., poaching, habitat loss), there is a risk of biasing our understanding of what is characteristic of the species or the whole, distributed population. For example, prior to comparative field work on gorilla populations, studies of the Virunga mountain gorilla served as a poor standard, in retrospect, for wild and captive gorilla behavior and social organization generally. Studies during the past 10 years on the nearby subspecies revealed various differences, including habitat use (Bwindi gorillas; Robbins and McNeilage, 2003) and prevalence of infanticide (Kahuzi-Biega gorillas; Yamagiwa and Kahekwa, 2001). Field studies on Western gorillas have shown, for example, that multi-male groups are rare, whereas they are common among mountain gorillas. While there are ecological differences between these populations of gorillas, the unique history of poaching and habitat encroachment in the Virungas may also contribute to these group demographic differences (Parnell, 2002). Similarly, in the Virungas the gorilla population growth rate evident for the subpopulation of research groups is significantly different (higher) than that for the Virunga population as a whole (Robbins and Robbins, 2004), which is likely due to local differences in degree of protection from poaching as well as abundance of food resources. Given our current understanding of the influence of ecology (including human influences) on population life history and sociodemographic characteristics, such variability among spatially distributed populations are to be expected if not predictable. The implication for long-term research is that its many benefits derived from a continuous, intense focus on one population must be weighed against the need and ability to broaden the research scope to other sites and/or populations.

A second potential cost to science concerns the subtle drift of a study population from a wild to a managed one. By "wild" we mean that there is little or no human interference in demographic processes. Again, the Virunga mountain gorilla is a good case in point. As we pointed out earlier, the combined scientific, conservation, and economic interests understandably have led to a high level of protection and health monitoring efforts. In the context of the latter, medical interventions, such as snare removal, vaccination, and antibiotic treatment of bite wounds, are not uncommon (MGVP/WCS, Chapter 2, this volume). Such often-life-saving interventions, however, raise serious questions for scientists about the degree to which the population can still be considered wild in terms of it sociodemographic characteristics. This is a case where pure scientific interests may conflict with conservation and economic ones.

Close range habituation, usually necessary for individual-based research, has the potential cost of disease transmission which can be a serious threat to the survival of the population (Wallis and Lee, 1999). This is particularly true for apes that are susceptible to many human pathogens and at the same time are already threatened. For mountain gorillas, PHVA models have shown that a human transmitted disease poses the most serious extinction risk (Miller et al., 1998). Indeed, in the late 1980s a suspected human-introduced measles outbreak required extensive vaccination of the habituated subpopulation in order to avert further morbidity and mortality (Hastings et al., 1991). Despite the subsequent imposition of rules designed to prevent cross-transmission by researchers and tourists, there is evidence for human-gorilla transmission of pathogens (Mudakikwa *et al.,* 2001). At present, 70% (269 of 380, Gray *et al.,* 2005) of the Virunga gorillas are habituated either for research or tourism, leaving a worryingly small reservoir of unhabituated gorillas that represents the population recovery potential following a disease epidemic (Butynski and Kalina, 1998).

In cases where different subpopulations or groups are separately used for research and tourism, there is a potential economic cost of long-term research if the revenues from tourism are perceived to be greater than the economic benefits from research. In the case of the Rwandan mountain gorillas, groups historically designated and used for research only have from time to time also been used, at the national park authority's request, for tourism, even though this incurred a cost of disrupting research. In addition, there has been pressure to habituate more groups for tourism or to consider the "conversion" of research groups for tourism entirely. Such conflict between scientific and economic interests is avoided in cases where the same subpopulation is used for research and tourism, such as the chimpanzees of Gombe Stream Reserve or lemurs of Ranomafana National Park, although some logistical conflict may still arise.

There may well be other costs that we have failed to consider that nevertheless need to be weighed before firmly concluding that, overall, the benefits of long-term research outweigh its costs, and, indeed, a more systematic cost-benefit analysis may be in order. However, in light of the ones we have considered, it seems warranted to conclude that the benefits of such research do indeed outweigh its costs, largely because long-term research provides many scientific benefits (e.g., documentation of rare events, acquisition of accurate life history information, and facilitation and stimulation of further research) alongside significant conservation and economic ones. At the same time, the costs we have drawn attention to are ones that can be managed, overcome, or at least minimized.

If this is a just conclusion, we might wonder why there are fewer long-term research projects than we should expect on the basis of such over-arching benefits—ones duly recognized by the National Science Foundation, for example (Collins, 2001). One likely reason for why there are few (but growing!) long-term research projects is the monetary cost involved and the requirement for consistent funding, while other reasons may concern significant logistical or political problems.

Financial costs and logistical challenges may make long-term serial research more feasible than long-term continuous research, especially since the former can provide many of the same benefits as the latter, though to a lesser degree (see Table 6.2). From a scientific standpoint, the decision about which type of field study to pursue necessarily depends on the goals of the study (see Table 6.2), as some goals can not be achieved with one or the other type of study. As we have suggested, whenever possible, the establishment of long-term research sites, either serial or continuous, should be encouraged in light of the broad benefits.

Lastly, we might ask a question that often justifiably arises concerning long-term research: How long should long-term research be? Nonscientists, in particular, frequently are puzzled by scientists' implicit assumption that there is no end to long-term research, that science can never have a complete understanding of a dynamic animal population, that indeed there is no end to any scientific pursuit. More concretely, the near 40 years of observation on mountain gorillas have not in the least completed our understanding of life history characteristics (e.g., mortality data for dispersing males), nor exhausted our questions about the adaptive capacities of this long-lived primate in a changing environment. Continuing long-term study will endlessly enrich our scientific understanding while also guiding our best conservation practices and generate significant economic benefits.

Acknowledgments. We are grateful to the many DFGFI-Karisoke researchers, assistants, and staff whose work has enriched our understanding of the biology of mountain gorillas and its ecosystem, and whose knowledge has contributed to capacity building. We particularly thank Martha Robbins for helpful comments on the manuscript. A large debt and ongoing gratitude is owed to the Rwandan Karisoke staff, who, for over nearly four decades, have given their hearts, minds, and, sadly, sometimes their lives, to make possible scientific research, and to insure the continuous protection of this population of mountain gorillas. We also thank the parks authorities of Rwanda (ORTPN), Democratic Republic of Congo (ICCN), and Uganda (UWA) for their protection efforts, support of, and permission to conduct long-term research in their national parks.

References

Alberts, S.C., and Altmann, J. (2003). Matrix models for primate life history analysis. In: Kappeler, P., and Pereira, M.E. (eds.) *Primate Life History and Socioecology.* University of Chicago Press: Chicago. pp. 66–102.

Alberts, S.C., Hollister-Smith, J, Mututua, R.S., Sayialel, S.N., Muruthi, P.M., Warutere, J.K., Altmann, J. (2005). Seasonality and logn-term change in a savannah environment. In: Brockman, D.K., and van Schaik, C.P. (eds.) *Seasonality in Primates: Studies of Living and Extinct Human and Non-Human Primates.* Cambridge University Press, pp. 157–196.

Altmann, S.A., and Altmann, J. (1970). *Baboon Ecology.* University of Chicago Press, Chicago.

Altmann, S.A., and Altmann, J. (1979). Demographic constraints on behavior and social organization. In: Bernstein, I.S., and Smith, E.O. (eds.), *Primate Ecology and Human Origins.* Garland Press, New York, pp. 47–63.

Butynski, T.M., and Kalina, J. (1998). Gorilla tourism: a critical look. In: Milner-Gulland, E.J., and Mace, R. (eds.) *Conservation of Biological Resources.* Blackwell Scientific Publications, Oxford, pp. 280–300.

Campbell, B. (2000). *The Taming of the Gorillas.* Minerva Press, London.

Carpenter, C.R. (1940). A field study in Siam of the behavior and social relations of the gibbon (Hylobates lar). *Comparative Psychology Monographs* 16(5):1–212.

Carpenter, C.R. (1964). *Naturalistic Behavior of Nonhuman Primates*. Pennsylvania State University Press, University Park, PA.

Caswell, H. (2001). Matrix Population Models (2nd Ed.). Sinauer Associates, Sunderland, MA.

Clutton-Brock, T.H. (ed.) (1987). *Reproductive Success*. University Chicago Press, Chicago.

Collins, S.L. (2001). Long-term research and the dynamics of bird populations and communities. *The Auk* 118(3):583–588.

Cowlishaw, G., and Dunbar, R. (2000). *Primate Conservation Biology*. University of Chicago Press, Chicago.

Czekala, N., and Sicotte, P. (2000). Reproductive monitoring of free-ranging female mountain gorillas by urinary hormone analysis. *American Journal of Primatology* 51:209–215.

Dittus, W.P.J. (2004). Demography: a window to social evolution. In: Thierry, B., Singh, M., and Kaumanns, W. (eds.) *Macaque Societies: A Model for the Study of Social Organization*. Cambridge University Press, Cambridge, pp. 87–116.

Dobson, A.P., and Lyles, A.M. (1989). The population dynamics and conservation of primate populations. *Conservation Biology* 3:362–380.

Fedigan, L.M., and Asquith, P.J. (eds.) (1991). *The Monkeys of Arashiyama: Thirty Five Years of Research in Japan and the West*. State University of New York Press, Albany, NY.

Fossey, D. (1983). *Gorillas in the Mist*. Hodder and Stoughton, London.

Fossey, D. (1984). Infanticide in mountain gorillas (*Gorilla gorilla beringei*). In: Hausfater, G. and Hrdy, S.B. (eds.) *Infanticide: Comparative and Evolutionary Perspectives*. Aldine Press, New York, pp. 217–235.

Franklin, I.R. (1980). Evolutionary change in small populations. In: Soule, M.E., and Wilcox, B.A. (eds.) *Conservation Biology: An Evolutionary-Ecological Perspective*. Sinauer Associates, MA, pp. 135–149.

Gerald, N. (1995). *Demography of the Virunga Mountain Gorilla* (Gorilla gorilla beringei). MSc. Thesis, Princeton University, Princeton, NJ.

Grant, P.R. (1989). *Evolutionary dynamics of a natural population*. Princeton University Press, Princeton, NJ.

Gray, M., McNeilage, A., Fawcett, K., Robbins, M.M., Ssebide, B., Mbula, D. & Uwingeli, P. (2005). Virunga Volcanoes Range Mountain Gorilla Census, 2003. Joint Organisers' report, UWA/ORTPN/ICCN.

Goodall, J. (1986). *The Chimpanzees of Gombe: Patterns of Behavior*. Harvard University Press, Cambridge, MA.

Harcourt, A.H., and Fossey, D. (1981). The Virunga gorilla: decline of an island population. *African Journal of Ecology* 19:83–97.

Harcourt, A.H., and Stewart, K.J. (1989). Functions of alliances in contests within wild gorilla groups. *Behaviour* 109:176–190.

Harcourt, A.H., Fossey, D., and Sabater Pi, J. (1981). Demography of gorilla. *Journal of Zoology* 195:215–233.

Hastings, B.E., Kenney, D., Lowenstine, L.J., and Foster, J. (1991). Mountain gorillas and measles: ontogeny of a wildlife vaccination program. In: *Proceedings of the American Association of Zoo Veterinarians Annual Meeting*, Calgary, pp. 98–205.

IUCN (1996). *IUCN Red List of Threatened Animals*. IUCN, Gland, Switzerland.

Jernvall, J., and Wright, P.C. (1998). Diversity components of impending primate extinctions. *Proc. Nat. Acad. Sci. USA*, 95:11279–11283.

Kalpers, J., Williamson, E.A., Robbins, M.M., McNeilage, A., Nzamurambaho, A., Lola, N., and Mugiri, G. (2003). Gorillas in the crossfire: population dynamics of the Virunga mountain gorillas over the past three decades. *Oryx* 37(3):326–337.

Likens, G.E. (ed.) (1989). *Long-term Studies in Ecology: Approaches and Alternatives*. Springer-Verlag, New York.

McNeilage, A.J. (1995). *Mountain Gorillas in the Virunga Volcanoes: Ecology and Carrying Capacity*. Ph.D. Thesis, University of Bristol, Bristol, UK.

Miller, P., Babaasa, D., Gerald-Steklis, N., Robbins, M., Ryder, O., and Steklis, D. (1998). Population biology and simulation modeling working group report. In: Werikhe, S., Macfie, L., Rosen, N., and Miller, P. (eds.) *Can the Mountain Gorilla Survive? Population and Habitat Viability Assessment for Gorilla gorilla beringei*. Apple Valley MN: IUCN SSC Conservation Breeding Specialist Group.

Morbeck, M.E. (1997). Life history, the individual and evolution. In: Morbeck, M.E., Galloway, A., and Zihlman, A.L. (eds.). *The Evolving Female: A Life-History Perspective*. Princeton University Press, Princeton, NJ, pp. 3–14.

Moss, C. (1988). *Elephant Memories*. Elm Tree Books, London.

Nsengimana, S.J., Fawcett, K., and Steklis, N. (2004). Biodiversity education and mountain gorilla conservation: the role of the Karisoke Research Center, DFGFI [Abstract]. *Folia Primatologica* 75(S1):37–38.

Mudakikwa, A.B., Cranfield, M.R., Sleeman, J.M., and Eilenberger, U. (2001). Clinical medicine, preventive health care and research on mountain gorillas in the Virunga Volcanoes region. In: Robbins, M.M., Sicotte, P. and Stewart, K.J. (eds.) *Mountain Gorillas: Three Decades of Research at Karisoke*. Cambridge University Press, Cambridge, UK., pp. 391–412.

Packer, C. (1986). The ecology of sociality in felids. In: Rubenstein, D.I. and Wrangham, R.W. (eds.). Ecological Aspects of Social Evolution. Princeton University Press, Princeton, NJ, pp. 429–451.

Plumptre, A.J. (1996). Modelling the impact of large herbivores on the food supply of mountain gorillas and implications for management. *Biological Conservation* 75:147–155.

Plumptre, A.J., and Williamson, E.A. (2001). Conservation-oriented research in the Virunga region. In: Robbins, M.M., Sicotte, P. and Stewart, K.J. (eds.) *Mountain Gorillas: Three Decades of Research at Karisoke*. Cambridge University Press, New York, pp. 362–389.

Pusey, A.E., Williams, J.M., and Goodall, J. (1997). The influence of dominance rank on the reproductive success of female chimpanzees. *Science* 277:828–831.

Pusey, A.E., Oehlert, G.W., Williams, J.M., and Goodall, J. (2005). Influence of ecology and social factors on body mass of wild chimpanzees. *International Journal of Primatology* 26(1):3–31.

Robbins, M.M. (1995). A demographic analysis of male life history and social structure of mountain gorillas. *Behaviour* 132:21–47.

Robbins, M.M., and McNeilage, A. (2003). Home range and frugivory patterns of mountain gorillas in Bwindi Impenetrable National Park. *International Journal of Primatology* 26:467–491.

Robbins, M.M., and Robbins, A.M. (2004). Simulation of the population dynamics and social structure of the Virunga Mountain Gorillas. *American Journal of Primatology* 63:201–223.

Robbins, M.M., and Robbins, A.M. (2005). Fitness consequences of dispersal decisions for male mountain gorillas (*Gorilla beringei beringei*). *Behavioural Ecology & Sociobiology* 58:295–309.

Robbins, M.M., Robbins, A.M., Gerald-Steklis, N., and Steklis H.D. (2005). Long-term dominance relationships in female mountain gorillas: strength, stability, and determinants of rank. *Behaviour* 142:779–809.

Robbins, M.M., Robbins, A.M., Gerald-Steklis, N., and Steklis, H.D. (2006). Age-related patterns of reproductive success among female mountain gorillas. *American Journal of Physical Anthropology* 131:511–521.

Robbins, M.M., Robbins, A.M., Gerald-Steklis, N., and Steklis, H.D. (2007). Sociological influences on the reproductive success of female mountain gorillas. *Behavioral Ecology and Sociobiology* 61:919–931.

Robbins, M.M., Sicotte, P., and Stewart, K.J. (eds.) (2001). *Mountain Gorillas: Three Decades of Research at Karisoke*. Cambridge University Press, Cambridge, MA.

Rubenstein, D.I., and Wrangham, R.W. (1986). Socioecology: origins and trends. In: Rubenstein, D.I. and Wrangham, R.W. (eds.), *Ecological Aspects of Social Evolution*. Princeton University Press, Princeton, NJ, pp. 3–17.

Sicotte, P. (1993). Inter-group encounters and female transfer in mountain gorillas: influence of group composition on male behavior. *American Journal of Primatology* 30:21–36.

Stearns, S.C. (1992). *The evolution of life histories*. Oxford University Press, Oxford, UK.

Steklis, H.D., and Gerald-Steklis, N. (2001). Status of the Virunga mountain gorilla population. In: Robbins, M.M., Sicotte, P., and Stewart, K.J. (eds.) *Mountain Gorillas: Three Decades of Research at Karisoke*. Cambridge University Press, Cambridge, UK., pp. 391–412.

Sterck, E.H.M., Watts, D., and van Schaik. C.P. (1997). The evolution of female social relationships in nonhuman primates *Behavioral Ecology & Sociobiology* 41:291–309.

Stewart, K.J., and Harcourt, A.H. (1987). Gorillas: Variation in female relationships. In: Smuts, B.B., Cheney, D.L., Seyfarth, R.M., Wrangham, R.W., and Struhsaker, T. (eds.) *Primate Societies*. University of Chicago Press, Chicago, IL., pp. 155–164.

Strier, K.B., Alberts, S., Wright, P.C., Altmann, J., and Zeitlyn, D. (2006). Primate life-history databank: Setting the agenda. *Evolutionary Anthropology* 15:44–46.

Strum, S.C. (1986). A role for long-term primate field research in source countries. In: Else, J.G. and Lee, P.C. (eds.) *Primate Ecology and Conservation,* Vol. 2. Cambridge University Press, Cambridge, MA., pp. 215–220.

Taylor, A.B. and Goldsmith, M.L. (eds.) (2003). *Gorilla Biology: A Multidisciplinary Perspective*. Cambridge University Press, Cambridge, UK.

Vedder, A.L. (1989). *Feeding Ecology and Conservation of the Mountain Gorilla* (Gorilla gorilla beringei). Ph.D. Dissertation, University of Wisconsin–Madison.

Wallace, J., and Lee, D.R. (1999). Primate conservation: the prevention of disease transmission. *International Journal of Primatology* 20:803–826.

Watts, D.P. (1984). Composition and variability of mountain gorilla diets in the central Virungas. *American Journal of Primatology* 7:325–356.

Watts, D.P. (1985). Relations between group size and composition and feeding competition in mountain gorilla groups. *Animal Behavior* 33:72–85.

Watts, D.P. (1988). Environmental influences on mountain gorilla time budgets. *American Journal of Primatology* 15:295–312.

Watts, D.P. (1989). Infanticide in mountain gorillas: New cases and a reconsideration of the evidence. *Ethology* 81:1–18.

Watts, D.P. (1990). Mountain gorilla life histories, reproductive competition, and some sociosexual behavior and some implications for captive husbandry. *Zoo Biology* 9:185–200.

Watts, D.P. (2000). Causes and consequences of variation in the number of males in mountain gorilla groups. In: Kappeler, P. (ed.), *Primate Males: Causes and Consequences of Variation in Group Composition*. Cambridge, Cambridge University Press, pp. 169–179.

Weber, A.W., and Vedder, A. (1983). Population dynamics of the Virunga gorillas: 1959–1978. *Biological Conservation* 26:341–366.

Williams, J.M., Hsien-Yang, L., and Pusey, A.E. (2002). Costs and benefits of grouping for female chimpanzee at Gombe. In: Boesch, C., Hohmann, G., and Marchant, L.F. (eds.), *Behavioural Diversity in Chimpanzees and Bonobos*. Cambridge, Cambridge University Press, pp. 192–203.

Wright, P.C., and Andriamihaja, B. (2003). The conservation value of long-term research: A case study from the Parc National de Ranamafana. In: Goodman, S.M. and Benstead, J.P. (eds) *The Natural History of Madagascar*. Chicago, University of Chicago Press, pp. 1485–1488.

Yamagiwa, Y. and Kahekwa, J. (2001). Dispersal patterns, group structure, and reproductive parameters of eastern lowland gorillas. In: Robbins, M.M., Sicotte, P. and Stewart, K.J. (eds.) Mountain Gorillas: Three Decades of Research at Karisoke. Cambridge, Cambridge University Press, pp. 89–122.

Yamagiwa, J., Kanyunyi, B., Kiswele, K., and Takakazu, Y. (2003). Within-group feeding competition and socioecological factors influencing social organization of gorillas in the Kahuzi-Biega National Park, Democratic Republic of Congo. In: Taylor, A.B., and Goldsmith, M.L. (eds.), *Gorilla Biology*. Cambridge, Cambridge University Press, pp. 328–357.

Chapter 7
The Art and Zen of Camera Trapping

Jim Sanderson

1. Introduction

Habitat destruction and illegal hunting have led to increased concern regarding the status of wildlife populations both inside and outside protected areas. However, the lack of baseline data and absence of accurate estimates of population trends have prevented conservationists and wildlife managers from identifying, quantifying, and addressing suspected negative impacts. To assess wildlife population trends, scientifically based monitoring programs must be implemented (Karanth and Nichols, 1998, Mackenzie *et al.*, 2006).

A common method for monitoring birds or primates is to walk through a forest, stop at each station in sequence, and record every bird or primate species observed or heard. This is the direct method of monitoring: the observer records detailed observations of what is seen or heard. Direct methods work well when the subject is easily observed or heard: most birds are comparatively easily observed or heard, many primates are vocal or move about in the canopy, and herding animals in savannas or woodlands have few places to hide. Monitoring prey populations is typically easier than monitoring predators simply because prey numbers are so much greater, and predators are, by their nature, often stealthy.

Most often, however, direct observations of animals cannot be made. For instance, to estimate the population of mountain gorillas (*Gorilla beringei beringei*) and orangutans (*Pongo pygmaeus*), nest counts are made. During some surveys, direct observations are never made. Moreover, monitoring cryptic wildlife species, such as top carnivores, is often difficult or impossible. Such animals are rarely observed in natural settings. Carnivores, particularly those in tropical forests, are usually elusive and are not easily observed by humans. Some are nocturnal or move about the landscape using dense cover. Typically, carnivores range widely and occur infrequently over large parts of their home range. Their population densities are usually low, making direct observation methodologies unreliable or impossible. The basic ecology of carnivores makes their populations inherently difficult to monitor. Population trends of some small carnivores are even more difficult to estimate effectively.

Sometimes the animals' presence is completely unknown. Thus, major challenges must be overcome to monitor carnivores and other shy species.

In spite of these difficulties, scientific monitoring of rare and elusive animals must be undertaken. Knowledge of their populations is an essential requirement to objectively evaluate the effectiveness of management decisions. Most often, baseline data on population numbers must first be established. Continuous population monitoring can then be used to assess outcomes of management strategies. As knowledge of various management strategies and their effect on a population increases, managers can develop predictive capacity that allows them to deal with new and unexpected situations.

2. Camera Traps

A new generation of camera traps and the use of well-developed capture-recapture models have led to an increase in the use of remote surveying and monitoring methodologies for terrestrial species. Camera traps can be used to make more accurate estimates of species diversity and richness, total mammalian biomass, the spatial and seasonal variation of some species, activity patterns, and seasonal changes in populations, and can be used to determine the presence of very rare species, even those that are highly arboreal. With long-term use, camera traps enable monitoring of many species. To aid law enforcement activities, camera traps have been used to identify individual humans committing illegal acts in protected areas. For instance, fully one quarter of all photographs taken by camera traps in a Malaysian national park were of humans (Sunquist, personal communication). In Cambodia, the use of camera trapping has increased enforcement capabilities.

Population estimates can now be made for individually identifiable species, and occupancy rate indices can be calculated for other species whose individuals cannot be identified. Year-to-year occupancy rates can be compared even when individuals cannot be identified. For instance, Karanth (2000) estimated tiger densities in four national parks in India, and Trolle and Kéry (2003) estimated ocelot densities in an area of the Pantanal. Carbone *et al.* (2001) suggested camera traps could be used to estimate densities of animals that cannot be individually identified, however, this remains controversial. For those species whose individuals can be identified uniquely, the methodology to estimate population densities using camera traps was given by Karanth and Nichols (1998). For species that cannot be uniquely identified, following occupancy rates is an effective monitoring tool (Mackenzie *et al.*, 2006).

Both closed (no immigration or emigration) and open statistical models can be used to estimate survival and recruitment in mammals that can be individually identified with camera traps (Pollock *et al.*, 1990). With sustained use, camera traps can be used as an early warning system to detect changes in number, composition, and relative abundance beyond background noise.

There are currently at least 22 manufacturers of camera traps. Most are single units that use off-the-shelf 35-mm or digital cameras and batteries. Some units are fully programmable, while others offer on-off toggle-switch choices. Most units can be programmed to operate continuously, during the day only, or at night only. All units have a wait or delay period between successive photographs, that is, after a photograph is taken, the camera must wait a certain time interval before another photograph can be taken. In this way, multiple pictures of the same subject are minimized.

Active camera traps, such as those manufactured by TrailMaster®, require an invisible beam to be broken so that a photograph can be taken. In this case, a sending unit and a receiving unit are paired, and a camera is cabled to the receiving unit. The sender and receiver both use batteries, and the camera also requires its own smaller battery. Preventing theft of these units is difficult since each piece must be secured. TrailMaster® units are fully programmable and thus offer complete flexibility in the hands of an experienced user.

Passive camera traps, such as those manufactured by Trapa-Camera, a Brazilian company, are self-contained single units. The sensor detects heat-in-motion that triggers the camera to take a photograph.

All camera traps use a recording device to photograph whatever walks in front of them. Thus, it is likely that in the future digital recording devices will replace film cameras in most camera traps. Because storage capacity is not limited to 36 photographs, as with 35-mm cameras, digital recording devices offer the possibility of replacing snapshots with video during daylight hours.

Imprinting information on each photograph is useful and depends on the camera's functionality. Most 35-mm cameras allow the day and time to be imprinted on the photograph. The month must be inferred. Digital recording devices record the full date and time on the file. Often, enabling the time stamp on the camera or programming the digital camera or camera trap is the most difficult mechanical hurdle that must be overcome. Anyone can attach a camera trap to a tree, but getting a good photograph is a technical challenge (Figure 7.1).

3. Site Selection

Time in place and complete area coverage are the two most important factors in camera trap data collection. As photographs accumulate, comparisons can be made between sites. Site selection is important in targeting specific species. Since some animals follow seasonal patterns, leaving cameras in place for a year or more might be required. Because human presence disturbs wildlife, leaving the phototraps unattended for as long as possible enables better opportunities to photograph wildlife.

As with humans, many animals use forest trails to move about the landscape. Other places where animals visit, such as water holes, salt licks, food sources such as fruiting trees, or patches of preferred vegetation, attract

FIGURE 7.1. A CamTrak Phototrap.

certain species. Obvious animal trails are often sites for possible camera trap deployment. Small species, such as the smaller carnivores, rarely use trails, however. Camera traps restricted to trails are thus less likely to record these elusive predators. One phototrapping campaign in Sumatra captured more tigers (*Panthera tigris*) on film than leopard cats (*Prionailurus bengalensis*), which probably occur more commonly in the forest (Holden, 2001). This might be because most of the camera trapping occurred on trails known to have been frequented by tigers. Perhaps the much smaller leopard cats are more likely to avoid these trails.

Visual aids and chemical lures are often used to attract wildlife to camera traps. Important questions arise with respect to data analysis when lures are used. Of primary importance is that data in the form of photographs must be collected, that is, if there are no photographs there is little data analysis that can be done (though this in itself should tell us something!). If the statistical model assumes that each individual animal has its own capture probability (the probability of getting a photograph of that particular individual), then the use of lures and attractants should increase the number of photographs.

Mark-recapture models assume that camera traps cover an area entirely and approximately uniformly. An important consideration is to ensure coverage of the entire sample area without leaving holes or gaps that are sufficiently large to contain a target species' movements during a sampling period and within which one of the target individuals has *zero chance* of being photographed. That is, the sample area must have *no holes* where the target species can hide during a trapping occasion. An *occasion* is the number of contiguous trap-days a set of camera traps operates in a sample area. One occasion might be 10 days, and 18 occasions might be used to sample an area.

A hexagonal grid minimizes both the size of the uncovered area and the overlap between camera sites. The distance between camera trap sites depends on the home range of the target species. For the mountain gorillas, we placed camera traps in a hexagonal pattern approximately 500 m apart. Similar hexagons were placed in different habitat types. Typically, opposing cameras are placed at each site in the hexagonal grid (Figure 7.2). Importantly, we did not place each camera trap precisely at the exact geographic coordinate selected by the GIS software. Instead, we placed the camera traps in the vicinity of the GPS coordinate we thought gorillas were more likely to frequent.

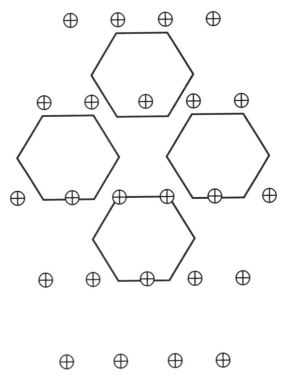

FIGURE 7.2. An Idealized Coverage Using 24 Sites. When placed in an approximate hexagon, each camera site is equidistant from all nearest neighboring sites. ⊕ represents a camera site with two opposing cameras.

In reality, covering an area is often a challenging task. Each location offers unique hurdles. Moreover, area coverage depends on the target species' home range and behavioral characteristics. For instance, determining the population density of sympatric ocelots (*Leopardus pardalis*) and jaguars (*Panthera onca*) is a challenging task because of the home range occupied by individuals of each species. Ocelots have smaller home ranges than jaguars, perhaps by two orders of magnitude. Ideally, an individual of the target species should be photographed in more than one camera trap location. If this does not happen, then the sites might be too far apart.

Habitat also plays a role in determining the distance between camera trap sites. Most often, "favorable habitat" is defined as habitat that supports more individuals. If more individuals occupy an area, then the distance between camera trap sites must be decreased to ensure that each individual is photographed at more than one site. The area covered by each site is smaller in favorable habitat. Thus, the area sampled by each camera trap depends on the home range occupied by members of the target species and on the relative quality of the habitat. Generally, the distance between sites should be proportional to the body mass of the target species and inversely proportional to the quality of the habitat.

Using the ocelot/jaguar example again, determining the population density of ocelots in an area requires a higher density of camera traps than does determining the population density of jaguars. However, the same number of camera trap sites can be used to determine the population density of each species, provided the studies take place on different spatial scales.

The challenge with monitoring mountain gorillas is that they sit in a huge salad bowl surrounded by their preferred foods. Certainly no cameras traps need be placed in areas where there is no food, despite what a GIS might indicate. Thus, a monitoring program for gorillas means covering an area used by gorillas.

4. Protocol

A camera trapping protocol for mountain gorillas has not yet been described and tested. However, following a general protocol offers a starting point that can be refined as the monitoring program progresses. Areas should be selected for camera trap deployment. Pairs of camera traps can be placed to enable the identification of individuals when individuals are uniquely marked in some way. Gorilla faces are distinct and enable individual identification. Generally, sites will be a roughly a constant distance apart. The distance between sites is related to the home-range size of the group of gorillas being monitored.

Recording devices should be set to record the date and time on the photograph. In the case of 35-mm cameras, the day and time should be printed on the film. The camera traps should be placed about 1 m above ground level so as to increase the chance of obtaining a full frontal photograph. If the camera traps are placed on narrow trails, perhaps two opposite facing camera traps are best, because in this way one will likely record a full frontal photograph.

As with all protocols, mentioning the obvious is required: record the GPS location of the camera traps so that they can be relocated, and the date and time the camera traps were enabled for data analysis purposes. Noting a description of the area that includes some information about the local vegetation, trail condition, or other useful information, such as if a fruiting tree is present, aids in data analysis. An attempt should be made to answer the question: why is this camera trap being placed here?

Most camera traps run for 30 days on a set of batteries. My suggestion is to let them do just that and inspect them only when necessary. I believe that the number of photographs is inversely proportional to the number of visits made to the camera trap. Thus, set the camera trap properly the first time and let it do its job.

5. Data Analysis

5.1. Statistical Method

Capture histories can be developed for each adult that has been photographed. The capture history of individual i consists of a row vector of J entries, where J denotes the number of occasions for the particular sample area. Each entry, denoted as X_{ij} for individual i on occasion j, assumes a value of either "0" if individual i was not photographed or "1" if individual i was photographed on the particular occasion. A camera trapping occasion might be 10 days, for instance. The matrix of such t-dimensional row vectors for all M individuals photographed during the sampling is often referred to as the X matrix, and these matrices comprise the data from which target species abundances are estimated. The X matrix is analyzed using a standard mark-recapture computer program such as MARK (White *et al.*, 1982).

For example, if 5 individuals were photographed multiple times during $J = 6$ occasions, then X might look like:

individual 1: 100101
individual 2: 001110
individual 3: 110010
individual 4: 000100
individual 5: 101010

The computer software will then be executed to compute a population estimate—the number of individuals populating the area. The area covered by the camera traps can also be estimated (Karanth and Nichols, 1998). Area estimates take into account the maximum distance traveled by individuals, so the area sampled by the camera traps is greater than the area calculated from the polygon circumscribed by the outermost camera traps (in mathematical jargon, the so-called *convex hull*). From the estimated abundance and the estimated area, an estimated density is calculated.

5.2. Occupancy Rates of Nontarget Species

Estimating population densities of species that cannot be identified as individuals is difficult (Karanth and Nichols, 1998) and remains controversial (Carbone *et al.*, 2002). Many use line-transect methods. For surveying large-bodied herbivores in fairly open country, line-transect methods work well. Line transects did not work well in Cambodia or Guatemala (this author), Taman Negara (Kawanishi, personal communication), or Venezuela (Sunquist and Sunquist, personal communication), where dense undergrowth obstructs views and where species are hunted. However, estimating occupancy rates is a preferred substitute for estimating population densities.

For even modest areas, estimating changes in absolute or even relative abundances of animals that cannot be identified as individuals is nearly impossible, especially in tropical forests. However, estimating the probability that a species is recorded at a random site is a far more tractable problem. By sampling a number of sites over a large area, occupancy rates for each species can be estimated (MacKenzie *et al.*, 2006).

6. Camera Trapping in the Virunga Volcanoes Region

In July 2003, the first camera trap study was begun in the only known habitat of the mountain gorillas. Five areas (one in the Democratic Republic of the Congo and four in Rwanda) were chosen for camera trap monitoring. In each area my collaborators and I placed one hexagonal grid, each with seven camera trap sites approximately 500 m apart. The habitat type of each area was recorded. At least one area was known to overlap with the home range of a single gorilla group. The camera trap protocols previously given were followed.

7. Common Problems and their Solutions

A camera trap is to a wildlife ecologist as a hammer is to a carpenter. Though anyone can use a hammer, few of us can construct a house. Thus, anyone can put a camera trap on a tree, but getting a good photograph requires much more. There are two common problems that can and do repeatedly occur and are most often "pilot error," the fault of the user:

1) *Animals are too close to the camera.* The camera trap may have been placed too close to the trail or allowed the subject to pass too close to

the lens. To prevent this from happening, side access to the front of the camera must be made more difficult or impossible. Often I select a tree that is off the trail 2–3 m and between two other trees located closer to the trail. Ideally, a 15-kg animal should be 3 m from the front of the lens, but closer if smaller, or further if larger. I use sticks or brush piled on either side of the camera that forces the subject to walk in front of the camera at a more optimal distance from the lens. I also prefer to use a rock or tree background that further channels the subject in front of the camera. Typically, after an approximate site is located, about an hour is required to precisely locate, clean, prepare, and set up a camera trap. Ideally, the area is free of vegetation and will remain that way for at least 30 days.

2) *"Ghosts" are photographed.* When a camera trap takes a picture without a subject, apparently a ghost has been photographed. This is a common problem with passive sensor camera traps. There are several ways this can happen. First, ground conduction can trigger the sensor. This happens when direct sunlight heats the ground, causing the air to heat and rise. This is "heat-in-motion" and can trigger the sensor. The solution is to relocate the camera trap, remembering to place it under a canopy offering constant shade. Secondly, in some units when the batteries have lost most of their power, weak voltage can also cause the sensor to trigger the camera. If this happens, photographs will be taken after the delay or wait period has expired, say every 20 seconds, until the film is spent. For instance, the last series of photographs will have the same or a close time stamp on them, and the film will be automatically rewound. This happens in some units when the camera traps are left unattended after the batteries have lost their useful life.

8. Art and Zen

Continued use of camera traps and study of the resulting photographs increases the skill and usefulness of this powerful technology. Placement and deployment will become an art, with site selection taking on a new meaning. Often a backdrop will grow in importance. Patience is required and, if not innate, will be learned. Most often the results of our efforts remain a secret until the camera trap is revisited for servicing. However, every new batch of photographs brings new surprises—the length of the wait time only increases the anticipation. I've never been unhappy with the results. I recall one experience in Guyana where my crew and I collected the film after a two-week campaign. One roll of film had only three photographs and another had a remarkable 29 photographs. Unfortunately, the 29 were all of the same three terrestrial birds that managed to stand in front of the camera trap as if inspecting it. But all three photographs on the other film were different species of cats: an ocelot, a margay, and a puma. Similarly, we required three years of camera trapping in Cambodia before recording our first clouded leopard, demonstrating that surprises can and do occur with sustained use. Finally, time has become my greatest ally. Time invested in placing the camera trap properly is usually rewarded handsomely. I can pleasurably invest an hour or more at each site setting and adjusting the camera trap and increasing the opportunities for a great photograph by what I refer to as *site improvements*. After all, the camera trap is going to work 24/7, so it's worth investing some time to get it right.

Acknowledgments. The author wishes to thank Dieter Steklis and Netzin Gerald-Steklis for the opportunity to train the Dian Fossey Gorilla Fund International's Congolese and Rwandan field crews and to assist them in deploying camera traps in the Virunga volcanoes. Thanks also for their comments and suggestions here. I also thank anonymous reviewers for their suggestions.

References

Carbone, C., Christie, S., Coulson, T., Franklin, N., Ginsberg, J., Griffiths, M., Holden, J., Kawanishi, K., Kinnard, M., Laidlaw, R., Lynam, A., Macdonald, D.W., Martyr, D., McDougal, C., Nath, L., O'Brien, T., Seidensticker, J., Smith, D., Sunquist, M., Tilson, R., and Wan Shahruddin, W.N. (2001). The use of photographic rates to estimate densities of tigers and other cryptic mammals. *Animal Conservation* 4:75–79.

El Alqamy, H., Wacher, T., Hamada, A., and Rashad, S. (2003). Personal Communication.

Holden, J. (2001). Small cats in Kerinci Seblat National Park, Sumatra, Indonesia. Cat News.

Karanth, K.U., and Nichols, J.D. (2002). *Monitoring tigers and their prey*. Centre for Wildlife Studies, Bangalore, India.

Karanth, K.U., and Nichols, J.D. (1998). Estimation of tiger densities in India using photographic captures and recaptures. *Ecology* 79(8):2852–2862.

MacKenzie, D.I., Nichols, J.D., Royle, J.A., Pollock, K.H., Bailey, L.L., and Hines, J.E. (2006). Occupancy estimation and modeling. Elsevier Academic Press, Burlington, MA.

Pollock, K.H. (1982). A capture-recapture design robust to unequal probability of capture. *Journal of Wildlife Management* 46:757–760.

Pollock, K.H., Nichols, J.D., Brownie, C., and Hines, J.E. (1990). Statistical inference for capture-recapture experiments. *Wildlife Monographs* 107.

Trolle, M., and Kéry, M. (2003). Estimation of ocelot density in the Pantanal using capture-recapture analysis of camera trapping data. *Journal of Mammalogy* 84:607–614.

White, G.C., Anderson, D.R., Burnham, K.P., and Otis, D.L. (1982). Capture-recapture and removal methods for sampling closed populations. Los Alamos National Laboratory, Los Alamos, NM.

Section 3
Approaches—Tools

Chapter 8
An Experiment in Managing the Human Animal: The PHVA Process and Its Role in Conservation Decision-Making

Philip S. Miller, Frances R. Westley, Ann P. Byers, and Robert C. Lacy

1. Introduction

An alarming proportion of the world's catalog of biological diversity appears to be in decline (Wilson, 1992; Purvis and Hector, 2000), and the steady losses of species may have serious or even catastrophic impacts on the stability and functioning of ecosystems (Tilman and Downing, 1994; McGrady-Steed *et al.*, 1997; Naeem and Li, 1997; McCann, 2000). Consequently, many of the services and benefits that humans derive from the natural world may be dangerously diminished (Chapin *et al.*, 2000; Tilman, 2000). The primary causes of the decline of nearly all endangered species can be directly related to the activities of human populations, both urban and rural (Caughley, 1994): wildlife populations are over-harvested; landscapes are polluted with the infusion of toxins into the air, water, and soil through industrial activity; exotic competitors, predators, parasites, and diseases are introduced into naïve communities that lack the proper defenses to combat these new invaders; wild habitat is converted to agricultural land; and recent evidence suggests that local and now even global climates are substantially modified by the actions of humans (e.g., Walthier *et al.*, 2002). Sadly, we have likely reached a point in time for much our world's biodiversity where these agents of decline will be difficult to reverse. Even if the original forces are relaxed, a remnant isolated wildlife population becomes vulnerable to other forces, intrinsic to the dynamics of small populations, which may drive the population to extinction despite our best attempts at scientifically based species and habitat management (Shaffer, 1981; Soulé, 1987).

It should be clear, then, that the responsibility for this global biodiversity crisis should be shared by all humanity. Stated another way, achieving meaningful and practical solutions to the problem should not be perceived as solely within the domain of the traditional biological sciences. In practical terms, implementing any strategy for biodiversity conservation demands an integration of both biological science and social science, expert and local knowledge, and even economic and conservation imperatives. This desperately needed synergy has been extremely difficult to achieve, but progress is being made.

This chapter describes a workshop process developed by the IUCN's Conservation Breeding Specialist Group (CBSG) that has been remarkably successful in leading the way towards this integration—the Population and Habitat Viability Assessment, or PHVA. Following a general discussion of the process and its key elements, we describe the PHVA workshop conducted for the mountain gorilla in Kampala, Uganda, in December 1997 and explain how this workshop represented a true landmark in our group's way of thinking about organizing and conducting these interactive and dynamic collaborative processes.

2. A Brief History of the PHVA Workshop

One of the cornerstones of applied conservation biology is the technique of population viability analysis, or PVA. PVA is a tool used to estimate the probabilities of wildlife population decline or extinction by analyses that integrate basic demographic and ecological data for a given species with identifiable threats to population survival. This integration is typically achieved through the use of computer simulation models that project the fate of a given population under a defined set of biological and environmental conditions (Burgman et al., 1993; Beissinger and McCullough, 2002; Miller and Lacy, 2003a). Simulation models are very adept at incorporating a large number of processes that can threaten the persistence of wildlife populations and, even more importantly, the interactions that can arise between them (e.g., inbreeding depression and resistance to disease). Since the first formal PVA on grizzly bears in western North America was completed in the late 1970s, a dizzying number of papers demonstrating the use of this tool—on everything from Minnesota moonworts to Wyoming toads to Sumatran tigers—have been published in conservation biology and ecology journals around the world (see Miller and Lacy, 2003b). To use the words of Michael Soulé, PVA has become conservation biology's "flagship industry."

Despite the general acceptance of PVA as a tool to assist conservation planning, this purely analytical process suffers from some fundamental flaws in its design. In short, most population viability analyses are conducted by mathematical ecologists and are typically intended to be read by other mathematical ecologists. The analyses focus very tightly on the biological issues surrounding population endangerment and recovery, with little to no recognition of the human social context within which the population became endangered in the first place. Sophisticated models are constructed, output data are often subjected to rigorous statistical tests, and (sometimes) recommendations for optimal biological management of the population are made to the relevant authorities. However, in a traditional PVA, those human groups responsible for both the causes of endangerment and implementation of the optimal management scenarios are almost never involved in the collection and/or synthesis of biological data or the development of meaningful and achievable management strategies that stem from the analyses. This disconnect between the practitioners of PVA and those who are most acutely impacted by its results often leads to a considerable degree of apathy or even mistrust among the latter domain toward the PVA process. If certain stakeholder domains are to be held responsible for a species' decline toward extinction, then those same stakeholders must be involved in the analysis of relevant biological and social information and the generation of solutions that all parties can live with. To date, PVA has not achieved this level of integration, now often referred to as "transdisciplinarity" (Westley, 2003).

In order to bridge this daunting gap, the late Ulysses Seal, Chairman of the IUCN's Conservation Breeding Specialist Group (CBSG), developed in the late 1980s a workshop process that quickly came to be known as a Population and Habitat Viability Assessment, or PHVA. The PHVA workshop is a highly participatory and dynamic species risk assessment process involving participation by all interested parties showing a stake in the development of management plans for the species or population in question. The workshop balances integrating the biological information required to evaluate the probability of species persistence with integrating, or at least connecting, the individuals from different disciplines and sectors who are centrally concerned with the conservation of the species. The objective is to create a realignment of priorities among individual stakeholder groups to take into account the needs, views, and initiatives of other groups. In this way, the PHVA workshop represents a broadening of the traditional PVA methodology to incorporate as much information as possible on the focal species, its habitat, and the ways in which local human populations impact this focal species and its surroundings (Miller and Lacy, 2003a).

Central to this workshop process is the use of a PVA simulation modeling approach. Our most common software of choice is VORTEX, a package written by Bob Lacy of the Chicago Zoological Society and JP Pollak of Cornell University (Miller and Lacy, 2003b). VORTEX serves as an exceptionally valuable tool to help stimulate discussion around population data collection and the assumptions built into that process, to integrate diverse biological and even social science-based data sets, and to evaluate—without judgment or bias—a set of proposed management alternatives. In this way, the software unites PHVA workshop participants in a common activity, leading to a greater degree of buy-in to the process among participating stakeholders and, consequently, a greater likelihood for positive action following the meeting.

VORTEX effectively simulates the "extinction vortex" of Gilpin and Soulé (1986), in which random events affecting external environmental conditions (e.g., weather, predator/competitor densities) or internal species biological processes (e.g., birth and death rates, offspring sex ratios) can dramatically influence the stability of small, isolated wildlife populations. The population simulation—consisting of mate selection, reproduction, mortality, increment of age by one time step (usually a year), dispersal among subpopulations, removal (harvest) of individuals, supplementation, and population limitation due to finite habitat availability (ecological "carrying capacity") as appropriate to the situation of interest—is repeated many times to generate the distribution of fates that the population might experience. The software is described in detail in Lacy (2000) and Miller and Lacy (2003b) and is available at http://www.cbsg.org.

PHVAs, however, are more than VORTEX-based scientific analysis—more than just a PVA. Over the past decade, considerable thought and experimentation has gone into the process design component of PHVA workshops: the design of the flow of human and task interactions that makes such interdisciplinary collaboration possible. As developed by CBSG, PHVA workshops are highly participatory processes, deliberately designed to combine optimal sophistication with optimal deliberation. Workshops are always conducted in the species' range country, at the direct invitation of the local wildlife management authority. The overall design allows for groups of 20 to 60 people to explore the implication of population dynamics, genetics, and a variety of threats to habitat and species persistence. Many of these people are wildlife management and academic professionals, but a considerable proportion of the total body of workshop participants lie far outside this sector of employment: social scientists, local and national government figures, and even private landowners are part of a typical mix of PHVA workshop attendees. For example, more than 60 people attended a PHVA workshop on the Houston toad (*Bufo houstonensis*) in Texas (USA) in 1994, but less than 20 of these people were biologists with expertise in the species or its conservation; the remainder included cattle ranchers, city mayors, real estate executives, and other concerned citizens (Seal, 1994). This seemingly chaotic mix of expertise and scientific experience was vitally important to the success of the workshop, since more than 95% of land in the state of Texas is privately owned. Management of this highly endangered species requires the direct participation of citizens across a wide range of sectors, so organization and implementation of the PHVA workshop must recognize this.

Participants work in small groups to identify and analyze risks and, ideally, to provide specific measures of such processes as habitat destruction and fragmentation or, if applicable, direct exploitation of the focal species. From the perspective of workshop design and facilitation, a PHVA workshop must provide encouragement for open and divergent expression of ideas as well as the tools necessary for convergence of these ideas and views in the interest of generating achievable action (Figure 8.1). The divergence phase allows for inclusion of a full range of data, views, and stakeholder needs, while the convergence phase allows for precision of analysis, risk assessment, and focused recommendations. Periods of small group work alternate with plenary

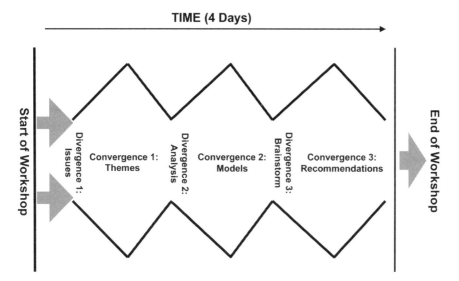

FIGURE 8.1. Diagrammatic representation of the flow of small-group work in a typical PHVA workshop. Adapted from Byers et al. (2003).

presentations, which allows all groups to comment on each other's analysis and recommendations. As more data are introduced and the complexity is increased, some level of acceptable consensus on recommended actions becomes more difficult to achieve. Some of the tools required to build consensus are the VORTEX model and the continual emphasis on prioritization and ultimately translation of analysis into specific plans to be implemented.

CBSG has conducted more than 100 PHVA workshops in nearly 50 countries, and the process has been recognized as an extremely effective vehicle for achieving meaningful decision-making for endangered species conservation (e.g., Conway, 1995; Westley and Miller, 2003). Even with this level of success, we continue to work to improve the process. Critical to this evolution has been the creation of a diverse group of experts devoted to the practical application of E.O. Wilson's concept of "consilience": the unity of knowledge between social and natural sciences as a means of addressing global environmental concerns (Wilson, 1998).

3. Expanding the PHVA Process: The Biodiversity Research Network

With funding from the Social Science and Humanities Research Council of Canada, a research network was formed in early 1997 to build interdisciplinary connections and facilitate exchange of information between specialists directly or indirectly involved with natural resource management. The work of this Biodiversity Research Network has focused on two primary avenues of research (Westley, 2003):

- *Expanding stakeholder inclusion and integrating expertise*—Can we find more effective ways to link social scientists with expertise in such fields as industrial geography, agricultural economics, human demography, and political science with conservation biologists? We must develop a better understanding of the dynamics of the human social system that, in part, defines species endangerment in order to engage those people with the proper expertise and bring them into conservation planning workshops like PHVAs more frequently.

- *Integrating tools for better risk assessment*—Many of the disciplines listed above have their own quantitative tools for data analysis and scenario evaluation. We postulate that output from these tools can be used as critical variables in tools like VORTEX to assess endangered population extinction risk. The task is to develop the appropriate interface to facilitate two or more models to successfully "talk" with one another so that a richer risk assessment can emerge (Figure 8.2).

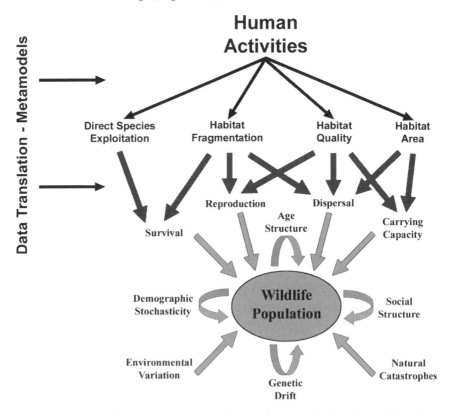

FIGURE 8.2. Major factors influencing the viability of threatened wildlife populations that need to be addressed in the development and application of expanded PVA models as envisioned by the Biodiversity Research Network. Adapted from Miller and Lacy (2003a).

Network members gathered at least twice each year in order to share expertise and to devise experiments around new and innovative approaches for expanding the traditional PHVA approach to stakeholder inclusion and analysis of diverse data sets. This type of transdisciplinary network is not always easy to manage: one of the first hurdles Network members faced was to develop a working understanding of the basic concepts underlying the represented disciplines: conservation biology, wildlife management, population genetics, interorganizational collaboration, human demography, political science, and business. Once it became a functional research unit, the Network quickly began to apply their diverse knowledge to the task of stretching the PHVA process beyond its traditional limits. Members immediately recognized that as the diversity of stakeholders invited to a PHVA is widened, the degree of divergence of ideas and viewpoints is likewise expanded and the task of subsequent convergence to action is made much more complicated. This was to be a major issue in the design and conduct of a revised workshop process.

Once the Network was confident in its conceptual foundation and had designed a revised PHVA workshop process to their satisfaction, it was time to "field test" the concept. In mid-1997, Network members defined the ideal characteristics of a workshop situation in which we could conduct this test: a diversity of available human demographic scenarios (defined primarily in terms of household-based fertility); dependency of local villages on local natural resources; and a well-defined distribution of the focal wildlife species. A workshop already scheduled for December 1997 appeared to be ideally suited for the Network's first case study: a Population and Habitat Viability Assessment for the mountain gorillas of eastern Africa.

4. The Mountain Gorilla PHVA Workshop

The Conservation Breeding Specialist Group, in collaboration with the IUCN Primate Specialist Group, was invited by the Director of the Uganda Wildlife Authority, the Office Rwandais de Tourisme et Parcs Nationaux, and the Institut Congolais pour la Conservation de la Nature to conduct a PHVA workshop for the mountain gorilla in December 1997 in Kampala, Uganda. Gorilla biologists saw considerable cause for optimism for the species' future based on the increase in the number of mountain gorillas over the previous two decades. However, the civil unrest and subsequent armed conflict in Rwanda and the Democratic Republic of Congo produced massive numbers of refugees seeking safety in protected areas such as the Parc des Volcans and Parc des Virunga regions. The potentially rapid rate of habitat destruction in the National Parks resulting from this crisis situation could result in a decline in mountain gorilla population size and a long-term reduction in the viability of the taxon. Local and international management agencies recognized the need for a systematic evaluation of species viability and the development of a regional management plan incorporating the needs of all relevant governmental and nongovernmental agencies as well as public and private stakeholders.

The Biodiversity Research Network saw this PHVA workshop as a critical opportunity to test three hypotheses that formed the foundation of their study (Byers *et al.*, 2003):

- Increased stakeholder participation would result in a richer result and a greater sense of ownership of both process and product;
- Incorporation of local human demographic data into the VORTEX-based modeling process would lead to a more informative picture of mountain gorilla population viability and, consequently, a more effective set of population management recommendations; and
- A firm institutional context including political stability, general social well-being, and the presence of effective government policy could influence the success of conservation initiatives.

In advance of the workshop, Network members collected information on the social, political, and demographic circumstances in the area surrounding the two mountain gorilla populations (Bwindi Impenetrable National Park and the Virunga Volcanoes region). For example, we obtained several articles on the ecological impact of refugee activities and the role of various nongovernmental organizations and other agencies in reducing that impact (Biswas and Tortajada-Quiroz, 1996; Pearce, 1996; UNHCR, 1996). A major challenge for the Network experiment was then to determine the best way in which these data could be successfully translated into input data for VORTEX through avenues such as reduction in habitat availability (carrying capacity) or indirect mortality. Additionally, Network members constructed a series of slide presentations designed to assist workshop participants understand the need to see species extinction risk in the context of definable and—more importantly from the standpoint of PVA—quantifiable consequences of human population growth.

Approximately 80 people, including biologists, researchers, governmental representatives and wildlife park managers, were in attendance on the workshop's first day, December 8, 1997. Although some individuals were unable to participate in the entire five-day event, the majority of these experts were committed to the intense discussions that became the defining element of the workshop. Twenty-six participants were from the three range states, and nearly 50 people had extensive expertise in working in these countries. Workshop sponsorship was generously provided by the Columbus Zoo (USA), International Gorilla Conservation Program, Dian Fossey Gorilla Funds Europe and International, Wildlife Conservation Society, Durrell Wildlife Conservation Trust, and Abercrombie and Kent. This diverse set of sponsors, including both *in situ* and *ex situ* conservation organizations, is a defining theme in PHVA workshop financial support.

The PHVA workshop began with overview presentations from mountain gorilla experts on the species' biology and past and present conservation activities, and from CBSG / Network members on general workshop process. In addition, two Network members gave detailed presentations on the intent of the expanded workshop "experiment" and the enhanced simulation modeling process with a focus on discussion of human demographic data from Uganda. Following this, and as a technique for surfacing issues around which the remainder of the workshop would proceed, the workshop facilitator led the participants through a problem-generation brainstorming exercise. More than 130 statements were recorded on flip charts and thereafter lumped into six categories: population biology and simulation modeling, local human population issues, park and protected area ecology and management, veterinary and health issues, revenue and economics, and political governance. These topics became the titles of six working groups that would stay together for the remainder of the workshop. Each working group was asked to examine their issues in the context of research, education and communication, building of local conservation capacity, and interorganizational collaboration.

As the workshop's first day drew to a close, we Network members readily saw that a major element of our experiment—increased diversity of stakeholder participation—was unsatisfactory. Despite discussions with local and international workshop organizers about the need to broaden the scope of participation, there was a glaring paucity of social science expertise in the room. We knew from the beginning, however, that this was not the fault of the organizers; the difficulty lay in our own ability to adequately explain to them and to other potential participants the vital role that experts outside the realm of traditional biological sciences can play in endangered species risk assessment. We quickly understood that, in order to secure their support and participation, we needed to more effectively speak their professional language and, more importantly, promote their own interests as stakeholders in the larger picture. Unfortunately, our failure to properly secure this broader participation suggested that some potentially critical human population information would not be available for analysis. Nevertheless, we had the world's experts on mountain gorilla conservation biology together for five days and we were very excited about the prospects for a successful workshop outcome.

Data for the PVA component of the workshop was based on nearly three decades of field data collected by a variety of researchers at the Karisoke Research Center and by those studying habituated groups visited by tourists (summarized in Gerald-Steklis, 1995; Steklis and Steklis, Chapter 6, this volume). This vast dataset allowed the modeling group to develop excellent estimates of long-term average demographic rates and, more importantly, the levels of annual variation in these parameters due to both demographic and environmental stochasticity (see Miller and Lacy, 2003b). The working group on local human population issues was tasked during the PHVA with providing the human demographic and land-use information to the simulation modeling group for incorporation into an expanded VORTEX-based PVA model for each of the two mountain gorilla populations. Despite a number of complexities revolving around working group dynamics, data availability, and working group structure (see Byers *et al.*, 2003 for a more detailed discussion), the participants made important progress in collecting and synthesizing a data set on the projected impacts of severe human civil unrest and war on mountain gorilla populations and their habitat. Specifically, in close collaboration with the population biology and simulation modeling working group, they proposed detailed scenarios in which a major event such as the Rwandan genocide of 1994 would occur on average every 30 years and have an average duration of 10 years. During the event, fewer adult females would produce offspring and mortality rates among adults and infants would increase (Figure 8.3). In addition, selected scenarios were extended to include a gradual and cumulative decrease in ecological carrying capacity of mountain gorilla habitat through the direct destruction of the habitat as well as indirectly through the encroachment of refugees and combatants into this habitat. These assumptions were based in part on direct observations of population demographic processes

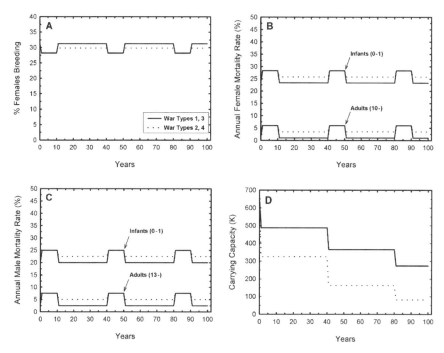

FIGURE 8.3. Simulated impacts of a war scenario in the Virunga Volcanoes region on local mountain gorilla population demographics and habitat ecology. Specific variables affected are (A) proportion of adult females breeding in a given year; (B) annual female mortality rate; (C) annual male mortality rate; and (D) habitat carrying capacity. War types 1 and 2 show full and partial return to normal demographic rates, respectively, in the time intervals between major civil unrest, while types 3 and 4 add either moderate or severe cumulative reductions in ecological carrying capacity on a schedule identical to the changes in population demographics. Adapted from Werikhe et al. (1998).

before and during the 1994 event, and also in part on expert judgment of the workshop participants. Unfortunately, the data needed to precisely quantify the demographic effects of major civil unrest on local mountain gorilla populations simply do not exist. Consequently, our computer simulations of gorilla population viability did not reach the level of sophistication to which Network members originally aspired. Despite this limitation, this is one of the first attempts to our knowledge at directly quantifying the anticipated population-level impacts of specific human activities on wildlife populations in the context of PVA. While some PVA "purists" may see this level of speculation as unproductive or perhaps even counterproductive, we feel strongly that ignoring such important human processes for the sake of scientific precision is even more unpalatable.

Our modeling efforts demonstrated the significant demographic impacts that periodic war could have on mountain gorilla populations (Figure 8.4).

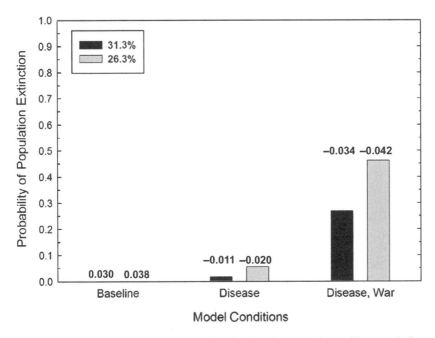

FIGURE 8.4. Extinction risk estimates for simulated mountain gorilla populations derived from VORTEX analyses conducted at the PHVA workshop. The baseline model incorporates the full set of demographic data collected over nearly three decades, while additional models include demographic impacts of possible disease epidemics and periodic civil unrest among the local human populations surrounding mountain gorilla habitat. Each pair of models shows projections for alternative levels of 31.3% or 26.3% adult female breeding success as part of a larger demographic sensitivity analysis conducted during the workshop. Numbers above each bar show the stochastic population growth rate calculated directly from the simulation. Adapted from Werikhe et al. (1998).

In addition to the inclusion of war in our models, we were able to work very closely with a group of gorilla veterinarians and health experts on the identification of a set of current and potential future disease threats and incorporate this potential for catastrophic outbreaks as simulation model elements. Under the combined effects of disease and severe human conflict, mountain gorilla populations in the Virunga Volcanoes region could face a major threat to their survival. It is important to note that while the extinction risk may not appear to be particularly high, especially in the disease scenarios, the negative population growth rates identified in Figure 8.4 indicate the simulated Virunga population is in decline and extinction risk will increase dramatically over a longer time period. Even though these sobering results may seem for some to be little more than plainly intuitive and, therefore, of little interest or value, an explicit and graphical depiction of the sometimes dramatic ways in which humans and wildlife interact on the landscape helps us to understand the nature of these relationships much more clearly and facilitates the successful communication of this understanding to decision makers. In this case, the enhanced PVA analysis and expanded PHVA workshop process helped stimulate new efforts among national and international conservation agencies to more carefully assess the impacts of the recent conflicts on local gorilla populations and their habitat.

As was discussed earlier in this chapter, a PHVA workshop is defined by a series of parallel discussions on many different topics that may or may not directly feed into a VORTEX–based PVA analysis. Lively discussions filled the full five days of the workshop on issues of gorilla management, research, institutional governance, revenue generation schemes, and regional and institutional collaboration. Based on the results of both the extended PVA modeling of war's impacts and the detailed discussions that define our expanded PHVA workshop process, a set of important workshop recommendations were created that included the following (for a full listing, see Werikhe et al., 1998):

• Work should be conducted with humanitarian agencies to ensure that their emergency plans fully address environmental conservation concerns. In addition, conservation agencies must prepare their own emergency plans that address identified critical interactions of humans with gorillas and their habitat.
• When human-gorilla population conflicts are slight or absent, it is important to recognize the potential for resilient growth of mountain gorilla populations. However, our PVA efforts clearly indicate that human population pressures resulting in severe loss of gorilla habitat and a reduction in gorilla survival require an even greater appreciation of the acute risks facing gorillas in order to minimize the risk of population or even subspecies extinction.
• Based on explicit disease risk assessments conducted at the PHVA, the existence of effective and sustainable national veterinary units, responsible for implementation of gorilla veterinary services, is critical to the conservation of the mountain gorilla (MGVP/WCS, Chapter 2, this volume).

- Lead conservation agencies must encourage range country ministers to meet and discuss legal issues relevant to mountain gorilla conservation.
- All relevant stakeholders should meet to discuss and develop appropriate revenue sources and revenue sourcing mechanisms based on an extensive list of alternatives developed at the workshop.
- Standardized park ranger–based monitoring should be developed and implemented throughout the Virunga Volcanoes and Bwindi regions to ensure more effective ecological data collection and analysis procedures (Lanjouw, Chapter 13, this volume).

Mountain gorilla conservation has been largely dominated by the work of international conservation nongovernmental organizations. A primary focus of the PHVA workshop was to develop improvements in the ways these organizations could more effectively collaborate. Following the completion of the workshop, the Mountain Gorilla Foundation (MOGOF) was formed in an attempt to bring together top management representatives annually to develop new cooperative mechanisms for implementing the many and varied action steps outlined in the workshop report. Considered by some to be the most significant outcome of the workshop, MOGOF had its first meeting in January 1999 in Rwanda. In addition, biological research priorities are being prioritized based on the recommendations produced at this workshop, tourist activities have been restructured so that guides adhere to the specified maximum number of visitors and that visitors maintain a required minimum distance from the animals, and broader ranger-based monitoring programs were implemented. All in all, the PHVA workshop had the desired outcome: to stimulate new ways of thinking about the difficult conservation problems facing the mountain gorilla in eastern Africa, and to spur people to action from different countries and different fields of expertise.

5. The Future of the Expanded PHVA Process

Members of the Biodiversity Research Network traveled to Uganda with the intent of conducting a species conservation workshop that builds upon the already respected PHVA process and pushes it further outside of the "box." While being rather pleased with the success of the workshop as a whole, we also realized that our Network experiment's first field test left ample room for improvement. As an example, we came to develop an even greater appreciation of the difficulties involved in integrating different quantitative datasets for use in a PVA. Achieving a successful synthesis requires careful and lengthy preparation in advance of the workshop. Additionally, we must put the same kinds of effort into generating a broader base of stakeholders among the pool of workshop invitees. This often involves extending our own network of contacts to local organizations with the required stakeholder-based expertise. For example, names of appropriate social scientists and academic researchers could perhaps be obtained through the IUCN's local Social Policy Program offices. Finally, we also learned extremely valuable lessons in Kampala about the complexities of designing and facilitating these kinds of diverse transdisciplinary processes. For example, we found that our own particular interests in implementing our experimental workshop designs may not be shared by other participants. If not handled very carefully, this can lead to rapid disaffection among skilled experts and their subsequent withdrawal from the discussions.

Learning these lessons proved invaluable in later implementations of similar field tests in PHVA workshops in Brazil, Papua New Guinea, and Canada (Westley and Miller, 2003). Knowledge gained from this experiment has since been applied in such diverse regions as Indonesia, Mexico, Bangladesh, and Colombia. For this reason, the mountain gorilla PHVA holds a very special place in the minds of those of us in the Network who strive to bring Wilson's notion of consilience to a more tangible reality.

Our Biodiversity Research Network continues to work toward achieving this goal. Specifically, our experiences have convinced us of the need to increase our understanding of both process and content in the field of biocomplexity. Without an appreciation of the mechanisms by which models—and modelers—can effectively communicate, how we can work across diverse disciplines, and how we can engage a wide range of stakeholder domains in complex discussions, we cannot adequately assist natural resource managers and decision makers dealing with difficult environmental problems. Toward this end, we are researching methods by which individual models can be physically linked together into "open-data metamodels" that are capable of passing data back and forth in a common data structure. Additionally, we are adapting techniques of scenario development and testing (Ringland, 1998; Gallopin, 2002) to our own workshop process as a mechanism to allow PHVA participants to use their judgment of likelihoods and their inherent sense of system structure to make predictions of particular events, which can then be translated into inputs to PVA tools like VORTEX. Through research like this, we hope to build upon the solid foundation provided to us by the late Chairman of CBSG, Ulysses Seal, in his persistent drive to get people talking and solving problems. Only by probing the ways we humans analyze and share knowledge will we be able to properly utilize the wisdom of those dedicated to preserving our wild species and spaces.

References

Beissinger, S.R., and McCullough, D.R. (eds.). (2002). *Population Viability Analysis*. Chicago: University of Chicago Press.

Biswas, A.K., and Tortajada-Quiroz, H.C. (1996). Environmental impacts of the Rwandan refugee crisis on Zaire. *Ambio,* 25(6):403–408.

Burgman, M., Ferson, S., and Akçakaya, H.R. (1993). *Risk Assessment in Conservation Biology*. New York: Chapman and Hall.

Byers, A.P., Miller, P.S., and Westley, F.R. (2003). Guns, Germs and Refugees: The Mountain Gorilla PHVA in Uganda. In: Westley, F.R., and Miller, P.S. (eds.). *Experiments in Consilience*. Washington, DC: Island Press, pp. 105–130.

Caughley, G. (1994). Directions in conservation biology. *Journal of Animal Ecology* 63:215–244.

Chapin, F.S. III, Zavaleta, E.S., Eviner, V.T., Naylor, R.L., Vitousek, P.M., Reynolds, H.L., Hooper, D.U., Lavorel, S., Sala, O.E., Hobbie, S.E., Mack, M.C. and Díaz, S. (2000). Consequences of changing biodiversity. *Nature* 405:234–242.

Conway, W. (1995). Wild and zoo animal interactive management and habitat conservation. *Biodiversity and Conservation* 4:573–594.

Gallopin, G.C. (2002). Scenarios, surprises and branch points. In: Gunderson, L. and Holling, C.S. (eds.). *Panarchy*. Washington, DC: Island Press, pp. 361–392.

Gerald-Steklis, C.N. (1995). *Demography of the Virunga Mountain Gorilla* (Gorilla gorilla berengei). Master's thesis, Princeton University. KRC Publication #139.

Gilpin, M.E., and Soulé, M.E. (1986). Minimum viable populations: Processes of species extinction. In: Soulé, M.E. (ed.). *Conservation Biology: The Science of Scarcity and Diversity*. Sunderland, MA: Sinauer Associates, pp. 361–392.

Lacy, R.C. (2000). Structure of the VORTEX simulation model for population viability analysis. *Ecological Bulletins* 48:191–203.

McCann, K.S. (2000). The diversity-stability debate. *Nature* 405:228–233.

McGrady-Steed, J., Harris, P., and Morin, J.P. (1997). Biodiversity regulates ecosystem predictability. *Nature* 390:162–165.

Miller, P.S., and Lacy, R.C. (2003a). Integrating the human dimension into endangered species risk assessment. In: Westley, F.R. and Miller, P.S. (eds). *Experiments in Consilience*. Washington, DC: Island Press, pp. 41–63.

Miller, P.S., and Lacy, R.C. (2003b). *VORTEX: A Stochastic Simulation of the Extinction Process. Version 9 User's Manual*. Apple Valley, MN: Conservation Breeding Specialist Group (SSC/IUCN).

Naeem, S., and Li, S. (1997). Biodiversity enhances ecosystem reliability. *Nature* 390:507–509.

Pearce, F. (1996). Soldiers lay waste to Africa's oldest park. *New Scientist* 3:4.

Purvis, A., and Hector, A. (2002). Getting the measure of biodiversity. *Nature* 405: 212–219.

Ringland, G. (1998). *Scenario Planning: Managing for the Future*. New York: John Wiley & Sons.

Seal, U.S. (ed.). (1994). *Houston Toad Population and Habitat Viability Assessment Report*. Apple Valley, MN: Conservation Breeding Specialist Group (SSC/IUCN).

Shaffer, M.L. (1981). Minimum population sizes for conservation. *Bioscience* 31:131–134.

Soulé, M.E. (ed.). (1987). *Viable Populations for Conservation*. Cambridge: Cambridge University Press.

Tilman, D. (2000). Causes, consequences and ethics of biodiversity. *Nature* 405:208–211.

Tilman, D., and Downing, J.A. (1994). Biodiversity and stability in grasslands. *Nature* 367:363–365.

UNHCR. (1996). *Executive Committee of the High Commissioner's Programme. Standing Committee, 3rd Meeting*. Update on regional developments in Africa. 31 May.

Walthier, G.R., Post, E., Convey, P., Menzel, A., Parmesan, C., Beebee, T.J.C., Fromentin, J.-M., Hoegh-Guldberg, O., and Bairlein, F. (2002). Ecological responses to recent climate change. *Nature* 416:389–395.

Werikhe, S., Macfie, L., Rosen, N., and Miller, P.S. (eds.). (1998). *Can the Mountain Gorilla Survive? Population and Habitat Viability Assessment Workshop for* Gorilla gorilla beringei. Apple Valley, MN: Conservation Breeding Specialist Group (SSC/IUCN).

Westley, F.R. (2003). The story of an experiment: Integrating social and scientific responses to facilitate conservation action. In: Westley, F.R., and Miller, P.S. (eds.). *Experiments in Consilience*. Washington, DC: Island Press, pp. 3–20.

Westley, F.R., and Miller, P.S. (eds.). (2003). *Experiments in Consilience: Integrating Social and Scientific Responses to Save Endangered Species*. Washington, DC: Island Press.

Wilson, E.O. (1992). *The Diversity of Life*. New York: Norton.

Wilson, E.O. (1998). *Consilience*. New York: Alfred A. Knopf.

Chapter 9
Approaches to Corridor Planning: Transitioning TAMARIN from Mata Atlantica to Madagascar

Karl Morrison, Charlotte Boyd, Keith Alger, and Miroslav Honzák

1. Introduction

Effective long-term species conservation requires a conservation approach targeted at all scales at which biodiversity occurs, from the scale of species occurrences to the scale of populations and the ecological processes needed to sustain them (Noss, 2002). Protected area- and site-level initiatives have proven to be effective at protecting habitat, even when resources for effective management are lacking (Bruner *et al.,* 2001), but protected areas alone are often of insufficient size to sustain viable populations of the species they are designed to protect (see, for example, Newmark, 1995). The total area accessible to conservation target species can be increased by connecting protected areas through biological corridors and stepping stones of habitat (Beier and Noss, 1998). But even with large protected areas and effective connectivity networks, human population pressures and incompatible land and resource use in surrounding areas can compromise biodiversity conservation goals (Wiens, 1996). In order to achieve the effective conservation of species, populations and ecological processes, a regional-scale biodiversity conservation corridor approach is necessary.

Biodiversity conservation corridors have two objectives—the primary objective is the conservation or restoration of naturally functioning landscapes and the species diversity naturally present within the landscape; the secondary objective is the reconciliation of biodiversity conservation with the livelihood aspirations of human communities in the region and national development goals. Within biodiversity conservation corridors, irreplaceable biodiversity areas are put under strict protection, areas that can support both conservation and development goals through sustainable use and direct incentives for conservation are also identified, while economically important areas are targeted for more intensive development. A biodiversity conservation corridor is therefore a landscape in which land use, incentives, and policies are designed to achieve conservation objectives while contributing to economic development (Sanderson *et al.,* 2003).

This regional scale approach to conservation planning generally provides sufficient flexibility to identify areas that will contribute to biodiversity conservation at the same time as economic development or with minimum economic cost. Development planning has long undervalued the services and resources provided by biodiversity and functioning ecosystems. This has resulted in infrastructure and policy developments that have not recognized the importance of maintaining economically important ecosystem services or capitalized on the value of biodiversity and so have undermined long-term economic development potential. Biodiversity conservation corridors generally provide sufficient scale and perspective to anticipate the effects of proposed development, infrastructure, and policy decisions on biodiversity and ecosystem services, and so to develop scenarios that meet the needs of both human and biological communities.

Biodiversity conservation efforts continue to be hampered by the lack of integration of conservation goals into regional development planning. For conservationists to work effectively with development authorities, a spatially explicit plan for biodiversity conservation that brings together biodiversity and economic information at a regional scale is often considerably more valuable than site-scale studies. Since there are many ways to design a landscape that will achieve both conservation and socioeconomic goals, enabling stakeholders to evaluate a range of specific landscape scenarios against explicit objectives, assessing the trade-offs between the location of development projects and maintenance of economically important ecosystem services is a more constructive approach than focusing attention on areas of conflict such as proposed protected areas or roads.

This chapter provides an overview of a planning approach that combines clear, simple, and defensible conservation targets (based on the principles of representation, viability, resilience, and redundancy) with economic information that permits the evaluation of the economic benefits and costs of different landscape scenarios. A key aspect of this approach is the Toolbox of Applied Metrics and Analysis of Regional Incentives (TAMARIN) (Stoms *et al.*, 2004) and, for convenience sake, we will refer to the planning approach presented here as the TAMARIN approach.

The chapter presents aspects of case studies from the Atlantic Forest in Brazil and the eastern Malagasy humid forests in Madagascar to show how this approach is being applied in different conservation contexts. The first section provides an overview of the TAMARIN planning approach, followed by a description of these two conservation regions in Section 2, highlighting the biophysical and socioeconomic differences that imply different questions for analysis. The next section provides an introduction to TAMARIN conservation planning software, which is designed to support this approach, followed by an account of how biodiversity conservation targets were set in the two case study regions and a detailed explanation of the construction of two of the five key GIS layers required for TAMARIN (current and future land cover/use).

Finally, we provide a brief overview of the initial results of planning processes using the TAMARIN approach in Brazil and Madagascar.

2. The TAMARIN Approach to Planning

To assist with conservation planning, a number of GIS-based spatial decision support systems have been developed (Figure 9.1). One of the innovative aspects of the planning approach being presented here is that it centers around the development and analysis of five spatial layers that allow planners to integrate economic opportunity cost and biodiversity targets into the development of future land use scenarios. The technical approach is

---

For more information on these software, please see the websites listed below:
SITES (Spexan/Marxan) - selecting reserve sites explicitly incorporating spatial design criteria into the site selection process. MARXAN finds reasonably efficient solutions to the problem of selecting a system of spatially cohesive sites that meet a suite of biodiversity targets
http://www.biogeog.ucsb.edu/projects/tnc/toolbox.html
or
http://www.ecology.uq.edu.au/marxan.htm

C-Plan
C-Plan is designed around the concept of a decision-support system. Together with a geographic information system (GIS) it:
➢ maps the options for achieving an explicit conservation goal in a region
➢ allows users to decide which sites (areas of land or water) should be placed under some form of conservation management
accepts and displays these decisions, and then lays out the new pattern of options that result
http://www.uq.edu.au/~uqmwatts/cplan.html

ResNet
ResNet 1.2, first outlined in the late 1980s, presumes a given area that has been divided into cells (based perhaps on geographical coordinates or ecosystem boundaries), a record for every cell of the presence or absence of every defined surrogate, and a definite target for representation, such as the minimum number of times any surrogate must be represented by the selected cells. The procedure is described in Aggarwal et al. (2002).
http://uts.cc.utexas.edu/~consbio/Cons/Labframeset.html

CODA (Conservation Options & Decisions Analysis)
CODA assists in the design of networks of nature reserves or protected areas. It allows you to define selection units, conservation features, land suitability units, units of costs and definition of conservation objectives.
http://members.ozemail.com.au/~mbedward/coda/overview.html

TAMARIN: The Toolbox of Applied Metrics and Analysis of Regional Incentives
TAMARIN, is a planning support system developed to assist in regional conservation planning. It is a customized ArcView project with scripts originally written specifically for the Central Atlantic Forest Corridor project in Bahia, Brazil. The tool was designed to test various strategies and assumptions about future land use against a set of descriptors of forest landscape configuration believed adequate to meet biodiversity conservation objectives.
www.tamarinmodel.org

FIGURE 9.1. Spatial planning tools.

embodied in the TAMARIN software, which can be downloaded for free at www.tamarinmodel.org. The software is designed to integrate the analysis of biodiversity and economic spatial layers to enable the design of different land use scenarios and assess trade-offs between biodiversity and economic development goals. The software can produce multiple scenarios, which can then be evaluated and scored against predefined conservation targets, for example, whether the proposed protected area system includes all endemic and threatened species to ensure representation. To assess viability, the software can evaluate a protected area system and habitat matrix in a corridor based on whether it provides the size and/or connectivity to prevent extinction of umbrella species with reference to Population and Habitat Viability Analysis. Resiliency targets are incorporated through assessments of core-edge ratios, and redundancy targets require replication of core habitat areas in order to mitigate stochastic risk factors. Finally, landscape scenarios can be evaluated based on the financial and economic costs and benefits of the proposed conservation plan. The software was developed through a partnership of local and international organizations to assist biodiversity conservation planning for the Central Atlantic Forest Corridor of Brazil, but has recently been redesigned so that it can be adapted to other regions of the world.

2.1. Inputs

The TAMARIN approach relies on five distinct layers of information to frame and evaluate corridor scenarios. A current land cover/use layer characterizes current land uses and landscape features. The bioregion layer identifies unique biodiversity regions within a landscape that should be the basis for biodiversity conservation targets for representation and viability. The threat layer characterizes the current threats to species and habitat within the corridor. The opportunity cost/land value layer demonstrates the value of land across the landscape. And finally, the "business as usual" layer represents a projection of land cover/use 20 years into the future if current trends and investments in conservation remain the same. The analysis behind each of these spatial layers can vary widely as long as the end results conform to the required technical format (Stoms *et al.*, 2002).

2.2. Outputs

From the information and analyses in the underlying information layers and the criteria and spatial design specified by the users of the program (development planners, decision makers, etc.), TAMARIN produces a map that depicts a future landscape scenario and summary statistics to show how well this scenario will meet both biodiversity and development goals. Key statistics generated include the number of protected areas of sufficient size and location to conserve threatened species, core to edge ratios, and the cost of conserving the areas selected. The program also provides supplementary analyses and can be adapted to provide a range of other statistics depending on the users needs (see TAMARIN Manual). These features allow users to test commonly held assumptions about what it would take to conserve bio-diversity and test them against economic and biodiversity criteria.

The information layers that form the basis of the analysis within TAMARIN are also key outputs of the process in themselves. Maps of priority areas for conservation, land cover, projected land use, threats, and opportunity costs or land values all have uses outside the framework.

2.3. Process

TAMARIN is designed as an aid to planning and should not usurp the role of policy- and decision-makers—its most useful aspect is its contribution to participatory conservation planning processes and its ability to move stake-holders toward a common vision for the landscape. While TAMARIN is capable of generating a map of an optimal corridor scenario based on bio-diversity conservation targets and economic criteria, the program is designed to facilitate participation in the design and assessment of different landscape scenarios. Scenarios can be designed interactively and iteratively, with rapid processing time allowing immediate evaluation of conservation targets and economic criteria. In a workshop setting with multiple stakeholders repre-sented, and multiple groups formulating possible scenarios, TAMARIN can foster the development of a shared land use vision for a region.

3. Characterizing the Context and Corridor Planning Goals: Identifying the Appropriate Questions

The TAMARIN approach is currently being used in a number of regions around the globe, including the Central Atlantic Forest Corridor of Brazil and the Zahamena-Mantadia Corridor in the eastern rainforests of Madagascar. Both of these regions are contained within areas of very high endemism and less than 10% of the original forest cover remaining (Mittermeier et al., 1999).

The Atlantic Forest is by far the most threatened major ecosystem in Brazil, with less than 8% of its original area remaining. Conservation International places it among the five highest-priority habitats for conserva-tion on the planet (Myers et al., 2000). Spanning a region equivalent in range and extent to the U.S. coastal states from Maine to Florida, the Brazilian Atlantic Forest is a mosaic of forest remnants harboring enormous biologi-cal diversity even as it accommodates more than 80% of Brazil's human pop-ulation and industry. The Central Corridor region of the Atlantic Forest is approximately 600 km long, includes 12 of the areas of highest biological importance identified in the 1999 Conservation Priority-Setting Workshop, possesses one of the highest indices of tree diversity in the world, and hosts a great number of endemic species.[1] It is also the region with the largest amount of remnant Atlantic Forest in the northeast, offering potential for the establishment of additional protected areas.

The eastern Malagasy humid forest is the richest habitat in Madagascar for endemic biodiversity. It occurs as a ribbon of forest along a scarp from north to south, over 1,000 km in length, but is often very narrow (less than 5 km) and is already broken into around 10 large fragments. Forest extends from sea level to about 2,800 m, being richest and most threatened in the lowlands. Forest is under considerable threat, especially from slash-and-burn agriculture in the lowlands and increasing forest exploitation in the mid-altitude regions. Hunting of primates is locally very important, as is collection of reptiles and palms and the bark of *Prunus africana* for commercial purposes. Mining for gold, rubies, and other precious stones has become locally important recently. The Zahamena-Mantadia Corridor represents one of the largest intact fragments of the eastern rainforests and is in itself one of the high priority areas for conservation identified in the 1995 priority setting workshop for Madagascar and again as a high priority in the 2001 Total Biodiversity Coverage Workshop.

While the two regions are similar in that they are forest biomes, contain high numbers of endemic species, and are under severe threat, the conditions in these regions have some fundamental differences that have shaped the way the TAMARIN approach is applied. In both biodiversity conservation corridors, the purpose of analyses and planning is to build a viable conservation corridor by 1) ensuring that the protected areas are of a sufficient size to sustain viable populations of species of conservation concern, 2) ensuring that the number and location of protected areas are sufficient to conserve all of the species of conservation concern within the region and the ecological processes that sustain them, and 3) minimizing the opportunity costs of achieving these conservation goals. But, because of different biophysical characteristics and policy and development agendas within the local, government and donor communities in the two regions, very different tactics are being adopted in the two corridors.

3.1. Biophysical Context

The Central Corridor of the Brazilian Atlantic Forest consists of a very fragmented landscape of small forest patches surrounded by a variety of encroaching agricultural land uses, while the Zahamena-Mantadia Corridor

---

[1] This region contains arguably the largest concentration of endangered and endemic taxa of the Atlantic Forest, including 19 mammal species, 32 bird species, and the highest tree diversity per hectare (over 450) of the region and one of the highest in the world. Notable endemic animal taxa include the golden-headed lion tamarin (*Leontopithecus chrysomelas*), Kuhli's marmoset (*Callithrix kuhlii*), the capuchin (*Cebus apella robustus*), spider monkey, white-winged cotinga, acrobat bird (*Acrobatornis fonsecai*), banded cotinga, and Geoffroy's marmoset. (Experts workshop for ecoregional priority setting, 10–14 August 1999, Atibaia, Sao Paulo, Brazil.)

consists of a large relatively intact forest threatened with fragmentation from encroaching agriculture. In both cases, the goal is to conserve the remaining primary forests and ensure connectivity between forest patches so as to avoid species extinctions, but, in the former, this requires the restoration of connectivity whereas in the latter the focus is on preventing the loss of existing connectivity.

3.2. Economic Development and Poverty Reduction Agenda

In the case of the Central Atlantic Forest Corridor, the remaining forest is under threat from the encroaching agricultural frontier, illegal logging, and resettlement of displaced peoples. Unemployment and rural poverty have stimulated the occupation of large landholdings by the landless, sometimes in areas with important forest fragments. The policy and development agenda is being driven by the region's attempt to diversify its economy from dependence on cacao cultivation for jobs and income, to other forms of agriculture and tourism. However, policy reform to increase agricultural assistance to small farmers, restrict unsustainable logging, and finance land acquisition for the landless on more productive soils has faced cutbacks in administrative capacity. With respect to the conservation agenda, important concerns within government relate to whether conservation of the remaining biodiversity is possible in such a fragmented landscape, the potentially prohibitive costs of achieving conservation within the region, and questions of how conservation goals can be achieved in practice. Biodiversity conservation corridor planning therefore needs to demonstrate that biodiversity conservation can be achieved in this context, without jeopardizing resettlement initiatives and development goals. The focus of economic analysis in the Central Atlantic Forest Corridor is therefore on minimizing the economic opportunity costs of successful conservation.

In the case of the Zahamena-Mantadia Corridor, the remaining forest is under severe threat from slash-and-burn agriculture and illegal logging, mainly by subsistence farmers living at or below the poverty line. Poverty reduction dominates the national agenda and hence analyses to demonstrate the economic value of biodiversity and ecosystem services are critical for conservation planning at the corridor scale. In Zahamena-Mantadia, therefore, economic analysis is focused on demonstrating what will be maintained or gained, in terms of watershed management and ecotourism values, through investment in successful conservation that prevents further forest degradation and fragmentation. Assessments are underway to determine the economic value of protected areas, the watershed value of the remaining primary forests, their ecotourism value, and the value being generated by the transfer of management of forests to local communities (*transfert de gestion*), and to compare these to the value of land if converted to agricultural production. This analysis also needs to take into account the important questions about the distribution of certain benefits, in particular watershed values.

Equally important is the use of TAMARIN results to assist the zoning process being led by the Madagscar Ministry of Water and Forests (MEEF) in the region. The forest zoning process is classifying forests according to three functions: production, regulation and conservation. For each function different management options are being proposed (e.g., community-based forest management, creation of site de conservation, etc.) to achieve the objectives of valorization of forest products, biodiversity conservation, and watershed protection. The people (MEEF, communities, local authorities) involved in the forest zoning need to know what activities or management options are best for each watershed unit and TAMARIN results can serve as an aide in determining effective resource allocation by providing information on the economic and environmental costs and benefits of different scenarios for comparison.

3.3. Land Tenure and Conservation Mechanisms

The state of land tenure and markets for land also shapes corridor strategies and analyses. Secure land tenure and developed markets for land allows the financial value of land and resources to be ascertained relatively straight-forwardly. Where markets for land and resource use do not exist or are distorted by barriers to entry and exit, inadequate access to credit, perverse subsidies, or other barriers to competitive markets, generating reliable estimates of the economic value of land is challenging.

Individual land ownership predominates the region of the Central Atlantic Forest corridor and a well-developed market for land exists, making it relatively easy to ascertain the financial value of a particular piece of land. By sampling the sale prices of lands throughout the region over a period of two years and then performing a regression analysis to identify significant explanatory variables such as distance to road, slope, soil quality, etc., planners have been able to project land prices across the landscape. These data enable conservation planners to design direct financial incentives for conserving biodiversity, such as conservation easements negotiated with large landowners, and provides a proxy for the economic opportunity costs of conservation. (It is assumed that land prices provide a reasonable signal of the economic value of land, including its potential value for resettlement.) (Chomitz et al., 2005).

In Madagascar, on the other hand, remaining primary forest is largely owned and managed by the government or parastatals (agencies owned or partly owned by the government) as protected areas or forest reserves. The market for land in surrounding areas is very limited. In a recent study, Minten and Razafindraibe (2003) found that almost three quarters of agricultural lands were acquired through inheritance and traditional customs and only 13% were purchased through some form of land market. Of the lands that were sold, the majority were sold out of great need and under dire conditions that do not represent an accurate economic value of the land. The land tenure system in Madagascar is one in which traditional and modern property rights coexist. Under customary law land can be allocated communally or individually. Access to land is also limited for some groups based on caste. In addition, credit markets are not well developed, imposing further barriers to competitive markets for land. Furthermore, representative land prices are unlikely to be a reliable signal for the economic value of land that includes watershed and biodiversity values. (Madagascar has been relatively successful at realizing biodiversity values through international funding of biodiversity conservation.) In this context, reorienting the management of government-owned forests from production to watershed protection and conservation and community-based forest management are more promising tactics than direct financial incentives to multiple small landholders.

4. Setting Biodiversity Conservation Targets

Species and area targets provide clear and objective indicators to measure progress towards conservation goals in both design and implementation of a biodiversity conservation corridor. A first step in the TAMARIN approach is therefore the identification conservation targets based on the principles of representation, viability, resilience, and redundancy.

In the case of the Central Atlantic Forest corridor, coarse scale representation targets were identified first as the basis for the development of finer scale targets. To ensure representation of the species diversity of the corridor, unique bioregions within the planning region were delineated based on natural breaks in the distribution of species associated with biogeographic barriers. Within each bioregion, different assemblages of plants, determined by climactic, altitudinal, and interfluvial zones bounded by major river systems, were identified and classified as separate vegetation types. A decision was made to focus conservation planning targets on primary forest vegetation, leaving mangrove and restinga vegetation for separate analysis. Adequate representation of each of the unique bioregions was then included as a conservation target.

The next step was to define adequacy based on assessments of population viability and the minimum habitat necessary for "umbrella" species (in this case, species known to require large areas of relatively intact vegetation). Population viability habitat analyses were conducted for *Cebus xantosthernos* and *Cebus robustus*, two critically endangered capuchin monkeys, each endemic to one bioregion in Southern Bahia. Based on these analyses, adequate representation was defined as a minimum contiguous area of strictly protected primary forest of 20,000 ha.

In forested hotspots, edge effects are a major threat to the resilience of even large and well-enforced protected areas (Gascon *et al.*, 2000), so resiliency targets were based on calculations of core-edge ratios.

Finally, in accordance with the principle of redundancy, the target was doubled to require two areas of 20,000 ha in each bioregion.

In the case of Zahamena-Mantadia Corridor, the approach is being formulated through a mix of expert workshop and primary research. Because species distribution within the corridor varies according to altitude, rather than north south or east west geographic barriers as in the Central Atlantic Forest Corridor, initial bioregions are delineated along altitudinal gradients, and viability, redundancy, and resiliency targets are being determined for each zone. An initial very crude layer was constructed using elevation data as a proxy for species distribution data. A contiguous digital elevation model was divided into four discreet elevation levels using 400 m increment: 1) the east coast lowlands (0–400 m), 2) the east side mountain slopes (400–800 m), 3) the central high plateau region (800–1200 m), 4) the mountain tops of the central high plateau region (1,200–1,600 m). Other supplemental data, such as environmental variables data and species distribution information, would significantly improve the accuracy of this layer.

5. Developing the Information Layers

Each layer of information that feeds into TAMARIN can be developed through a variety of means and analyses requiring varying levels of effort. The following explanation of how the land cover/use and business as usual layers demonstrates how development of the layers can be adapted to the available data.

5.1. Land Cover/Use layer

The land cover/use map functions as a reference layer that guides TAMARIN users in selecting planning units and supports the business as usual analysis. Despite the high demand for land cover information for developing countries, few up-to-date maps and digital databases are available globally. Those that are produced often lack standardization of scales and legends, making comparisons among land cover maps extremely challenging. This situation applied to both the Central Atlantic Forest Corridor and the Zahamena-Mantadia Corridor. To overcome this challenge, new standardized land cover maps were produced using different techniques in the two regions.

For the southern part of Bahia, which includes the Central Atlantic Forest Corridor, a land cover map construction was based on an interpretation of eight adjacent Landsat Thematic Mapper images acquired within a five-month period between 1996 and 1997 (Landau *et al.*, 2003). The approach used to produce a land cover map for Zahamena-Mantadia Corridor differed slightly from the one adopted in Bahia. Two existing maps of different origin were combined: 1) Conservation International's Landsat forest change map derived from satellite images acquired in the early to late 1990s (CI Deforestation Map Reference, 2002), and 2) the BD500 land cover/use map of Madagascar (1:500,000 scale) published by the Foiben-Taosarintanin'i Madagasikara (FTM) Institute (1997). This approach was adopted despite the different scales of the data in order to ensure that additional land cover categories, required by the TAMARIN program that were not included in Conservation International's forest change map, would be included in this map (Conservation International's map consists only of categories describing change between the two time periods of forest, nonforest, water, and cloud cover).

The key aspects of the land cover layer, which determine the analyses possible with TAMARIN and the relevance of the results, include the availability and resolution of data, the number of different land cover/use classes, and the accuracy of classification. With improved quality and availability of Landsat imagery and aerial photography, the resolution of land cover layers is becoming less and less of an obstacle to mapping the landscape and performing useful analyses. TAMARIN can accommodate any number of classes during customization of the model to a region, but defining and choosing classes should be based on what is relevant to planners to make informed decisions. Too many classes may unnecessarily complicate the analyses, while too few makes meaningful analysis difficult. Finally, any analysis is only as good as the data inputted into it. Ground-truthing is a key element to the creation of a useful land cover map.

5.2. Business as Usual (BAU) Layer

The business as usual (BAU) layer, as used in TAMARIN, represents the type of land cover that would be expected to exist approximately 20–30 years from today if no new conservation initiatives were undertaken. The methods for generating the BAU layer range from very simple to very sophisticated analyses. The two examples—the Central Atlantic Forest Corridor and the Zahamena-Mantadia Corridor—illustrate two approaches from the range of options. They illustrate the implications of these choices on the analyses and products produced by TAMARIN.

In general, construction of the BAU layer requires multiple sources of information including the land cover/use map, socioeconomic data, population data, road networks, environmental variables like soil, climate, historical records, etc. In the Central Atlantic Forest Corridor, the conversion/deforestation of primary forest into other land cover categories is driven by cacao production, whereas in the Zahamena-Mantadia Corridor, slash-and-burn (*tavy*) practices are the leading cause of deforestation. Different approaches using distinct variables were therefore required to predict the future configurations of the landscapes in these different regions.

In the absence of quantitative information on socioeconomic drivers of future land use change in Central Atlantic Forest Corridor, authors of TAMARIN employed a simple deterministic model of change, based on the following assumption: the recently observed land use trends will continue over the next two decades if conservation interventions are not applied. In particular, primary forest will no longer be converted to other uses because it primarily occupies marginal lands and is legally protected. However, it will continue to be degraded into secondary forest through firewood gathering and other resource extraction, hunting, and other human-related impacts. Secondary forest will be permanently converted to pasture or agriculture except in areas where it is near primary forest; in such situations, it will remain secondary forest. Cabruca, a traditional form of cacao cultivation that retains overstory trees for shade, will be replaced by other forms of agriculture, including sun-grown cacao, coffee, crops, or pasture, except in areas where previously established on soils with the highest agricultural production capacity, on steep terrain (larger than 70% of slope), or on fertile flood plains. Pasture and agriculture will generally remain unchanged, as we assume no spontaneous abandonment and regeneration of farmland. Bare land is assumed to be a temporary state of agricultural land that is reclassified as agriculture/pasture for the future. Urban land uses and other habitat types (such as mangrove, wetlands, and water bodies) cannot be converted into forest and are assumed to remain in their present condition. Similarly, we did not anticipate changes for restinga and caatinga. The BAU layer for the Central Atlantic Forest Corridor was constructed to reflect this situation by reclassifying the current land cover layer and slope and agriculture suitability maps.

The projection of future land cover under BAU scenario for the Central Atlantic Forest Corridor was based on some simple assumptions based derived from secondary research, consultation with experts, market trends, and local knowledge of the region. A more sophisticated approach to predict the future landscape was adopted in the Zahamena-Mantadia Corridor. Projected deforestation of primary forest, approximately 25 years into the future, was produced by CI's Center for Applied Biodiversity Science (CABS) using a prediction procedure based on Markov chain analysis and a cellular automata algorithm (Eastman, 2002). In particular, distance from towns, roads, and rivers and the degree of slope were used to derive probability data about forest cover to agriculture changes (vulnerability map of forest to agriculture conversion). The prediction procedure then utilized the probability layer and an element of spatial contiguity of cellular automata to predict deforestation in the future. Similar to the case of the land cover/use map construction, the resultant future forest/nonforest map was combined with other land cover categories derived from the BD500 land cover/use map of Madagascar. The following modifications were made to account for the future scenario: area of deforestation, secondary forest, and grassy savanna were assumed to become agriculture/pasture; woody savanna was expected to degrade to grassy savanna. All other land cover/use categories were assumed not to change over time.

These assumptions and modeling produced a business as usual scenario for the Central Corridor with very small areas of primary forests remaining in existing protected areas and the majority of the land reverting to agriculture/pasture use.

In both cases, the BAU layer was produced using a limited set of variables and only a few generic rules that describe the mechanisms governing forest to agriculture conversion. With increased knowledge about the land cover change processes and supplemental data adequately describing such processes (such as location of mines, bridges, soil topography, population, planned roads, micro-watersheds, and conservation layer), we may have a chance to improve the accuracy of predictions. Other more realistic and/or empirical models may also provide significant assistance.

6. Presenting the Results

The key aspect of the TAMARIN approach is the integration of biodiversity and economic analyses into regional scale planning. The spatial nature of the analyses also helps to engage a variety of stakeholders and facilitates a transparent decision-making process.

In the case of the Central Atlantic Forest Corridor, the goal is to rebuild a highly fragmented landscape and assessing opportunity costs at a relatively fine scale was comparatively straightforward. It was therefore possible to break opportunity costs down to a relatively fine level (990 m^2 cells) and assess a multitude of scenarios based on their contribution to biodiversity and economic goals.

In the case of the Zahamena-Mantadia Corridor, the initial aim is to show the value of maintaining the large patch of primary forest. Economic analysis is focused on benefits associated with large forest blocks, and initial analysis is based on a watershed scale. In order to facilitate participatory review of regional planning choices, the results will be presented as different landscape scenarios with associated biodiversity and economic development values (Figure 9.2).

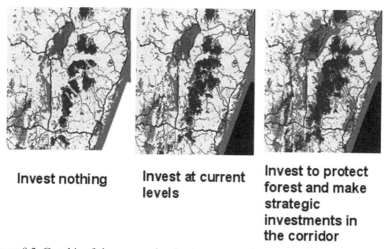

Invest nothing   **Invest at current levels**   **Invest to protect forest and make strategic investments in the corridor**

FIGURE 9.2. Graphic of three scenarios: business as usual, conservation, and investment.

7. Conclusions

In both the Central Atlantic Forest Corridor and the Zahamena-Mantadia Corridor, the TAMARIN approach is proving a useful framework to assist decision makers in designing sustainable landscapes to achieve the two objectives of biodiversity conservation and economic development. By combining the concepts of representation, viability, resilience, and redundancy and with assessments of economic benefits and opportunity costs, TAMARIN is providing an opportunity for mainstreaming conservation into regional development planning.

In both corridors, the development of biodiversity targets based on threatened and restricted range species, their habitat needs, population distribution, and viability criteria is essential for identifying conservation priorities. These targets provide objective measurable indicators that allow us to measure progress toward conservation outcomes. While economic benefits and opportunity costs are being used to inform conservation strategies and regional planning in both corridors, the different conservation goals, poverty reduction agendas, and land tenure have led to the adoption of different planning solutions in the two regions.

In the Central Atlantic Forest Corridor, the TAMARIN approach and software have been used to shed light on commonly held misconceptions about conservation scenarios and their costs. TAMARIN has been able to help demonstrate the affordability and achievability of conservation within the region. It has also helped move key decision-makers and stakeholders toward a common vision for the landscape and has provided a means of engaging the development community in conservation and integrating conservation goals into regional planning. It has provided decision-makers with a planning tool that can be used to improve and refine the corridor design as new information becomes available.

In Madagascar, the TAMARIN approach is being used in the Zahamena-Mantadia Corridor for many of the same purposes. It is being used to demonstrate the economic value of conserving the remaining forests and biodiversity in the corridor and providing economic arguments that conservation is necessary and a prerequisite for any poverty reduction plan for the region. It can also provide a tool for government and aid agencies to help target investment towards the most productive ends while conserving the natural heritage of the country for future generations.

By analyzing and incorporating both biodiversity priorities and economic values within the same framework and representing these values spatially, TAMARIN is proving to be a useful tool to engage governments, regional authorities and multilateral/bilateral international institutions responsible for development planning and investment. By defining conservation goals and economic benefits and costs spatially, conservationists and development planners are able to identify opportunities for win-win situations, areas where biodiversity conservation or development can be achieved at relatively low opportunity cost and conflict areas, then use this information to design sustainable landscapes and negotiate tradeoffs. While the TAMARIN software may not be appropriate for all planning efforts within biodiversity conservation corridors, the TAMARIN approach of setting specific biodiversity targets and incorporating opportunity costs into regional planning is sound and should prove to be useful in a wide variety of regional contexts.

References

Beier, P., and Noss, R. (1998). Do habitat corridors really provide connectivity? *Conservation Biology* 12:1241–1252.

Binswanger, H.P. (1989). Brazilian policies that encourage deforestation in the Amazon. The World Bank. Environment Department Working Paper No.16.

Eastman, J.R. (2002). Idrisi for Windows user's guide version 32. Clark labs for cartography technology and geographic analysis. Worcester, Clark University.

Bruner, A.G., Gullison, Rice, R.E., and da Fonseca, G.A.B. (2001). Effectiveness of parks in protecting tropical biodiversity. *Science* 291:125–128.

CABS/IESB (2000). Designing Sustainable Landscapes. Center for Applied Biodiversity Science (CABS), Conservation International, Washington, DC, and Instituto de Estudos Sócio-Ambientais do Sul da Bahia (IESB), Ilhéus, BA, Brazil.

Chomitz, K.M., Alger, K., Thomas, T.S., Orlando, H., Vila Nova, P. (2003) Opportunity costs of conservation in a biodiversity hotspot, The case of southern Bahia. *Environment and Development Economics* 10:293–312.

Gascon, C., Williamson, G., and da Fonseca, G. (2000). Receding forest edges and vanishing reserves. *Science* 288:1356–1358.

Landau, E.C., Hirsch, A., Musinsky, J. (2003). Cobertura Vegetal e Uso do Solo, escala 1:100 000, data dod dados:1996-97 (mapa em formato digital). In: Prado, P.I., Landau, E.C., Moura, R.T., Pinto, L.P.S., Fonseca, G.A.B., and Alger, K. (orgs.) Corredores de Biodiversidade na Mata Atlantica do Sul da Bahia. Publicacao em CD-ROM, Ilheus, IESB/CI/CABS/UFMG/UNICAMP.

Minten, B., and Razafindraibe, R. (2003). Relations Terres Agricoles–Pauvrete a Madagascar. FOFIFA, Cornell University, Madagascar.

Mittermeier, R.A., Myers, N., Robles Gil, P., and Mittermeier, C.G. (1999) *Hotspots*. Mexico City, Mexico, Cemex.

Myers, N., Mittermeier, R.A., Mittermeier, C.G., da Fonseca, G.A.B., and Kent, J. (2000). Biodiversity hotspots for conservation priorities. *Nature* 403:853–858.

Newmark, W.D. (1995). Extinction of mammal populations in western NorthAmerican national parks. *Conservation Biology* 9:512–526.

Noss, R.F. (2002). Context Matters: Considerations for large-scale conservation. *Conservation in Practice* 3(iii):9–19.

Ostrom, E. (1999). Coping with tragedies of the commons. Workshop in Political Theory and Policy Analysis Center for the Study of Institutions, Population, and Environmental Change, Annual Review of Political Science.

Stoms, D., Chomitz, K.M., and Davis, F.W. (2004). TAMARIN: A landscape framework for evaluating economic incentives for rainforest restoration. *Landscape Urban Plan* 68(1):95–108.

Stoms, D.M., Davis, F.W., Church, R.L., and Gerard, R.A. (2002). Economic Instruments for Habitat Conservation: Final Report to The World Bank, Washington, DC.

Wiens, J. (1996). Wildlife in patchy environments: metapopulations, mosaics and management. In: McCullough, D.R. (ed.). *Metapopulations and Wildlife Conservation.* Island Press, Washington, DC.

World Bank. (2003). Sustainable Development in a Dynamic World: Transforming Institutions, Growth, and Quality of Life. World Bank, Washington, DC.

Chapter 10
Linking the Community Options Analysis and Investment Toolkit (COAIT), Consen_sys_® and Payment for Environmental Services (PES): A Model to Promote Sustainability in African Gorilla Conservation

Michael Brown, Jean Martial Bonis-Charancle, Zephyrin Mogba, Rachna Sundararajan, and Rees Warne*

Most approaches to gorilla conservation have been top-down national park approaches that have included some limited form of community participation.[1] The top-down approaches have worked relatively well in Uganda and Rwanda; as Adams and Infield (2001, p. 146) put it, "the patient is stabilized, but the harder tasks of surgery and post-operative recovery lie ahead, but they do not appear to have guaranteed sustainability." Eves and Bakarr (2001, p. 53) note meanwhile that "the maintenance of protected areas is an extremely costly and difficult process, and, despite tremendous concern and long-term efforts, most governments are hard-pressed to secure the human and financial resources necessary to monitor, manage and protect wildlife populations." Given the economic, social, political, and population pressures many communities face, frontline communities' neighboring parks could represent a serious medium- to long-term threat to gorilla conservation in the absence of innovative approaches to gorilla conservation. Considering this sobering reality, communities must at least accept, if not actively support, protection of gorillas and their habitat if gorillas are to have a chance at

---

*The authors all worked for Innovative Resources Management Inc. (IRM), a nonprofit organization specializing in sustainable development and biodiversity conservation (www.irmgt.com).

[1] The recently instituted Tayna community-managed nature reserve initiated by Dian Fossey Gorilla Fund International (DFGFI) with local communities is a notable exception (See Patrick Mehlman's contributions in this volume).

survival into the next century. This chapter presents the use of tools developed by Innovative Resources Management (IRM) for community mobilization in landscape level biodiversity conservation [Community Options Analysis and Investment Toolkit (COAIT) and Consen*sys*®], along with the use of incentives that can be provided through Payments for Environmental Services (PES). When combined, these become a potential model for biodiversity and gorilla conservation.

Although park boundaries, management plans, and attention to local people's needs already are considered in gorilla conservation, solutions to the *root causes underpinning local people's pressures on protected resources* are not comprehensively addressed through gorilla conservation approaches. The strictures that get placed on people's livelihood activities by regulations prohibiting access to land and resources in protected areas where gorillas range (such as agricultural and grazing lands, water sources, firewood, timber, and nontimber forest products that were formerly harvested by local peoples as food, medicinals, housing materials, raw input for cottage industry and artisanry, and hunted bushmeat for either consumption or sale) have typically not been matched by adequate compensation or by new livelihood or economic opportunities that correspond to the value of the resources lost. Induced behavior change that can become sustainable of those that most directly threaten gorilla habitat is questionable so long as incentives for behavior change does not match perceived costs of such change. Despite years of outreach programming and infrastructure investments by the national park services and nongovernmental organizations (NGOs), including roads, water systems, and schools, communities living near protected areas often do not perceive "benefits" as commensurate with the costs they have been forced to bear, as these benefits most often are not negotiated but, rather, are imposed. Integrated Conservation and Development Projects (ICDPs) funded by NGOs or other external agencies often do provide limited short-term development benefits. These typically have not proven to be effective mechanisms for sustaining people's interest in conserving habitat by offering them adequate alternatives to habitat encroachment in the medium and long-term. Even with the increased attention given to "consultation"[2] with nearby communities, resentment of parks is still strong. The issue is whether protected area (PA) authorities responsible for gorilla conservation will be in the position to negotiate agreements with peripheral communities to gorilla PAs that will prove sustainable, or, will they be heading "back to the barriers" (Hutton *et al.,* 2005) with the result being an increasing "dialog of the deaf" (Redford and Robinson, 2006).

---

[2] We argue that "consultative" approaches are part of the problem. Consultation has come to be equated with "participation," acceptance, and even for some, bottom-up driven approaches. Most often consultation is no more than that: dissemination of information, responses to some questions, and limited discussions between planners and people who will be impacted by projects about which key decisions have already been made.

1. Importance of Communities for Gorilla Conservation

Community acceptance of, and cooperation with, conservation measures is especially important for the protection of gorillas and their habitat (Mehlman, Chapter 1, this volume). The agro-ecological context for African gorilla conservation contrasts significantly with that applying to other charismatic African mega-fauna, such as elephants, rhinoceros, or major predators. In areas of high human population density (some primate habitat in the Afromontane forests of Burundi and Rwanda occurs near areas with up to 1,200 people/km^2) communities will likely pose a threat to gorillas so long as the root causes of their poverty and pressure on land and natural resources have not been effectively addressed. In this region, anthropogenic pressures have reduced the land area of natural forests from approximately 30% of total land area at the turn of the century to 7% in 1997 (Mitchell, 1997). As such, opportunity costs for local subsistence farmers of not utilizing gorilla habitat for cultivation or for nontimber forest product (NTFP) extraction are very high. Conversely, lowland gorilla habitat generally occurs in areas of low human population density (often under five people/km^2). However, in these areas, opportunity costs of not respecting gorilla habitat through poaching of various wildlife for the bushmeat trade are also high, as local hunters have few options for revenue generation or protein consumption (Wilke, 1999). As long as viable economic incentives for local stakeholders in these landscapes remain limited, the opportunity costs of foregoing hunting bushmeat for personal consumption and sale will remain high. In order to preserve habitat integrity it is clear that safeguards coupled with incentives must occupy a primary place in current and future gorilla conservation.

For example, despite at least seven years of government and NGO funding for gorilla conservation at Mgahinga, communities in the peripheral area to the park still express strong resentment about the "inadequacy of the 'compensation' paid," and, more widely, the loss "of the park as a source of land for food production and (to a lesser extent) as a place of residence." (Adams and Infield, 2001, p. 146). Park neighbors simply do not see that the benefits they have received from the park and associated projects make up for the costs they bear, particularly since local discontent over designation changes from the original game and forest reserves in place in the 1930s led to recent evictions that have never been fully remediated. Unless communities have sufficient, long-term incentives to protect gorillas and their habitat, they are likely to represent a medium- to long-term risk to gorilla conservation. This will be true even though the area has enjoyed sustained financial and project support from the Global Environment Facility (GEF), United States Agency for International Development (USAID), CARE/International, and the Government of Uganda. Since projects *per se* have not proven to be the solution, we present a model for how communities could determine 1) what they objectively assess as needed to elicit sustained buy-in, 2) what "threat reduction assurances" they could negotiate to any linked incentives projects could offer, and 3) and how the linkage and negotiation between stakeholders could occur and be adaptively managed.

2. The Need for an Innovative Approach

Whether it is through top-down approaches to gorilla conservation via national parks or through more participatory techniques to gorilla conservation [such as community managed nature reserves similar to Tayna-Maiko in the Democratic Republic of Congo (DRC)], new methods must be employed to achieve sustainability. As Eves and Bakarr (2001, p. 53) put it, "we are only

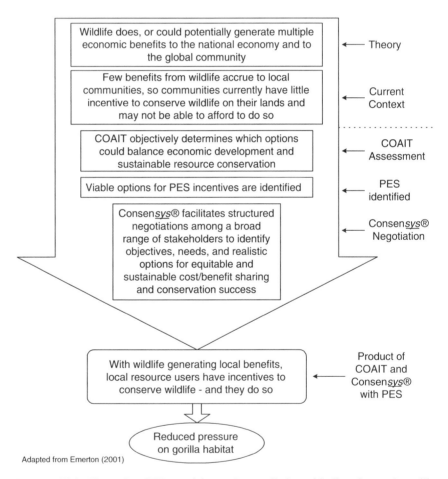

FIGURE 10.1. How the CCP model can be applied to biodiversity and gorilla conservation.

beginning to engage in debates that will develop clear roles that local communities, national and international experts, universities and national and international governments and conservation agencies will play in building management and schemes to sustain both human and wildlife." The question, then, is: *What are the key next steps?*

In this chapter, we outline what we feel are the key next steps to sustainable gorilla conservation. We propose tools[3] to facilitate these steps, describing two toolkits that IRM has developed and is currently using in biodiversity conservation and rural development contexts in the Democratic Republic of the Congo under funding from the Central Africa regional Program for the Environment (CARPE) and Congo Basin Forest Partnership (CBFP) that we feel are pertinent to gorilla conservation - COAIT and Consen*sys*®. We then present how COAIT and Consen*sys*® can be linked to the provision of incentives through PES[4] in a structured approach we label "the CCP model" [COAIT, Consen*sys*® and PES]. Figure 10.1 provides a conceptual overview of how these tools can be linked in gorilla conservation.

3. The CCP Model

In protected area management, there are three principles that we feel have not been adequately addressed in PA planning to promote sustainability. These three principles underpin the CCP model we present in this chapter. These principles address the following:

1. In the establishment and management of any kind of protected area, and especially for endangered species such as gorillas, the potential benefits, costs and decision making options faced by "frontline communities" in gorilla conservation must be very clearly elaborated and understood by community members themselves to successfully elicit their buy-in.
2. In addition to regulations and prohibitions, communities must have both short- and long-term incentives to protect both gorilla habitat and gorillas themselves. These incentives must be directly linked to the quality of the protected resources, and to the costs that community members bear from the strictures placed on them by protection mechanisms.
3. Finally, to ensure the feasibility and the sustainability of gorilla conservation, several conditions must be guaranted: a) institutionalized relationships among frontline communities, government agencies, NGOs, and the private sector must be forged and maintained; b) mutual accountability of

---

[3] See www.irmgt.com for documents on the COAIT and Consen*sys*® toolkits.
[4] PES in our usage can encompass any form of incentive provided in exchange for conservation services rendered. These can range from cash to in-kind payments to other potential instruments such as conservation concessions in which stipulation of benefit distribution to various holders of resource ownership *and* usufruct use rights are made.

partners, together with compliance in any agreements, must be assured; and c) conflict management mechanisms should integrated (and if need be used) from the start.

To secure local buy-in for gorilla conservation, we propose that these three principles be addressed through the use of the CCP model. The elements of this model are defined here.

1. COAIT is a participatory tool designed to enable communities to collect and analyze economic, ecological, and social data. COAIT helps communities to identify appropriate development and resource management pathways by injecting objective business principles of cost-benefit and feasibility analysis into community deliberations.
2. Measures restricting resource use identified through COAIT should be complemented with *incentives* (such as PES) that are directly linked to the continuing achievement of long-term conservation goals and are specifically designed to compensate for short-term costs local people bear due to the restrictions.
3. Consen*sys*® facilitates the creation of the viable working relationships and institutional arrangements needed to carry out sustainable gorilla conservation.

When used in conjunction, the CCP model *can address* the current methodological challenges in gorilla conservation, adding further innovation to what has previously been labeled as the "new conservation."[5,6] This three-part CCP model can mitigate or remove critical threats by targeting neglected community capacities, and internalizing costs and benefits in mutually advantageous ways. CCP is not premised on a traditional project-based model, which some argue is the wrong mechanism for achieving biodiversity conservation (see Kiss, 1999). Rather, each of the tools within the model contributes to the social capital that communities must draw upon for *any* conservation, or sustainable development activity, to be achieved.

4. How COAIT can Contribute to Gorilla Conservation

COAIT helps communities make objective decisions based on their analysis of short- and long-term costs and benefits. As such, COAIT is a methodology for mobilizing communities in sustainable development and conservation programming. COAIT helps answer the question conservationists pose regarding *"how"* to conserve (see McShane, 2003). McShane refers to this

---

[5] See Hulme, D. and Murphree, M. (2001) for an assessment of "new conservation."
[6] See "Integrating COAIT, Consen*sys*® and Payment for Environmental Services (PES) as an approach to landscape-level conservation of protected areas" by Brown, M., Bonis-Charancle, J.M., Mogba, Z., Sundararajan, R. and Warne, R. (forthcoming) from Innovative Resources Management, Washington, D.C.

central problem: conservation organizations are historically strong at spatial analysis of biological resources, but weak at the social, economic, socio-political and human capacity building aspects of conservation particularly outside of protected areas where major threats to conservation reside. In the absence of these capacities, it is challenging to see how conservationist missions are to be achieved. COAIT, as the cornerstone of the CCP model, enables conservationists to collaborate with communities to generate the ecological, economic and social data required to answer the following questions: 1) "what do we want to achieve?" and 2) "what is objectively feasible?" Without this information generated by communities themselves, conservation must continue to rely on the tools of exclusion and enforcement. This however will only lead to further shrillness in the increasingly heated debate about people and parks, as it is clear that "protected areas of all types will not survive without people – inside them, using them in sensible ways, or outside them, respecting and defending them" (Redford and Robinson, 2006). CCP is a means to move toward negotiated multiple land use planning involving people and parks, in this case people living peripherally to parks where gorillas and their habitat exist.

COAIT falls at the "empowerment" end of Barrow and Murphree's (2001) spectrum of community conservation approaches. COAIT builds on popular methods for promoting participation, such as participatory rural appraisal (PRA), which have clearly proven to be necessary components of conservation planning.[7] COAIT is a set of tools[8] that responds to the lesson learned by development planers that it is not enough to run successful computer models for rates of return on investment for successfully implementing complex projects. Rather, where community impacts are concerned, community members themselves must have the opportunity to evaluate different options (including assessing short- and long-term implications of resource use choices) and to identify what will work for them from an investment *and* impact standpoint. In the case of gorilla conservation, community buy-in is clearly needed, and obtaining this buy-in *must be* a component of any management plan that seeks sustainability.

COAIT was originally designed to maximize the potential for self-mobilization and empowerment at community levels in Congo Basin forest conservation. By considering the range of full ethnic, class and resource user diversity and agro-ecological complexity characterizing these communities, COAIT addresses technical and institutional issues that determine how community conservation can be designed and scaled-up to landscape levels. A key feature of COAIT is that it specifically strengthens the capacity of communities to negotiate outcomes with state and private sector agencies in

---

[7] While necessary in conservation and development planning, PRA does not address technical feasibility issues as COAIT does. Technical feasibility issues are key for viable planning and for developing fundable and implementable projects.
[8] See http://www.irmgt.com/html/papers.php#coait for a full listing of all COAIT manuals and documents.

areas where biodiversity values are high and where local behavior changes are crucial for conservation success. Tables 10.1 and 10.2 present an overview of what COAIT does and the steps and tools employed.

IRM has applied COAIT in the Congo Basin in partnership with communities and technical partners including the Center for International Forestry

TABLE 10.1. What does COAIT do?

Landscape	What does COAIT do?	Outcomes
High biodiversity value outside of protected areas	Help communities assess natural resources management and sustainable development options in an integrated and structured manner	• Communities likely favor more sustainable options • Community choices are expressed in local management plans, prospectuses, good practices manuals and proposals
In and around protected areas	Helps communities determine how they can deal with the external constraints to maximize advantages and minimize disadvantages	• Communities and conservation agencies achieve a better level of collaboration

TABLE 10.2. The three phases of COAIT.[a]

Phase	Steps	Tools employed
Community mobilization and generation of baseline data	• Study of local forest resource management systems • Resource mapping • Resource inventories	• PRA tools that focus on links between resources practices and social organization • Landscape-level Participatory Mapping[b] • Participatory Resource Inventorying • Nontimber forest product analysis
Analysis and comparison of options	• Participatory cost-benefit and risk analysis of options (PCBRA) • Determination of criteria for comparing options • Data synthesis • Comparison of options	• Capacity building • Testing of options • Focus groups and representative and full community meetings • Multi-criteria analysis and tools for environmental impact assessment
Capitalization	• Prospectus preparation including: business plans, management plans, local codes and standards for sustainable resource management, project identification, pilot projects, developing partnerships, and lobbying for enabling environment	• Internal rate of return for business plans • Geographical Information Systems (GIS) for management plans • Planning by Objectives for pilot projects • Consensys® for partnership and alliance building

[a] This table reflects the work of IRM through USAID-CARPE funding.
[b] For more information on Landscape-level Participatory Mapping, see http://ag.arizona.edu/OAL/ALN/alm48/brown&hutchinson.html

(CIFOR) and the *Centre de Coopération Internationale en Recherche Agronomique pour le Développement* (CIRAD), along with the World Wildlife Fund (WWF) in the DRC. Results to date demonstrate that through COAIT communities will actively and enthusiastically participate in determining how they can feasibly participate in activities by balancing economic benefits against any conservation costs they may bear. Results from COAIT work in Southern Cameroon (Brown, 2001a & b) show that when communities as a whole (not only the elite or leaders) are better informed and actively participate in generating data for analysis pertaining to natural resources management options, they will be likely to opt for the more sustainable choices which balance environment and development objectives. Communities require a high level of confidence in data and analysis to make credible and durable decisions at levels that represent a "community vision."

In the following example from a COAIT exercise held in Cameroon, communities summarized the five conditions most essential for achieving sustainable development in response to the current threats and opportunities they faced. These conditions are summed up in the local dictum as "*Homme Bien/Forêt Bien*"— "Good for People/Good for Forest." Within these five (out of a total of 39 conditions identified overall), the participants incorporated two "conservationist" criteria (italicized):

1. Better access to health services
2. *Low-impact logging*
3. *Long-term presence of all species of flora and fauna*
4. Secure access to community forests
5. Better access to information

These criteria are now the framework for a conservation and sustainable development program in a 1,200 km^2 area of Southern Cameroon. While these criteria may be broader than those defined by the conservation community, they were generated internally, agreed upon by community members, and can serve as the legitimate basis for conservation and development programming in this region.

In this same region, gorilla conservation has been seriously discussed in the context of a recently legislated protected area that will have significant impact on several communities, with COAIT part of this local discussion. Community-level facilitators in Djoum have recently requested that IRM work with them to design a COAIT process *specifically* addressing gorilla conservation in the area on the periphery of the Mengamé Gorilla Sanctuary. They have written to us:

Game meat is a big part of our food intake. Any activity that concerns wildlife touches at the heart of our society, particularly if this is done in the absence of awareness raising. This is what our decision-makers are doing to us with the creation of Mengamé Gorilla Sanctuary. This is why, we, local facilitators trained by IRM, have taken the initiative to prepare to carry out a sociological, ecological, economic analysis of this kind for the protected area.

For this COAIT work communities proposed the following objectives: 1) highlight the impacts of the intended sanctuary on current natural resource use patterns of communities; 2) identify the interactions that would be created between the sanctuary and the communities; and 3) define how best to prepare the communities for the creation of the sanctuary. They have requested that the COAIT work focus on: a) gathering and disseminating information on protected area legislation; b) gathering information on the planned activities of the Mengamé Gorilla Sanctuary; c) sociological, ecological and economic analysis of the impact of the sanctuary; d) integration of the results of the analysis; analysis of the conclusions with the communities; and e) definition and prioritization of feasible conservation and development options.

Local analysis of options would focus on the following:

- Economic consequences: a) the number of local jobs that could be created or lost; b) possible revenue generated (and equitable revenue sharing) by the development of ecotourism or other tourism service activities; c) revenue (and protein source) lost from diminished hunting activities; d) impacts of loss of access to firewood, building and artisanry materials, medicinal plants, wild foods, and other non-timber forest products; and e) crop losses due to raiding by wild animals
- Ecological consequences: a) benefits of gorilla habitat conservation to the natural resource base as a whole and b) analysis of the ecological impacts of limited logging activities on the resource base and gorilla populations
- Social consequences: a) gorilla conservation cost/benefit analysis for different stakeholder groups (including distribution of costs and benefits within those groups); b) governance issues for sanctuary management and tourism revenue sharing; and c) utility and feasibility of creating a code of ethics to shape sustainable interaction between humans and gorillas (*code de déontologie*)
- Plans: definition of the required steps for developing prospectuses, partnerships, and management and implementation plans for transforming gorilla conservation in the Mengamé Sanctuary into a viable option for local populations.

At the end of the COAIT process, local communities should be in a position to make informed, transparent and locally enforceable decisions regarding sustainable natural resource use and gorilla conservation.

5. Incentives: Why Payment for Environmental Services (PES) is Needed and How it Fits into the CCP Model

To be effective, gorilla conservation must go beyond regulations and prohibitions. As described for Mengamé, frontline communities must endorse the protection of gorillas and their habitat. They must have incentives to do so, and these incentives must be compelling in the short and long term. Incentives must have two central characteristics: 1) they must be tied to the achievement and maintenance of conservation objectives, and 2) they must compensate people for the costs of conservation that they actually sustain. So too, they must extend beyond this to provide enough additional benefits so that people clearly buy-in to gorilla conservation objectives because *it is* in their self-interest to do so. Simply put, in the absence of perfect enforcement (which will be increasingly difficult to secure in much gorilla habitat over time), if people are not just as well (or better) off after the imposition of restrictions to protect gorillas, why will they engage to significantly change their behavior?

In neo-classical economics, the benefits of a proposed change must be, in total, greater than the costs for these to be accepted. In hypothetical compensation analysis, there must be enough benefit to potentially compensate the losers for their losses. There is increasing recognition that to achieve effective community involvement in protecting resources (such as gorillas and their habitat), frontline communities that bear the costs of living near protected areas should *actually* be provided compensation for their losses. In past practice, this "compensation" has taken the form of schools, roads (which were often actually built for tourism or other private sector and/or government purposes), health posts, rural development projects, or scholarships. While these are all useful to communities, they do not, for the most part, address the actual losses sustained by the communities in question. Nor do they address core development concerns for moving forward. Generally, negotiation has not occurred over "compensation packages."

Through a first generation of ICDPs and then Community-Based Natural Resource Management (CBNRM) projects (and more recently adaptive co-management projects), local communities over the past 20 years have been encouraged to conserve their own resources and at the same time, take advantage of livelihood improvements offered. However, the benefits from ICDPs and/or CBNRM approaches have tended to not be immediate, if indeed there are any (Ferraro and Kiss, 2002). According to Wells *et al.* (1999), there has been a notable lack of success from these models and few convincing cases where people's development needs have been reconciled with conservation. While we would argue that this has been due to poor ICDP design to begin with,[9] results of ICDPs as implemented have, from a conservation standpoint, been equivocal as the development incentives offered have not been securely linked to changes in local conservation practices, nor have they been sufficient in the mind of frontline communities bearing most of the costs. Gullison *et al.* (2000) state that most efforts to promote more sustainable use of natural resources have in general failed for one reason—they have not provided direct incentives to conservation. Our conclusion: the types and levels

---

[9] See Brown, M. and Wyckoff-Baird, B. (1995) for a description of the ideal approach suggested for ICDP design.

of magnitude of incentives, appropriate methods to provide incentives, and clear linkages between incentives and conservation results have been lacking in conservation programming and need to be strengthened.

5.1. *Payments for Environmental Services*

Perhaps the most relevant type of incentives for conservation fall under the umbrella term of "payments for environmental services" (PES)—especially payments linked to forests and water quality. PES in the form of direct payments, forest concessions, land leases and easements have been popular in developed countries. In developing countries, projects and governments have begun to work with PES financed through various mechanisms including carbon sequestration and offset sales, upstream/downstream payments, taxes on urban, hydroelectric and irrigation water users, taxes on tourists, and conservation concessions. Common types of PES have included payments given for preventing deforestation; bonuses paid if periodic surveys indicate the presence or increased levels of wildlife within an area; access to a certain portion of the land resources in exchange for complying with prohibitions on access to biodiversity-rich lands; and payments made to compensate people for crop losses caused by agricultural pests and wild animals (Ferraro, 2001). Successes achieved with PES in Costa Rica, Mexico, Brazil and El Salvador offer precedents for how incentives specifically linked to ecosystem health can be used to promote conservation in developing country contexts (Herrador and Dimas, 2000). PES in these countries places value on at least some of the environmental and social benefits that have previously gone unrecognized by both markets and concessional aid donors.

In gorilla conservation, listing gorillas on the Convention on International Trade in Endangered Species of Wild Flora and Fauna (CITES) endangered species list is of course only a beginning to gorilla conservation challenges. To ensure that trade or habitat exploitation does not threaten this extraordinary species under multiple threats, conservation of gorillas and gorilla habitat must be the *direct objective* of a policy and interventions, not just an expected by-product. Direct payments to communities and community members provide a safety net for protecting single species populations by reducing the risk of irreversible damage (such as loss of the species itself) posed by continued exploitation (see Gullison *et al.*, 2000). PES offers new possibilities to both resource "owners" (in Africa, generally the state) and "administrators" (usually communities) who at a minimum may have recognizable usufruct rights to land and forest resources. That said, we are convinced that direct payments by themselves will not be a sustainable solution either.

Conservation concessions are one form of incentive that has begun to be discussed in the African context. Conservation concessions are a form of renting rights to resource use as an incentive to compensate "resource owners" for not using them. For instance, payments might be made for the rights to keep a forest intact (rather than for the rights to cut trees down)

(Rice, 2002). Given the tenure regimes common in gorilla habitats, concessions might be instruments created with the stipulation that a significant portion of the purchase price go to the communities with objectively veritable usufruct rights. This arrangement could tie payments (for both the government and communities) to the continued health of the gorillas and their habitat, with compliance incumbent on both stakeholders. Considering the number of stakeholders involved, conservation concessions will require considerable negotiation to be successful.

5.2. Mgahinga Gorilla National Park: A Case of Insufficient Match Between Costs and Incentives

In a concrete example of the standard that PES will need to reach to achieve local buy-in, we refer to Adams and Infield (2001), who provide a ledger of negative impacts perceived by people in communities within the Mgahinga Gorilla National Park. These are contrasted with community benefits provided to Mgahinga communities from a variety of sources (see Tables 10.3 and 10.4). This case is particularly interesting because park engagement with local communities has been high, and considerable investments were made to provide viable benefits for people in peripheral communities to the park.

There are several notable characteristics of these ledgers. First, only benefits A, C, G and J were provided by entities related to the park itself, while all the others were provided by international agencies. Second, only benefits C, E, F, H and J have anything to do with the negative impacts the local people say they suffered as a result of the existence of the park—and many of these benefits were incipient as of 1998 and of very limited scale. Third, six of the benefits are projects, presumably with fixed time frames—*they are not benefits explicitly linked to the medium or long-term existence and success of the park*. In any case, two of these projects (reliable water sources and agricultural development) are NGO projects that could have been provided even if the park had not existed. Fourth, and perhaps most importantly, only three specific losses cited by the community were addressed by these interventions: #6 (poverty), #7 (crop raiding), and #10 (no water); in none of these cases

TABLE 10.3. Negative impacts cited by people in the Mgahinga area (in order of prevalence).

1. Eviction/famine/hunger	7. Crop raiding
2. Inadequate compensation for losses	8. No resources
3. Shortage of land	9. Loss of grazing land
4. Eviction from land	10. No water
5. Presence of Interahamwe (Rwandan Hutus who were involved in the genocide of the Tutsis and who are cited by residents as being armed bandits)	11. Eviction from homes
	12. No firewood
	13. Emigration (and death)
	14. Harassment by rangers
6. Poverty	

TABLE 10.4. Benefits being provided to local people based on the existence of the park.

A. Conservation education (Community Conservation Rangers—supported by the Uganda Wildlife Authority)
B. Participatory needs assessment exercises (CARE)
C. Compensation paid for loss of physical improvements after eviction (loss of land was not compensated)—this ranged from US$ 6 to US$ 1200 with an average one time payment of US$ 27/person
D. Support for local projects (funded by USAID and the Dutch Government)
E. A lava block wall to prevent the incursion into local agricultural fields of the increasing buffalo populations in the park (CARE)
F. Establishment of some secure water sources (CARE)
G. Revenue sharing—the only example provided was the construction of classroom blocks in each of three parishes. (It is important to note that revenue sharing funds are not being efficiently captured and that total funds available have declined over the life of the park.)
H. Enhancement of productivity and sustainability of agriculture in the park area through support for agricultural extension, tree planting, agroforestry, composting, seed multiplication, potato stores, and other local projects (CARE)
I. Teachers for open air schools (provided by the Adventist Alliance Development Association)
J. Limited extractive activities: controlled bamboo rhizomes collection has been organized, and some opportunities for honey collection and the extraction of specific NTFPs and medicinal plants were being planned (Park Management)

did the intervention provide benefits to *all* who suffered the loss. It is not surprising that Adams and Infield (2001, p. 144) report that when they asked participants at parish meetings about the positive impacts of the park "the question usually evoked much noisy incredulity."

5.3. Central Challenges in Adapting PES for Use in Gorilla Conservation

The Mgahinga case, as documented by Adams and Infield, illustrates the frequent mismatch between benefits provided, and the incentives that will *actually be needed* to secure protection of valuable resources. There, the framework and mechanisms to elicit local buy-in were insufficient to overcome the negative impacts local communities felt that they bore. We believe that it could be possible in both Mgahinga and Mengamé to build an incentive framework with mechanisms that would elicit community level buy-in. The incentives will, however, need to be *much* clearer, and we feel that PES experiences elsewhere do offer lessons for how this could be structured for gorilla conservation in Africa.

The form that PES can take depends largely on local agro-ecological conditions, on available institutional structures, and on specific conservation objectives. That said, it will be no simple task to design appropriate PES and the institutional mechanisms to implement them for gorilla conservation. Conservation initiatives using PES in developed countries are established based on strong and reliable institutions for channeling payments and checking compliance. Reliable institutions are notably absent in many developing countries, particularly in Central Africa. While PES payments in developed countries are usually made directly to the private owners of land or of resource use rights, in most African countries, governments retain title to all land resources in the absence of locally held titles (an extreme rarity in rural areas) and resources are used by local people under a variety of local traditional access/use arrangements. Given all this, the central challenges to PES in gorilla range countries are 1) the design and maintenance of effective institutional arrangements for transferring payments; 2) a means of firmly linking payments to objectively monitored performance; and 3) equitable and effective means for distributing the benefits in ways that reduce the pressures on gorillas and their habitat.

We argue that for PES-type incentives to work in gorilla conservation (and we believe they can), mechanisms for communities to directly benefit from the concessions will need to be *negotiated* with the communities, the governments, and the NGOs or any private sector agencies involved. Since local resource users will be the ones whose performances will be crucial to the success of concessions, they must be central to the identification of appropriate incentives and forms of local distribution. Governments must take a role in providing policy and contexts that facilitate PES. NGOs or private firms or other entities that provide the funds for the PES must be closely involved in negotiations as well. All must work together to create stable institutional structures for realizing both payments and verification of conservation results. We posit Consen*sys*® as a model for both structuring equitable and effective negotiations and monitoring compliance with management plans and other agreements.

6. What is Consen*sys*® and How Does it Fit in the CCP Model?

Conflict is endemic to many regions where gorilla conservation occurs. The sources of conflict in gorilla range countries include political, economic, and ethnic dimensions. While resolution of these conflicts is not the mandate of conservationists, nor is it possible to address all conflicts through a conservation project, the existence of conflict must be clearly taken into account, and the strategic management of their impact on conservation programming must be prioritized. Consen*sys*®[10] offers a systematic way to achieve this.

Conflict may also be a by-product to the conservation of high-value species. Low-grade localized conflicts may be provoked by protection itself and may indirectly pose threats to gorilla survival if poorly handled. Localized conflicts

---

[10] For more information on Consen*sys*® see: Consen*sys*® FAQ sheet; Consen*sys*® Typology; Applying Consen*sys*®; A hypothetical case study; Consen*sys*® Conceptual Framework; and the Consen*sys*® process is available at www.irmgt.com.

can result when unacceptably high opportunity costs faced by local people with limited incentives and low levels of trust in existing relationships between government and themselves provoke community frustration leading to illegal behaviors such as poaching or illegal extraction of forest or nontimber forest products. While broad stakeholder participation is increasingly recognized as necessary in conservation, direct attention to conflict management is also necessary to minimize threats. We believe that it is possible for culturally, politically, and economically diverse groups to design the means to manage and reduce conflicts embroiling them and that, in turn, impact gorilla conservation. We also believe that these same groups can come to consensus on complex issues and cooperate to achieve shared goals and individual goals. These are not empty beliefs—they are based on our experience in environmental and biodiversity conservation conflict management in the DRC, Guatemala, Honduras, and the Dominican Republic as well as from lessons learned by other practitioners.

However, few systematic methods for achieving these results exist. We designed Consen*sys*® to fill this gap, using a strategic toolkit of proven tools and strategies adapted to local conditions to help communities, governments, NGOs, and private-sector actors work together effectively to address causes of conflict and to achieve complementary goals.

Consen*sys*® is an integrated set of tools, strategies, and facilitation services for strengthening multi-stakeholder decision-making processes, addressing conflicts productively, and enhancing project sustainability through culturally appropriate means at the relevant regional, national, and local levels. Consen*sys*® integrates best practices from alternative dispute resolution, participatory development, and local decision-making practices to create a systematic process for supporting conflict management and consensus-building among development agencies, government, the private sector, and civil society organizations in complex development and environmental conservation scenarios.

The community-level analysis facilitated by COAIT, while necessary, runs the risk of being an isolated local learning experience *if* there are no links created between communities and institutions that can help carry out decisions made through the community options analysis. Likewise, incentives are an excellent idea—in theory—but decision-making on *how*, and in *what form* to make benefits available to local people can be terribly complex (this is particularly true for the institutionalization of payments for environmental services over the long term). In order for COAIT and PES to have lasting impacts, reliable working relationships must be created among communities, government, NGOs and the private sector. Agreements must be made through *sufficient* consensus to enable their implementation. A joint stakeholder process must include clear representation and strong buy-in from all participants, and transparent joint monitoring of the implementation of decisions. Although complex issues like gorilla conservation involving multiple tasks and coordinated efforts require stakeholder groups to work together in order for these results to be achieved and sustained, few toolkits have been available that effectively facilitate multi-stakeholder team building. Consen*sys*® was designed to help key stakeholders work together to accomplish these tasks. The Consen*sys*® process builds trust, creates stable working relationships, and establishes and supports working groups made up of key stakeholder representatives. Figure 10.2 illustrates the stages of Consen*sys*®.

In most conservation and development processes, stakeholder participation, environmental and social assessment, and identification of projects for local people (essentially incentive and compensation mechanisms that pay at suboptimal levels for environmental services) are carried out as *ad hoc* activities. We hypothesize that this is the root of the disconnect between parks and surrounding communities, as in the case of Mgahinga discussed above. In the CCP model, Consen*sys*® strategically connects stakeholders and activities. In CCP, a coherent set of information gathering, negotiation, mediation, conflict-management, and consensus-building activities based on efficient collection, analysis, and dissemination of information are integrated. These can lead to identification of viable options for gorilla and other biodiversity conservation acceptable to diverse stakeholders. Effective management of diverse viewpoints leads to reduced conflict and to the progressive emergence of teams around common objectives.

From the very beginning, Consen*sys*® incorporates means of reducing and/or productively managing the stakeholder conflicts that inevitably arise. Within the CCP model for gorilla conservation, Consen*sys*® provides a structured approach for negotiating outcomes between conservation agencies, local government jurisdictions, communities, nongovernmental entities, tourism operators, loggers, and other stakeholders at national, regional, and local levels. Any of these stakeholder groups can offer to use Consen*sys*® to jumpstart the process. For example, the Fang or Baka Pygmy communities surrounding Mengamé in Cameroon could request the use of a Consen*sys*® process, or the Government of Cameroon through the Ministry of Environment, National Parks, or the WWF that is landscape lead in the Tri-National Park between Gabon, Central African Republic (CAR), and Cameroon, or the local government could do so. What is prerequisite is that the will to *negotiate outcomes* be present among the key stakeholders—this is critical to creating joint gains and achieving outcomes that work for all involved. Specifically, Consen*sys*® can be used to assist multi-stakeholder negotiations to:

- Identify effective and equitable payments for environmental services
- Define respective roles and responsibilities of different stakeholders in gorilla conservation
- Design decision-making process compatible with different stakeholder "corporate" culture and agendas
- Monitor compliance with agreements
- Develop a comprehensive plan for how stakeholders will work together around specific objectives

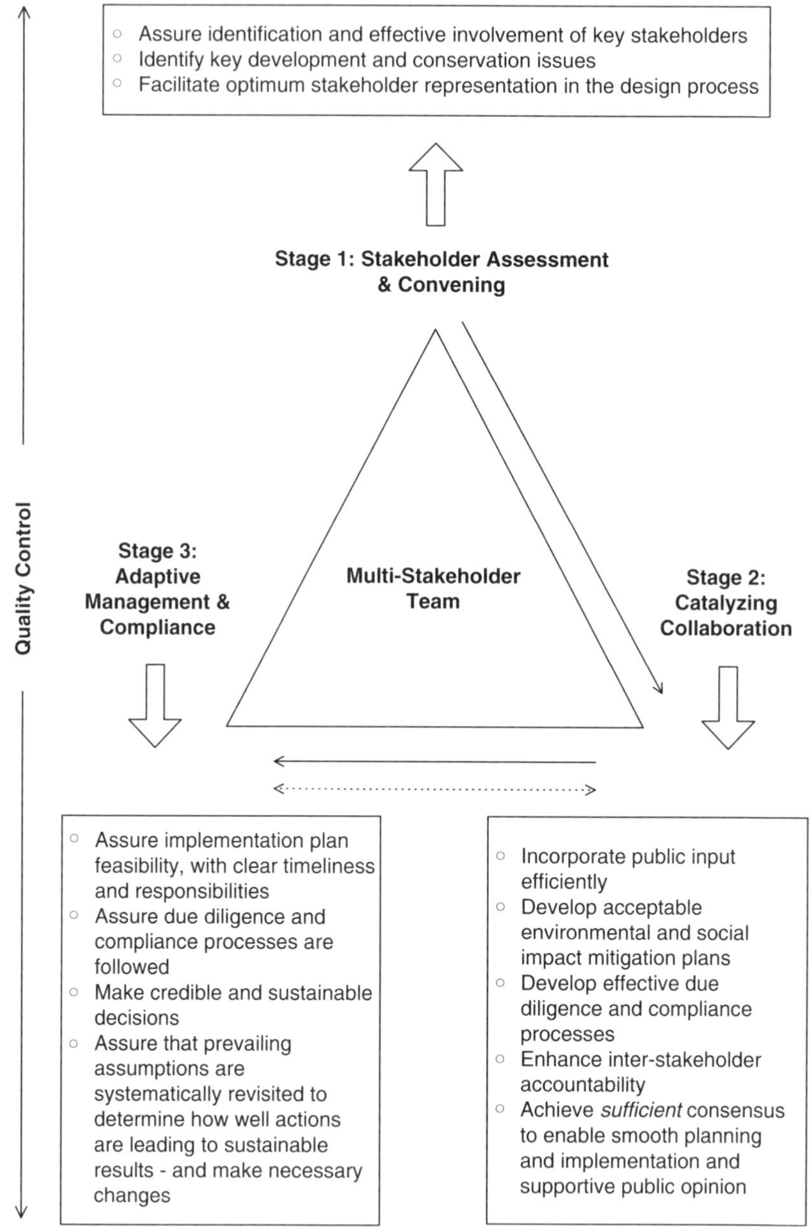

○ Assure identification and effective involvement of key stakeholders
○ Identify key development and conservation issues
○ Facilitate optimum stakeholder representation in the design process

Stage 1: Stakeholder Assessment & Convening

Quality Control

**Stage 3:
Adaptive
Management &
Compliance**

**Multi-Stakeholder
Team**

**Stage 2:
Catalyzing
Collaboration**

○ Assure implementation plan feasibility, with clear timeliness and responsibilities
○ Assure due diligence and compliance processes are followed
○ Make credible and sustainable decisions
○ Assure that prevailing assumptions are systematically revisited to determine how well actions are leading to sustainable results - and make necessary changes

○ Incorporate public input efficiently
○ Develop acceptable environmental and social impact mitigation plans
○ Develop effective due diligence and compliance processes
○ Enhance inter-stakeholder accountability
○ Achieve *sufficient* consensus to enable smooth planning and implementation and supportive public opinion

FIGURE 10.2. The Consen*sys*® Process.

Building capacity to mutually define objectives and processes while creating durable linkages between people in a variety of sectors is a key element of achieving conservation. We see this type of team-building as a central element of the CCP model. Consen*sys*® facilitates team-building by bringing data, analysis, training, facilitation, and decision-making tools together under a single purposive framework. Absent this, alliances are unlikely to emerge and crisis, conflicts, and the need to perpetually put out "fires" becomes the norm.

7. CCP: Integrating COAIT, PES and Consen*sys*® in Gorilla Conservation

CCP can be the driver behind feasible community participation in gorilla conservation. We believe that the CCP model can strengthen communities' capacities to effectively participate in effective negotiations that create functioning alliances with other key stakeholders. These broader alliances are the foundation for operational task-oriented teams. But that is just the first step. We also believe that CCP can assist a broad range of stakeholders to work together to increase the probability of achieving sustained gorilla conservation in either Afro-montane or lowland gorilla forested landscapes.

COAIT enables communities and conservationists to objectively identify *the minimum technical conditions* that will be required to elicit community buy-in to gorilla conservation. Different types of PES can be structured as incentives to be matched to the technical resource requirements, management, and sustainable development plans identified through the COAIT process. Consen*sys*® frames and facilitates negotiations among the government, local resource users, conservation practitioners, technical assistance providers, the private sector, and other gorilla conservation stakeholders. It does so by first incorporating the results of COAIT and the PES mechanisms identified, and then creating the conditions for achieving the conservation and development objectives these same institutions jointly identify. The linkage of COAIT, Consen*sys*®, and incentives through PES can achieve the community-level buy-in that has historically proven to be among the most challenging aspects of conservation. The integration of the elements of the CCP model is illustrated in Figures 10.3 and 10.4.

8. Conclusions

For biodiversity conservation to be achieved and sustained in Africa, we argue that a new approach to involving communities and other stakeholders is needed. For gorilla conservation (or any landscape level conservation) to be achieved, the innovative types of capacity building and effective decision-making frameworks provided by COAIT and Consen*sys*® in the CCP model are required. One of the reasons that full community participation in conservation and development planning and implementation is rare is that few

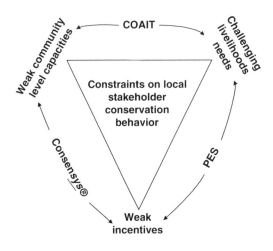

FIGURE 10.3. Conceptual overview: how the CCP addresses constraints in support of gorilla and biodiversity conservation.

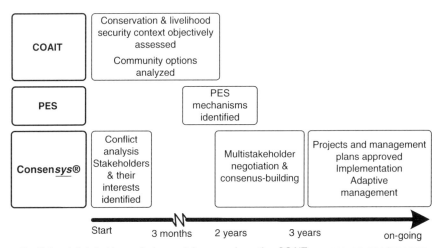

Conflict and stakeholder analysis may take several months. COAIT may occur over a two-year period. Negotiation and consensus-building may take a year or more. Implementation and adaptive management should have a medium-term time horizon.

FIGURE 10.4. Approximate timeline for CCP implementation in gorilla conservation.

communities have sufficient capacity to fully engage in these activities. In the absence of both COAIT and Consen*sys*®, few communities near gorilla habitat (and not many other stakeholders for that matter) have the level of capacity in cost-benefit analysis, decision-making, and negotiation that would make them strong partners and enable them to negotiate appropriate conservation outcomes that will cement their sustained participation. Without enhanced capacities in these areas, local stakeholder groups will likely *never* be well positioned to assume greater responsibility over resource management in landscapes where conservation values and threats to conservation (such as in the case of gorillas) are very high. Assuming greater responsibility at the community level is possible with capacity building and is needed for conservation momentum to be gained and sustained.

This chapter has presented a model that promotes 1) integration of participatory data generation and situational analysis (COAIT); 2) capacity strengthening in negotiation, consensus-building, and conflict management skills for a broad spectrum of stakeholders (Consen*sys*®); and 3) clear incentives linked to both conservation costs and outcomes (PES). The CCP model is based on the past 10 years of IRM's work under USAID supported CARPE funding, and is further informed by over 30 years of natural resource management and development experience of IRM staff and partner organizations in Africa and other parts of the developing world.

The CCP model is applicable to gorilla conservation in both well-demarcated protected areas and broader landscapes managed under government protected area agencies with full statutory authority, or by local management authorities such as the communities empowered through DFGFI's activities at the Tayna Nature Reserve in the eastern DRC (Mehlman, Chapter 1, this volume). Elements of CCP will need to be integrated over time to guarantee that the conditions for sustainability in conservation are met and adaptively managed in each of these cases. Land use planning, livelihood security, organizational capacity, equity, palpable incentives, decision-making, and conflict management processes are all brought to the forefront through the CCP model. We argue that to nibble *ad hoc* around the edges of these issues, as current conservation approaches do, is to endanger the biodiversity we all seek to protect. Furthermore, to revert back to the barrier approach to conservation will be even worse. Gorillas need more than this if they are survive into the 22nd and 23rd century.

9. List of Acronyms

CAR	Central African Republic
CARPE	Central African Regional Program for the Environment
CBFP	Congo Basin Forest Partnership
CBNRM	Community-Based Natural Resource Management
CIFOR	Centre for International Forestry Research
CIRAD	*Centre de Coopération Internationale en Recherche Agronomique pour le Développement*
CITES	Convention on International Trade in Endangered Species of Wild Flora and Fauna
COAIT	Community Options Analysis and Investment Toolkit
DFGFI	Dian Fossey Gorilla Fund International
DRC	Democratic Republic of Congo
GEF	Global Environment Facility

GIS   Geographic Information Systems
ICDPs   Integrated Conservation and Development Projects
IRM   Innovative Resources Management
NGO   Non-governmental Organization
NTFP   Non-timber Forest Product
PA   Protected Area
PCBRA   Participatory Cost-Benefit and Risk Analysis
PES   Payment for Environmental Services
PRA   Participatory Rural Appraisal
USAID   United States Agency for International Development
WWF   World Wildlife Fund

Acknowledgments. The authors thank Brian Greenberg and Christin Hutchinson, formerly of IRM, for their review and technical inputs.

References

Adams, W., and Infield, M. (2001). Park outreach and gorilla conservation: Mgahinga Gorilla National Park, Uganda. In: Hulme, D., and Murphree, M. (eds.), *African Wildlife and Livelihoods: The promise and performance of community conservation*, James Currey Ltd., Oxford, pp. 131–147.

Barrow, E., and Murphree, M. (2001). Community conservation: From concept to practice. In: Hulme, D., and Murphree, M. (eds.), *African Wildlife and Livelihoods: The promise and performance of community conservation*, James Currey Ltd., Oxford, pp. 24–37.

Brown, M. (2001a). Community management of forest resources: Moving from "Keep Out" to "Let's Collaborate!" In: Somé, L., Wilkie, D., and Oglethorpe, J. (eds.), *Congo Basin: Information Series: Taking Action to Manage and Conserve Forest Resources in the Congo Basin: Results and Lessons Learned from the First Phase (1996–2000)*, World Wildlife Fund, The Nature Conservancy, and World Resources Institute: The Biodiversity Support Program, Washington, D.C., Briefing Sheet # 17 (March 2001).

Brown, M. (2001b). Mobilizing communities to conserve forest resources: Cameroon case study. In: Somé, L., Wilkie, D., and Oglethorpe, J. (eds.) *Congo Basin Information Series: Taking Action to Manage and Conserve Forest Resources in the Congo Basin— Results and Lessons Learned from the First Phase (1996–2000)*, World Wildlife Fund, The Nature Conservancy, and World Resources Institute: The Biodiversity Support Program, Washington, D.C., Briefing Sheet # 20 (March 2001).

Brown, M., and Wyckoff-Baird, B. (1995). *Designing Integrated Conservation and Development Projects*, Revised edition. World Wildlife Fund, The Nature Conservancy, and World Resources Institute: The Biodiversity Support Program, Washington, D.C.

Emerton, L. (2001). The nature of benefits and the benefits of nature: Why wildlife conservation has not economically benefited communities in Africa. In: Hulme, D., and Murphree, M. (eds.), *African Wildlife and Livelihoods: The Promise and Performance of Community Conservation*. James Currey Ltd., Oxford, pp. 208–226.

Eves, H.E., and Bakarr, M.I. (2001). Impacts of bushmeat hunting on wildlife populations in West Africa's Upper Guinea forest ecosystem. In: Bakarr, M.I., da Fonseca, G.A.B., Mittermeier, R., Rylands, A.B., and Painemilla, K. W. (eds.). *Hunting and Bushmeat Utilization in the African Rain Forest: Perspectives Toward a Blueprint for Conservation Action*, Conservation International: Center for Applied Biodiversity Science, Washington, D.C., pp. 39–53.

Ferraro, P.J., and Simpson, D.R. (2002). The cost-effectiveness of conservation payments. *Land Economics* 78(3):339–353.

Ferraro, P.J. (2001). Global habitat protection: Limitations of development interventions and a role for conservation performance payments. *Conservation Biology* 15(4):990–1000.

Ferraro, P.J., and Kiss, A. (2002). Direct payments to conserve biodiversity. *Science* 298:1718–1719.

Gullison, R.E., Rice, R.E., and Blundell, A.G. (2000). 'Marketing' species conservation. *Nature* 404:923–924.

Herrador, D., and Dimas, L. (2000). Payment for environmental services in El Salvador. *Mountain Research and Development* 20(4):306–309.

Hulme, D., and Murphree, M. (2001). Community conservation in Africa: An introduction. In: Hulme, D., and Murphree, M. (eds.), *African Wildlife and Livelihoods: The Promise and Performance of Community Conservation*, James Currey Ltd., Oxford, pp. 1–23.

Hutton, J., Adams, W.M., and Murombedzi, J.C. (2005). Back to the barriers?: Changing narratives in biodiversity conservation. *Forum for Development Studies* 32(2):341–370.

Innovative Resources Management, Inc., Introduction to Consen*sys*®. 2002, Washington, D.C.; available at: http://www.irmgt.com

Kiss, A. (1999). Making community-based conservation work. Society for Conservation Biology Annual Meeting, College Park, MD. (June 1999).

McShane, T.O. (2003). The devil in the detail of biodiversity conservation. *Conservation Biology* 17(1):1–3.

Mitchell, T. International conflict and the environment: Case Name: Rwanda and Conflict. (1997). ICE Case Study # 23. In: Trade and Environment Database, (Spring 1997), available at: http://www.american.edu/TED/ice/rwanda.htm

Redford, K., and Robinson J. (2006). Parks as shibboleths. *Conservation Biology* 20(1):1–2.

Rice, R. (2002). Conservation concessions: Our experience to date. Society for Conservation Biology Annual Meeting. Canterbury, UK. (July 2002).

Wells, M., Guggenheim, S., Khan, A., Wardojo, W., and Jepson, P. (1999). *Investing in Biodiversity: A review of Indonesia's integrated conservation and development projects*, The World Bank, Washington, D.C.

Wilke, D.S., and Carpenter, J.F. (1999). Bushmeat hunting in the Congo Basin: An assessment of impacts and options for mitigation. *Biodiversity Conservation* 8:927–955.

Chapter 11
An Integrated Geomatics Research Program for Mountain Gorilla Behavior and Conservation

H. Dieter Steklis, Scott Madry, Nick Faust, Netzin Gerald Steklis, and Eugene Kayijamahe

1. Background

1.1. Introduction

The mountain gorilla (*Gorilla beringei beringei*) of the Virunga volcanoes region of Central-East Africa has been the focus of much research and conservation activities since Schaller's pioneering field study in the late 1950s. Despite large declines in this population during the 1970s and early 1980s due to poaching and habitat loss, active research, protection, and a successful ecotourism program resulted in the population's recovery by the mid-1980s to a size of about 380 individuals by late 2003 (Kalpers *et al.*, 2003; Fawcett, pers. comm.). Nevertheless, the Virunga mountain gorilla population is small from the standpoint of an effective breeding population size (Steklis and Gerald Steklis, 2001), is isolated, and is surrounded by a dense, growing human population. Its long-term survival continues to be threatened by the introduction of human disease, habitat loss and degradation, and poaching. The IUCN Red Book classifies the population as "Critically Endangered."

The threatened status of this population calls for management strategies that, in addition to a high level of protection and monitoring, rely on innovative research that addresses fundamental problems such as: 1) the dynamics of continued population growth in a limited habitat; 2) density and distribution of gorilla groups in relation to habitat quality; and 3) the effects of human use of the habitat on gorilla behavior and biology. Such management issues and their connected research agendas also concern other African ape populations (e.g., Walsh et al., 2003), and many primate species the world over are equally threatened with extinction (Jernvall and Wright, 1998), frequently due to the combined threats of disease, human population growth, and habitat loss. In response to this growing conservation crisis, conservation biologists have increasingly begun to rely on geographic information system (GIS) technologies , including Global Positioning System (GPS), satellite imaging, and other remote sensing tools. This technology, developed during the 1970s and 1980s, has come to play a vital role in providing quantitative, visually based tools for both conservation assessment and modeling of impacts and outcomes over large temporal (e.g., decades) and spatial scales (e.g., entire nations, ecological zones). We turned to this technology in the early 1990s, as civil war began to plague the Virunga region, and field work in this international border area became dangerous and at times impossible, as a means initially for habitat assessment and monitoring over the entire Virunga region. More generally, our objective has grown to understand the dynamic relationship between the Virunga gorillas, their habitat, and human neighbors and to apply this knowledge to the sustained, effective management of the gorilla population and its habitat.

Our purpose in this chapter is to review this corpus of work, employing what we prefer to call a "Geomatics" approach (see below) to mountain gorilla conservation, in the hope that both our methods and results, and "lessons learned," will be of use to other primatologists and conservationists faced with similar conservation challenges.

1.2. Study Site and Subjects

The mountain gorilla habitat comprises the 430 km^2 area, commonly referred to as the Virunga Conservation Area (VCA), in the Western Rift Valley of Africa, where the borders of Rwanda, Uganda, and Democratic Republic of Congo join together (Figure 11.1). The VCA consists of three national parks: the Parc National des Volcans in Rwanda, the Parc National des Virunga in the Democratic Republic of Congo, and the Mgahinga Gorilla National Park in Uganda. Prior to its destruction in the aftermath of the 1994 civil war, the Karisoke Research Center was located at 3100 m in Rwanda's Parc

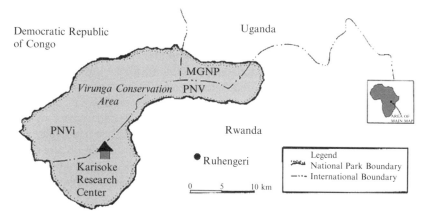

FIGURE 11.1. The Virunga Conservation Area, showing the international borders and location of the now destroyed Karisoke Research Center.

National des Volcans (Figure 11.1), where it served as a base for researchers and conservationists, including members of our Geomatics research team. The center's cumulative database on the life history, demography, behavior, and ecology of mountain gorillas as well as human illegal activity (poaching) serves as a powerful resource for conservation biology (Steklis and Steklis, Chapter 6, this volume) and for multi-faceted research projects of the sort we are reviewing here.

Within a few decades of the mountain gorilla's scientific discovery in 1902, its imperiled status became clear as habitat loss and poaching took their toll. The human population encircling the VCA has grown dramatically since 1902, with a current staggering density of 400–600 individuals/km² (Weber, 1995). The majority of these are subsistence farmers that historically have relied on the forest as a source of construction materials, bushmeat, firewood, water, and land for livestock grazing. Population pressure and consequent need for agricultural lands soon led to the loss of gorilla habitat. The greatest loss of habitat occurred in the Rwandan portion of the VCA, where large tracts of forest were given to local farmers in the late 1960s, mostly for *Pyrethrum* cultivation, reducing the size of Rwanda's national park to less than half its original size. This massive habitat loss—much of it the most fertile lowlands (1600–2600 m)—by the mid-1970s had effected a drastic decline in the mountain gorilla population (Weber and Vedder, 1983). To this day, human population growth and habitat loss or degradation from encroachment persist as significant threats to gorilla survival (Plumptre and Williamson, 2001). Unlike the lowland forest habitat of gorilla populations found to the west of the Virungas, much of the Virunga habitat, because of its higher altitude, is classified as Afro-montane forest with very little closed canopy. It is mixed with patches of bamboo and herbaceous vegetation, and yields to sparse, low shrub sub-alpine and alpine vegetation zones above 3300 m (White, 1978; Fischer and Hinkel, 1992). The primarily folivorous diet of mountain gorillas is much less diverse than that of their lowland neighbors, consisting primarily of herbaceous vegetation found on the slopes of and saddles between volcanoes. Fruit is rare in their diet, as there is generally very little fruit availability, with the exception of the lower lying Mixed Forest zone (2000–2550 m) in the lower parts of the DRC, where gorilla groups may make greater use of fruit resources (McNeilage, 2001).

There is significant variation in abundance and distribution of food resources, human disturbance, and other factors that, collectively, comprise "suitable habitat" for the gorillas. While the potential effects of this habitat heterogeneity on gorilla diet, group size, structure, social organization, population density, ranging, and ultimately habitat carrying capacity have been recognized and begun to be addressed by research (McNeilage, 2001), they deserve significantly more research effort if we are to fully understand the effects of habitat change or human activity on gorilla biology and behavior.

1.3. Project Objectives

The overall aim of our research over the past decade, using GIS technology in combination with extensive field work, has been the identification and quantification of the variables that define "suitable gorilla habitat" in the Virungas, and to understand the effects of habitat change on gorilla behavior and population biology. Our intent is to make our results available and useful to government and park officials, rangers, and other field personnel who are charged with the protection and management of the mountain gorilla and its habitat. Our first objective was to treat the VCA as a single habitat, and to define as closely as possible the environmental features of this habitat. This is achieved in a GIS through a single geo-referenced map that contains (as separate GIS data layers) the major vegetation types, topographic (elevation) and geographic (streams and trails) features, and political-cultural features (park and international borders, local place names). Second, we wanted to bring into this GIS environment both historical and current data on gorilla group ranging and human illegal activity, so as to assess, both historically and presently, how habitat features and human activity influence gorilla habitat use.

1.4. A Geomatics Approach to Conservation

Geomatics is a recent concept, developed and advanced in Canada, and one that is particularly appropriate for this research project. It expands upon the tools of GIS and remote sensing, and provides a true multidisciplinary, multiperspective, and multitechnology environment for regional and environmental analysis. Geomatics is defined as the functional integration of GIS, GPS, remote sensing, simulation and visualization, databases, spatial statistics, and related technologies all conducted within an inherently multidisciplinary context. The Geomatics approach takes multiple advanced spatial technologies and integrates them so that the information can be interpreted by multiple specialists with different backgrounds, fields of expertise, and technical skills.

GIS is the heart of the Geomatics approach. GIS computer systems allow both mapping and spatial analysis and have developed rapidly over the past decades. Any information that can be represented on a map can be entered into a GIS. Typically, a GIS database includes human-derived data such as political boundaries, zoning and tax maps, roads, and socioeconomic information, as well as physical data such as hydrography, land use and land cover, soils, geology, wetlands, elevation, slope, and aspect. Data derived from satellite imagery, scanned aerial photographs, and historic maps can also be entered into the system. Electronic maps are thought of as "data layers" because they represent a single set of attributes or categories of a landscape. All data are co-registered to a common map projection, coordinate system, and datum, so that new combinations of data, even data derived from different sources at different scales, can be collated and their relationships analyzed. Maps can also be tied to existing tabular databases that contain data derived from many sources, including field work, GPS, etc. A GIS lends itself to the production of compelling graphics that can easily be tailored to particular needs. Output in the form of color maps at any scale, tables, charts, etc. can be created as needed. Three-dimensional representations showing any combination of data layers draped over the elevation can also be viewed on the screen or 3-D maps can be produced. Such systems also have the ability to include multimedia capabilities, such as the ability to incorporate video, field photographs, sound, and scanned documents into a single information environment.

Once we have the information from various sources in our system, we can conduct a wide variety of analyses. The ease of testing hypothetical scenarios encourages looking at multiple alternatives before we commit ourselves to a given course of action that carries with it financial, environmental, and political costs. GIS also promotes logical problem-solving methods and produces analyses that are easily replicable, based on quantitative methods, and that are defensible in public meetings and the political process. These applications require only the basic GIS tools. Beyond these basics, powerful spatial, statistical, and physical process modeling are all possible using modern GIS technology (e.g., see Scott et al., 2002).

GIS adds a cumulative perspective to long-term analyses. In the past, each researcher would conduct fieldwork and publish a paper with a few maps or graphics. These results would not be cumulative, however, as the next researcher would have to start essentially from scratch again. GIS allows us to create a long-term and integrative research system that builds upon previous work, facilitating the testing of new hypotheses over time. GIS tools have been used in a variety of ways to help define, model, and solve conservation issues, usually by bringing together into a digital, visual environment separate data layers, such as processed satellite images of vegetation cover and other land features, digitized topographic maps, aerial photos of terrain or habitat, and spatial distribution of wildlife (e.g., from GPS data). Commonly, ArcView or Arc/Info software packages, easily run on either a desktop or laptop computer, provide the GIS environment for the analyses. Most analyses are concerned with a quantitative assessment of the relationship between current, past, or projected habitat characteristics and a target species' distribution and abundance. By incorporating data layers across time, GIS analyses can be retrospective and/or prospective in modeling impacts of environmental change on wildlife populations.

2. Data Sources and Processing Methods

Our effort began in 1992 at Rutgers University with a small United States Agency for International Development (USAID) grant to the Dian Fossey Gorilla Fund International's (DFGFI) Karisoke Research Center that included the creation of a mountain gorilla habitat map. Although there had been some effort made in that direction in the late 1980s by John Kinneman and colleagues, it had not resulted in the sought-after habitat map and vegetation classification. As a result, one of us (HDS), then director of DFGFI's Karisoke Research Center in Rwanda, contacted his Rutgers colleague Scott Madry at Rutgers' Center for Remote Sensing and Spatial Analysis (CRSSA), to see whether we could produce a vegetation and topographic map for the VCA. What began as a seemingly simple project has grown into a complex, long-term, multidisciplinary collaboration among primatologists and field biologists, image-processing experts, and GIS specialists, who have by necessity joined forces in a Geomatics approach to the problem at hand.

In order to achieve our objectives, we required 1) cartographic, remote sensing, and other data to provide a detailed characterization of the gorilla habitat. Such a "base map," which includes basic environmental and cultural features (elevation contours, streams, trails, national boundaries, park boundaries, roads) would serve as our GIS foundation, to which all subsequently acquired data and analyses could be geo-referenced; and 2) data on gorilla ranging behavior and human activity in the VCA, so as to analyze gorilla ranging behavior in relation to both environmental features of the habitat and patterns of human activity.

2.1. Habitat GIS

Ideally, a GIS base map is digitized from existing maps of the region that contain the environmental and cultural features. While the layers of these features can be derived from any source, they must all be appropriately geo-referenced so that all data are in reference to the same geodetic datum and coordinate system. In our case, what might ordinarily be a relatively straightforward task was complicated by the fact that the VCA is comprised of three countries with different colonial mapping histories and traditions and different levels of access. In addition, the region is generally poorly mapped and cloud covered most of the year. For example, there were no useful maps of the DRC (then Zaire), as there was a civil war underway in the region, and possession of topographic maps or aerial photographs was politically sensitive. Unfortunately, the DRC area accounts for more than half of the VCA (Figure 11.1).

2.1.1. Cartographic Data

We did acquire a set of four 1:50,000 Rwandan topographic maps from the 1960s (courtesy of the Institut Géographique National, Paris, France) that cover Rwanda's part of the VCA and surrounding area. These did not include any portions of DRC or Uganda. We also acquired a 1:50,000 topographic map covering the Ugandan portion of the VCA, which was produced in the 1970s by the British Royal Ordnance Survey, using a different datum and map projection from the Rwandan maps. Using these maps, we manually digitized, as GIS data layers, the location of national park boundaries, the international boundaries, streams, and contours employing the GRASS GIS software, v. 4.2, as developed by the U.S. Army Corps of Engineers, Construction Engineering

Research Laboratory, Champaign, IL (Goren *et al.*, 1993), running on Sun UNIX computers. When the Rwandan and Ugandan data were corrected for both datum and map projection and displayed, it became clear that they did not correspond, even after extensive digital "massaging." Nothing matched—contours, the international boundary, streams—all were in different locations. This was clearly an unsatisfactory solution, as nearly one half of the VCA (DRC) was not available, and the available data were dated and questionable at best. While another source was needed for our base map, the environmental and cultural features were retained as data layers in our GIS.

In the mid-1990s, we initiated a search for a single topographic map-set that would cover the entire study area. Our best bet was to find some historic Belgian colonial maps of the region, as the VCA was part of the Belgian Congo. With the help of a Belgian colleague, a search of the Belgian Colonial Service records located a 1936–1938 set of four 1:100,000 topographic maps (including elevation contours, roads, streams, lakes, for the entire VCA) at the Belgian Royal Museum of Central Africa in Tervuren. The contours and other features were manually digitized from these maps, as before, using the GRASS GIS module *v.digit*.

A set of nine 1:50,000 maps from the same period was also acquired. These maps were without contours, but they did show the hydrology, political boundaries, roads, and major settlements in more detail. Hence, these features were also manually digitized as GIS data layers.

These historic maps are still, amazingly enough, the most recent comprehensive maps of the entire area (covering Rwanda, Uganda, and DRC portions of the VCA) with significant detail. In modern terms, they are poor in quality and control, and we have little information about how they were produced. Nevertheless, they provide an important "snapshot" of the entire region as it was before modern overpopulation, deforestation, and human development.

In 2004, we were fortunate to obtain from the Center for GIS and Remote Sensing of the National University of Rwanda a 1:50,000 scanned map of the Rwandan portion of the VCA. This was a mosaic of the four 1960s maps we had digitized earlier. Because this map was in a Gauss-Kruger projection (Clarke 1880 spheroid and ARC 1960 datum), we re-projected it into Universal Transverse Mercator (UTM), zone 35 S, WGS 84, so that it would be compatible with our GPS data as well as that of our collaborators. This provided us with our best approximation of spatial accuracy, given the sources at hand. All existing data were converted to this new coordinate system and map projection, which serves as the base map for all subsequent GIS analyses.

2.1.2. Global Positioning System (GPS) Data

A final but critical data source for our GIS base map consisted of ground truthing points (e.g., on the ground location of trails, roads, streams) collected with GPS units. We began training field staff in 1993, using Trimble GeoExplorer receivers, and later switched to Garmin GPS 12 units. The GPS units were also used to map park boundaries and other major features in the area for use in geo-referencing the subsequently acquired satellite imagery. The U.S. government's disabling of the Selective Availability (SA) feature on May 2, 2000, improved the GPS data precision from ± 100 m to ± 7 m (95% of readings).

All GPS field data are summarized in spreadsheets. This growing GPS database (near continuous from 1993 to 2003) provides us with the ability to include direct field observations into the GIS.

2.2. Digital Elevation Model (DEM)

Our next step was to create a three-dimensional map (i.e., Digital Elevation Model, or DEM) of the VCA, so that slope and aspect could be used in later analyses. Typically, a DEM is created from cartographic sources that contain elevation contours or, more recently, from stereo satellite imagery. As discussed later, appropriate satellite imagery was not available for this purpose, so cartographic maps were our only option.

We constructed the DEM based on the contours of the four 1936 1:100,000 historic maps. Mylar separates for the elevation contours were hand copied from these maps and manually digitized using the GRASS GIS module *v.digit*. The labeled contours were then interpolated into raster digital elevation data with a cell size of 30 m. Digital raster slope and aspect (compass direction of slope) data were created using the GRASS GIS module *r.slope.aspect*. The generation of raster elevation, slope, and aspect files also allows us to visualize the Virunga volcano region in three dimensions, overlay satellite images, and create visualizations.

Figure 11.2 shows the DEM of the VCA produced from the digitized 1936 contour maps. Various 3-D visualizations and "fly-throughs" of the region have also been created using the DEM and various overlays, including the Landsat and radar images. These are useful in showing the entire Virunga region as it exists in three dimensions and without cloud cover.

2.3. Vegetation Classification

We employed data from several sources to derive a habitat, or vegetation, classification for the VCA. These sources, described in this section, include satellite, radar, and hyper-spectral imagery, aerial photography, GPS data, and other visualization tools.

2.3.1. Satellite Imagery

Our objective was to utilize satellite remote sensing data to accurately map the Virunga region's environment and vegetation and the changes in these over time. Such a map would normally be produced from Landsat or SPOT multispectral satellite imagery, or aerial photography, but persistent cloud cover has made these unreliable sources for long-term environmental

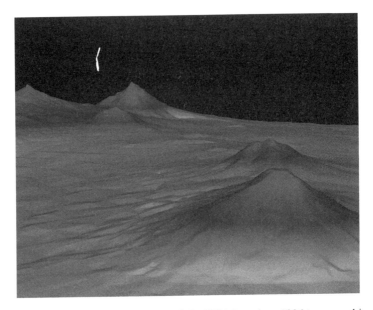

FIGURE 11.2. Digital Elevation Model of the VCA based on 1936 topographic maps.

monitoring in the Virunga area, because this region, due to its specific micro-climate, is often shrouded in mist and clouds. Indeed, a scene search of existing archives for Landsat and SPOT data back to 1984 was conducted, and only one, mostly cloud-free, Landsat 5 image from August 1987 of the entire region was available at the start of the project.

To analyze the 1987 Landsat multispectral data, several digital processes had to be performed. In these and later analyses of remote sensing data, we used ERDAS Imagine software. (Later, we converted all GIS data over to the ESRI ArcInfo, ArcView 3.X, and ArcGIS 8.X environments running under MS Windows, with some additional processing work conducted using Idrisi.) First, the imagery had to be geometrically corrected to be able to compare with existing map and GIS data sets. The Landsat system has crude data on the satellite orbit and sensor parameters that allow the image to be "calibrated" so that each pixel in the satellite image is related to a coordinate on the ground in a map projection appropriate for our study area. Unfortunately, the "calibration" only allows the image pixel to be located with an accuracy of 500 m to 1 km. This is not sufficiently accurate to correctly tie the image data to previously generated map data for vegetation or other GIS data that had been obtained as discussed above. GIS data were used to identify Ground Control Points (GCPs) that could be located both in the imagery and in the GIS data sets. This included road intersections, points at which streams crossed roads, and other prominent points. Using polynomial approximation, we were able from these data sets to increase the placement accuracy of the image data.

Once the image rectification process was completed, image enhancement functions were performed on the multi-spectral imagery to increase the contrast between vegetation types in the imagery. Using the enhanced imagery, a technique known as multi-spectral pattern recognition was performed to separate out vegetation classes within the multi-spectral imagery. Two types of pattern recognition techniques were attempted: Supervised and unsupervised classification. Supervised classification is a technique where, if a user can visually identify specific regions of different vegetation types in the image by manually drawing polygons on the imagery and extracting statistics, then the computer can find regions in the image that have the same or a similar spectral signature. Since at the time the only "ground truth" information that existed was McNeilage's (1995) vegetation map at a much coarser scale, his vegetation map was overlaid on the imagery and used to try to identify regions that corresponded to color variations in the Landsat. This technique was not as satisfactory as subsequent unsupervised classification techniques. In this technique, the computer analyzes all spectral bands of Landsat (six bands in the visible and near infrared) and automatically determines which pixels in the image "look alike" based on some parameters provided by the user. The user selects the number of classes to be separated, and based on spectral distance calculations, the computer algorithms take each pixel and categorize it into one of those classes. The initial classifications used 50 classes. Once the unsupervised classification had been performed, the user was required to identify (again from limited "ground truth" data) what each category represented in terms of vegetation. This initial classification showed more detail than McNeilage's vegetation map, but it had never been checked in the field with higher resolution data sets.

In January of 2003 we acquired a Landsat 7 (Enhanced Thematic Mapper, ETM+), almost cloud-free, scene over the Virunga area. The ETM+ has the six visible and near infrared bands similar to the 1987 Landsat 5 image and a lower resolution thermal band, but it also has a co-registered 15-m panchromatic band that gives significantly more spatial detail.

The 2003 Landsat imagery allowed us to revisit the vegetation classification with newer and more detailed data (especially GPS ground truth data of specific vegetation types, radar, and hyperspectral imaging data; see below). Several approaches were used to concentrate more on the vegetation within than outside the VCA. Using GPS boundary points for the VCA, we extracted the ETM+ data only for the areas within the VCA. We also excluded cloud and cloud shadow regions and used information from the 1987 Landsat data to determine vegetation class in those areas. We merged the visible and near infrared 30-m bands with the 15-m panchromatic bands and provided a classification of the merged data set at 15 m.

Figure 11.3 shows a side-by-side comparison of the 1987 and the 2003 Thematic Mapper data sets for the VCA. Note that neither scene is totally cloud free; however, for the most part, the clouds are in different places for the two dates. This allows the extraction of vegetation information for cloud and cloud shadow regions.

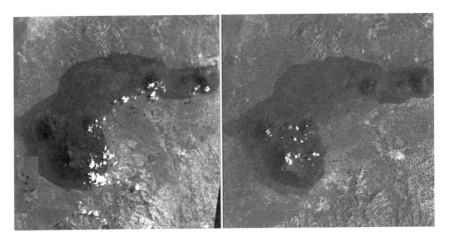

FIGURE 11.3. Landsat images from 1987 (on left) and 2003 showing the VCA with different patterns of cloud cover (in white).

2.3.2. Space Radar

A relatively new and powerful technology, space imaging radar, is now available from several civilian satellite systems, including advanced systems developed by NASA. RADAR stands for radio detection and ranging. It refers to electronic equipment that detects the presence, composition, direction, height, and distance of objects by using reflected electromagnetic energy. Unlike other environmental satellite systems like Landsat and SPOT, the electromagnetic energy wavelengths used in imaging radar can penetrate clouds (and even dry sand in some cases), and an imaging radar system can acquire data day or night. This is because radar satellites are "active" systems, meaning that they carry their own energy source that is transmitted to the ground. The reflected energy is recorded, stored, and digitally processed to make an image or map.

In 1994, NASA was flying its third research radar system (The Spaceborne Imaging Radar–Version C and the X-band Synthetic Aperture Radar, or SIR-C/X-SAR) on the Space Shuttle Endeavor (Jordan *et al.*, 1991; Stuhr, *et al.*, 1995). A colleague at NASA arranged for the VCA to be imaged during the April and October 1994 shuttle flights. Before the SIR-C/X-SAR mission, radar imaging satellites used a single band and polarization for each image. This meant that each dataset could only produce a black-and-white image. SIR-C/X-SAR allowed for multiple simultaneous bands (C, L, and X) and polarizations, which could be combined to create color images (see below).

The resulting data are very accurate with regard to locating features on the ground, and have a 15-m spatial resolution. However, radar imagery requires a great deal of processing and computer manipulation to produce an image, much more than passive remote sensing systems such as Landsat. The data, therefore, were pre-processed at the Jet Propulsion Laboratory, which included geometric correction of the data, and then sent to us in digital format for further analysis and integration with our other GIS data.

We first processed the radar data to produce various black-and-white images of individual bands, different color composite images, and thematic classifications showing different vegetation and land use zones. False color composites were then created by combining the three different bands (X and two C bands) and by assigning blue, green, and red to each of the grayscale images. While these color composites are "false" colors, and do not accurately reflect the colors of the landscape, they do provide good visual differentiation among the major vegetation zones, as can be seen in Figure 11.4.

One significant problem with radar systems is that they will produce large shadow areas on the side farthest from the instrument. There is also a significant problem with "layover" in mountainous regions (like the Virungas) where the mountains will appear to be tilted towards the radar receiver. The layover caused serious problems for us. Areas on the lee side of the volcanoes were in shadow, and not imaged at all. We attempted several thematic classifications of the radar data, seeking to create a single vegetation map of the region, but our attempts were less than satisfactory due to the extremely different illumination of the volcanic mountains. Some areas in shadow that were known to be the same vegetation types were classified as different vegetation cover from areas that were strongly illuminated. Some general vegetation data were derived, but a single, accurate vegetation map was not

FIGURE 11.4. The 1994 Space Shuttle radar image covers an area of 58 km × 178 km, at about 1.75 degrees south latitude and 29.5 degrees east longitude. The Virunga Conservation Area is to the right, divided by lava flows from the 1977 eruption (purple streaks) from the still active Mt. Nyiragongo to the left. (See Color Plate)

produced. Nevertheless, several areas within the VCA that contained accurate vegetation data proved helpful to our final vegetation classification.

2.3.4. Aerial Photography

Vertical aerial photography is the source of most cartographic products and maps. Such photographs are excellent sources of high-resolution vegetation and land use data. They can be used to assist in identifying vegetation categories, to provide information about changes over time, and to assist in classifying satellite data. Our search for aerial photos turned up several sets covering Rwanda at the Institut Géographique National (IGN) in Paris, but we could not get copies without permission of the Rwandan government, which at that time was in turmoil. McNeilage (1995), however, did have access to a second generation set of high-altitude 1950s black-and-white aerial photographs of the region. These photographs were 35-mm copies of the original 9 × 9 inch Rwandan mapping photos.

These aerial photographs were photo interpreted using stereo pairs and a zoom transfer scope by McNeilage (McNeilage, 1995) at Rutgers' CRSSA. A zoom transfer scope is a manual photo interpretation and mapping device that uses mirrors to allow the user to plot features on a photograph accurately onto a map. Aerial photographs are very useful, but they contain significant spatial error due to various types of displacement of features on the photograph caused by elevation differences, camera lens imperfections, and radial displacement out from the nadir point directly under the lens. In this case, these errors were worsened by the fact that the photographs are second generation, taken by a 35-mm camera held over the original mapping photos. Therefore, features on an uncontrolled aerial photograph cannot be directly transferred to a map without significant spatial error. The zoom transfer scope avoids this problem for the most part.

Based on these photos, the 1:100,000 topographic map, and his extensive field experience, McNeilage produced a vegetation classification of the VCA that showed eight habitat types. The aerial photos served to delimit each of the eight habitat types as polygons in a GIS. While this polygon classification method is relatively crude and does not likely reflect the heterogeneity within the polygons, it was nevertheless an important first step in a quantitative definition of the VCA's major vegetation zones. Moreover, this map served well in our subsequent satellite imagery classification process.

2.3.5. Hyperspectral Imagery

In August 1999, we acquired data of higher spatial (approximately 5 m) and spectral (hyper-spectral imagery) resolution over the Virunga area (courtesy of a collaboration between Earth Search Sciences Incorporated [ESSI], DFGFI, and the National Geographic Society). Hyper-spectral imagery may have as many as 512 spectral bands that cover the visible and near infrared parts of the spectrum. Computational analysis of this massive amount of data can provide more detailed discrimination of vegetation types than is possible either from panchromatic or multi-spectral data (e.g., Landsat or Spot).

The hyper-spectral data were acquired from a light plane equipped with the sensor (ESSI's Probe 1) and flown over the Virunga area over the course of two days. Numerous flight lines of data were acquired (about 45) in order to map as much as possible of the habitat. As is customary, the mountains were covered with clouds, and flights had to be modified to steer around or under (not always possible) clouds. Most of the flight lines recorded had significant problems with roll and yaw of the aircraft and the presence of haze and clouds that prevented a clear image of the surface below. The pilot tried to keep the plane at an altitude such that the pixel size on the ground was relatively consistent, but this was not always possible. Even though the Probe 1 instrument can record 128 spectral bands, only 64 of the bands contained meaningful data, possibly due to calibration or sensitivity issues. We decided to use the hyper-spectral data as a kind of "ground truth" for the limited areas in which we had coverage since the higher spatial resolution gives a more detailed understanding of the vegetation distribution.

2.3.6. GPS and Visualizations

As mentioned previously, we have accumulated a large database of GPS positions with associated habitat attributes. Since the mid-1990s, our Karisoke field staff have collected such attribute data when GPS positions are recorded for the location of gorilla nests and group locations at noon. Moreover, we have made a systematic effort since 1993 to collect GPS data for all vegetation types (following McNeilage and others, see below) during the frequent field trips by members of the Geomatics team. This database was a critical resource for deriving the final VCA vegetation classification (Section 3).

2.4. Gorilla Ranging

In this section, we discuss the data sources used in our analyses of gorilla ranging behavior. For our analysis of recent group ranges, these data include, principally, GPS field data of group movement. In order to bring a historical dimension to gorilla group ranging behavior, we also make use of archival ranging maps prepared during the 1980s.

2.4.1. GPS Data

Our field staff has been logging daily gorilla group movements since 1993. The daily gorilla group location data, for three to four groups in the Karisoke sector, consists of the previous night's nest location in the morning (i.e., the group's starting point for the day's travel at sunup) and group position at noon (essentially, the position of the group every six hours during the day). GPS data are also recorded for lone males as well as interactions between groups.

2.4.2. Fossey Maps

Our archival library includes the original, hand-drawn gorilla movement maps created by Fossey as part of her 1984–1985 research on gorilla group habitat use. These maps were created each month in the field for three to four groups, as well as lone males, in the Karisoke sector of the VCA, and we have a total of 23 monthly maps (September 1984 is missing). Fossey produced these by enlarging the area around the volcanoes Karisimbi and Visoke on the 1930s volcanology map (Belgian Geological Survey). This topographic map, described earlier, was the only map available to Fossey and other researchers at Karisoke at that time. Fossey used a commercial Xerox machine several times (with different amounts of magnification) to enlarge the area between the volcanoes where she followed gorillas. The daily locations and trails of the groups and the lone males, were drawn on a Mylar overlay for each day.

These maps permitted us to extend our analysis of gorilla ranging to the years prior to 1993, when we first began collecting gorilla ranging and poaching data with GPS technology. Although of great historical value, there were several problems with these maps. First, they are in very poor condition (e.g., stains, rips, and tears). More importantly, there are several watermarks where the patterns of gorilla movement (marked in non-waterproof ink!) are obliterated. It appears that several different Xerox background volcanology maps of different magnifications were used, so geo-referencing them was challenging. We used various volcanic lakes on the scanned modern maps as reference marks. Not all group locations are marked on all days, and some of the monthly movement patterns are in such a tight pattern that instead of showing a clear line, they more resemble a plate of spaghetti. In these cases, we utilized an automated GIS line thinning function with limited success. In the maps where we could not discern the movement pattern over some days, we created a polygon to represent the area of gorilla activity over those days.

The digitized Fossey maps, in combination with the ranging data from the 1990s onward, allows us to compare gorilla ranging behavior and examine patterns of continuity and change in the Karisoke sector over a 20-year time frame. The final GIS database can be queried, and individual days or months can be called up for individual groups. All groups can be viewed on a given day, or specific groups can be visually tracked over a given period of time. This makes possible our analysis of gorilla movement patterns in relation to various environmental and cultural data such as vegetation zones, distance to park boundary, elevation zones, slope, and poaching activity.

2.5. Human Activity Data Sources

2.5.1. Anti-poaching Records and GPS

The anti-poaching records kept by field staff go back to 1978, but prior to 1993, they do not contain GPS location data. For the present exploratory analyses, we used 2003 GPS data collected by the Karisoke anti-poaching patrols. The GPS data include attribute data for all evidence of poaching (e.g., snare traps, footprints, shelters, cut wood, weapons) encountered by the patrol on a given day.

3. Analysis and Results

In this section, we present our results to date on the Virunga habitat classification and analyses of gorilla ranging behavior in relation to human activity. Our gorilla ranging behavior analyses must be regarded as preliminary. We have selected some initial analyses more to convey the utility of the Geomatics approach than to reach any conclusions about the dynamics of gorilla ranging behavior.

3.1. Virunga Habitat GIS

Prior to conducting the GIS analyses, it was necessary to manipulate the data into the proper raster, vector, or point data format. Some raster data were then reclassified, re-combined, buffered, or otherwise manipulated in order to get the data into the proper categories and structure for the final analyses. This process is referred to as intermediate data processing. The final analysis and modeling routines are conducted using raster data for each of the individual study areas with a cell resolution of 30 m. This cell size was chosen because the digital elevation, slope, and Landsat-derived landcover data all have 30-m cell sizes. Each 30×30 m cell in each "layer" in the GIS contains an individual value for each layer in the database.

Figure 11.5 shows our vegetation classification. As stated earlier, this classification was derived from the 2003 Landsat data along with the various other sources described earlier, and the resulting classified Landsat image has been draped over the DEM developed earlier. Our classification consists of 12 habitat types, which, with the exception of the Heather/Lichen, Bare Earth, Bare Rock, and Agriculture classes follows the definitions provided by McNeilage (1995). The habitat GIS database, of course, contains much additional information, as separate data layers, such as streams and trails that can be utilized in later analyses.

As McNeilage (1995) provides the only GIS-based vegetation classification for the Virungas, Figure 11.5 also compares the percentages of each of eight of the classifications provided by us and McNeilage. McNeilage used subalpine (3300–3600 m) and alpine (3600 m and above) altitude classes. Since the latter classes are not habitat types per se, in our classification we derived habitat classes (e.g., Heather/Lichen) within the sub-alpine and alpine classes.

3.2. Gorilla Ranging Behavior

Our analysis of gorilla group ranging in the Virungas is not the first to examine the variables that determine gorilla ranging behavior or habitat use (e.g., Fossey, 1974; Fossey and Harcourt, 1977; Vedder, 1984; Watts, 1991;

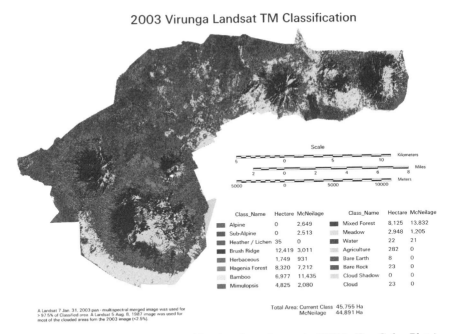

2003 Virunga Landsat TM Classification

Scale

Class_Name	Hectare	McNeilage		Class_Name	Hectare	McNeilage
Alpine	0	2,649		Mixed Forest	8,125	13,832
Sub-Alpine	0	2,513		Meadow	2,948	1,205
Heather / Lichen	35	0		Water	22	21
Brush Ridge	12,419	3,011		Agriculture	282	0
Herbaceous	1,749	931		Bare Earth	8	0
Hagenia Forest	8,320	7,212		Bare Rock	23	0
Bamboo	6,977	11,435		Cloud Shadow	0	0
Mimulopsis	4,825	2,080		Cloud	23	0

A Landsat 7 Jan. 31, 2003 pan - multispectral merged image was used for > 97.5% of Classified area. A Landsat 5 Aug. 8, 1987 image was used for most of the clouded areas form the 2003 image (<2.5%).

Total Area: Current Class 45,755 Ha
McNeilage 44,891 Ha

FIGURE 11.5. VCA vegetation classification draped over the DEM. (See Color Plate)

McNeilage, 1995). It is the first, however, to use complex GIS tools to do so, owing to both the growth in this technology since the earlier studies and our extensive historical digital database (described above) of imagery, cartography, gorilla behavior, and poaching activity. Only McNeilage (1995) employed GIS tools. This involved superimposing the outlines of minimum convex polygons (MCP) for group ranges derived from field location data hand recorded in 250-m grid on a 1:100,000 contour map, onto his digital habitat map. Enhanced tools and databases provide us with the opportunity to examine more accurately and quantitatively the relationship between gorilla behavior and habitat variables. It also puts us in a position of testing, for example, the accuracy of popular techniques (e.g., MCP) for home range estimates or for estimating daily travel lengths. Thus, unlike MCP methods, GIS analyses can include a third dimension (elevation), and further, we can compare actual travel distances or daily ranges (measured in the field) to GIS-generated ones. Finally, contrary to previous work, which provides one snapshot in time of gorilla habitat use, our historical database will allow us to compare over decades the patterns of gorilla group ranging and thus ultimately to disentangle the complex interrelationships among habitat features and gorilla behavior.

Our analyses rely on software developed by Phillip Hooge (Research Population Ecologist at the USGS-Alaska Science Center-Biological Science Office, Glacier Bay Field Station, Alaska). GIS software specifically developed

Color Plate

FIGURE 4. The 1994 Space Shuttle radar image covers an area of 58 km × 178 km, at about 1.75 degrees south latitude and 29.5 degrees east longitude. The Virunga Conservation Area is to the right, divided by lava flows from the 1977 eruption (purple streaks) from the still active Mt. Nyiragongo to the left.

2003 Virunga Landsat TM Classification

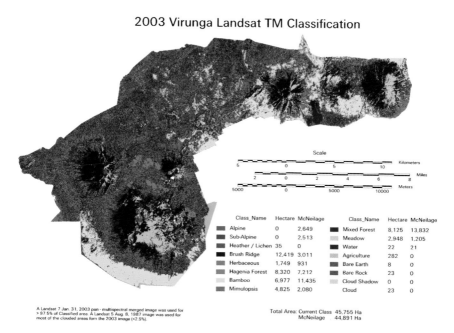

Class_Name	Hectare	McNeilage		Class_Name	Hectare	McNeilage
Alpine	0	2,649		Mixed Forest	8,125	13,832
Sub-Alpine	0	2,513		Meadow	2,948	1,205
Heather / Lichen	35	0		Water	22	21
Brush Ridge	12,419	3,011		Agriculture	282	0
Herbaceous	1,749	931		Bare Earth	8	0
Hagenia Forest	8,320	7,212		Bare Rock	23	0
Bamboo	6,977	11,435		Cloud Shadow	0	0
Mimulopsis	4,825	2,080		Cloud	23	0

A Landsat 7 Jan. 31, 2003 pan - multispectral merged image was used for > 97.5% of Classified area. A Landsat 5 Aug. 8, 1987 image was used for most of the clouded areas form the 2003 image (<2.5%).

Total Area: Current Class 45,755 Ha
McNeilage 44,891 Ha

FIGURE 5. VCA vegetation classification draped over the DEM.

for use in wildlife tracking and analysis is rare, and most generic GIS functions are not well suited for this analysis. Hooge had written a series of software extensions to the popular ArcView Spatial Analyst 2.x/3.X software that was specifically designed for the spatial analysis of animal movements (Hooge and Eichenlaub, 2000). His "Animal Movement" software contains more than 40 routines that are specifically designed to analyze animal movement patterns. Many of the functions are implementations of algorithms derived from reviews of the published scientific literature such as fixed kernel home range utilization distribution, minimum convex polygons, descriptive statistics of the animal location point patterns, and point in polygon analysis and histograms, to name a few. The software is written in the Avenue scripting language, and there are no plans to port the code to Visual Basic to run under ArcGIS. This necessitated our running both ArcView and ArcGIS environments, but the tailored capabilities of the software for this application make it worth the effort.

For our initial analyses, we generated several monthly and complete annual ranges for the three Karisoke groups during 2003 using the MCP function of the Animal Movement program. The results were used to 1) compare overlap in a sample of monthly group ranges to that of annual ranges; and 2) determine range size in relation to group size (obtained from our long-term demography database).

Figure 11.6 shows each of the three groups' ranges for April 2003 (indicated by the daily GPS locations) and their 2003 annual ranges (indicated by the polygons). Although a statistical comparison is not possible, differences in group size (i.e., Shinda, upper polygon is smaller than Pablo's) appear to correspond well with differences in annual range size, a finding consistent with previously established correlations between daily travel distance and group size (Watts, 1991) and between group biomass and range size (McNeilage, 1995).

In comparing the annual range of Beetsme group for 2003 to that plotted by McNeilage (1995), it is apparent that on McNeilage's map (representing 1991–1993 data) Beetsme extended its range far deeper into DRC than it did during 2003.

This difference in the extent of home range for Beetsme's group prompted a second analysis in which we explored the possible reasons for the apparent shift in Beetsme's range. One reason for the change might be the reported increased poaching activity in DRC (primarily buffalo) and refugee movement through the gorilla habitat during and following the 1994 civil war. One test is to compare the recent home range of Beetsme's group as established by Fossey in 1985 to that estimated by McNeilage and by us in both 2002 and 2003. Another test is to examine the 2003 group ranges in relation to locations of poaching activity from the same year entered into the GIS database.

Figure 11.7 provides a comparison of the ranges of Beetsme's group between 1985 and 2003, while Figure 11.8 shows the relationship between the daily location of poaching activity (i.e., snares set for antelope and hyrax), coverage

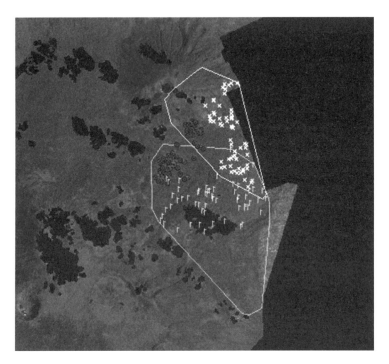

FIGURE 11.6. Polygons (in white) show 2003 annual ranges for two Karisoke groups: top polygon for Shinda (smallest group), lower polygon for Pablo (largest group). Symbols in white (x, cross) and dark circle symbols represent April 2003 GPS points for three Karisoke groups: Shinda, Pablo, and Beetsme.

by anti-poaching patrols, and gorilla group locations for all of 2003. As seen in Figure 11.7, between 1985 and 2003, Beetsme progressively shifted its range out of DRC into Rwanda. Further, as seen in Figure 11.8, gorilla groups generally exclude or avoid areas of high snare trap density, that is, poaching activity. While these results are consistent with high poaching in DRC as a potential cause of the shift, other factors (e.g., habitat quality, interactions with other groups) cannot be ruled out and will need to be controlled for with additional analyses.

4. Discussion

The principal result of our efforts has been the creation of a comprehensive geo-referenced GIS database for the Virunga Volcano region of Central-East Africa. This includes current and historic maps, GPS data, imagery from Landsat TM and ETM, historic aerial photographs, airborne hyperspectral imagery, and SIR-C radar satellite data. An integrated Geomatics approach has allowed us to combine all of the data in a manner that facilitates their exploration and the testing of hypotheses in ways that were not possible only

FIGURE 11.7. Shown are Beetsme's range in 1985 (far left polygon) compared with that plotted by McNeilage (1995) (middle, dark line polygon) and to its range in 2003 (right polygon).

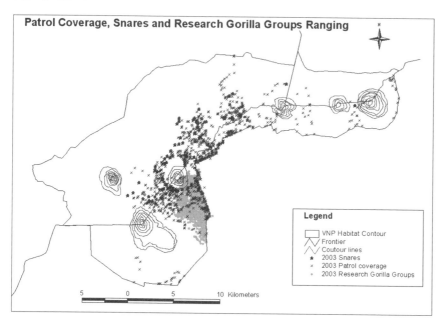

FIGURE 11.8. Daily location in 2003 of snares, anti-poaching patrols, and research gorilla groups.

a decade ago. A significant benefit of our GIS database lies in its long-term and cumulative nature, in that we can track patterns over time in ways that are not possible using traditional field notes and hand drawn maps. Our ultimate goal is to understand the dynamic relationship between mountain gorillas and their habitat over time. We seek patterns of gorilla ranging behavior, their relationship to the environment, and to quantify the impacts of poaching and human encroachment. For example, gorilla population censuses over time have noted that the Karisoke sector of the VCA has the highest concentration of gorilla groups and largest group sizes of the entire VCA (Steklis and Gerald Steklis, 2001). Indeed, the most recent census (Fawcett, pers. comm.) shows that the 17% growth in the population since the previous census in 1989 is entirely accounted for by growth in four groups in the Karisoke sector. There has been much speculation over the years as to the reasons for the greater density of gorillas in this sector. Some postulate that greater protection (from Karisoke anti-poaching patrols and daily presence of researchers) and richer vegetation are the primary causes (Vedder, 1986; McNeilage, 2001). A definitive answer to this important demographic question will require a comparison of ranging data for groups outside the Karisoke sector, as well as data on levels of protection or monitoring in other sectors of the VCA. Such data can be acquired, and in the future we will conduct these analyses using our GIS database and GIS tools.

Our growing understanding of the relationship between the mountain gorilla and its environment needs to be made accessible and usable by parks authorities and other agencies directly involved in mountain gorilla conservation. Publication and public presentations are important means of achieving this information transfer, but it needs to be supplemented with significant in-country capacity building and technology transfer. Indeed, an important aspect of our work has been technology transfer to the people of Rwanda and the region. For example, as a result of a grant from the Georgia Research Alliance and the USAID Gorilla Directive, DFGFI, in partnership with Georgia Tech, Clark Atlanta University and the National University of Rwanda helped establish a GIS Center in Rwanda, with emphasis on remote sensing technology. Scientists from the Center and DFGFI are using these new tools not only for mountain gorilla research and conservation but also for other applications in Rwanda. The GIS Center serves as a regional resource for GIS data and training, which makes it possible for the people of the region to harness these powerful tools to create a better future.

4.1. Future Directions

Key aspects of our future research on the mountain gorilla habitat will depend on access to timely, higher quality data. A collaborative effort is underway by The European Space Agency (ESA) and UNESCO (the United Nations Educational, Scientific, and Cultural Organization) to generate a habitat classification for the Virungas and nearby important biological regions. At the 52nd International Astronautical Congress in October 2001 in Toulouse, France, ESA and UNESCO presented their new initiative on the monitoring of World Heritage sites using space technology, including Earth observation satellites. The Virunga region is the first test case. A new satellite, ENVISat, will have a powerful radar imager, and ESA has committed to acquire numerous images of the region for several years. A major product will be a new 1:25,000 basemap of the VCA, to be generated collaboratively by our team and others. The overall goal is to create a new remote sensing unit inside the UNESCO World Heritage Center, to create a new virtual network of co-operating entities, and to increase the capabilities in the areas of remote sensing and monitoring within developing countries. DFGFI is participating in this new, international collaboration that will provide important new data to monitor natural and human-induced changes in the Virunga region.

New radar imagery recently acquired by NASA can also significantly improve the accuracy of our Virunga DEM. In 2000, NASA and the National Image and Mapping Agency (now National Geospatial Agency) of the U.S. Defense Department flew a third flight of the SIR-C radar. The SIR-C/X-SAR missions of 1994 proved that interferometry from space was practical using the system. Interferometry is the creation of DEMs using two different radar images taken from different angles. The Shuttle Radar Terrain Mapping (SRTM) Mission flew February 11–22, 2000, using a modified version of the same radar instrument that comprised the SIR-C/X-SAR in 1994. SRTM was designed to collect 3-D measurements of the Earth's surface (NASA, 2003). To collect the 3-D data, engineers added a 60-m (approximately 200-ft) mast, installed additional C-band and X-band antennas, and improved tracking and navigation devices. The mission is a cooperative project between NASA, the National Imagery and Mapping Agency (NIMA) of the U.S. Department of Defense (DOD), and the German and Italian space agencies. It is managed by NASA's Jet Propulsion Laboratory (JPL), Pasadena, California, for NASA's Earth Science Enterprise, Washington, D.C.

Once these data are made available, they will give us a superbly accurate 90-m resolution DEM of the entire region for analysis (or possibly 30 m if it will be released to us). The data have already been processed at JPL, but have not been released outside of NASA and DOD. Once released, the imagery will be a great improvement over the original DEM produced from 1930s contour maps, as described earlier. We will compare the two datasets, and will have to re-run our analyses if there are significant differences.

A future direction of our research program concerns the historical development of the VCA. This includes the changes in the park boundaries, human population and encroachment patterns, and vegetation changes over time going back to the origins of the Albert National Park. The incorporation of historical cartographic data into GIS analysis is a fairly recent development (Knowles, 2002), but one that has significant potential benefit to our research.

We have acquired historic cartographic and photographic data through visits to the Royal Museum of Central Africa (Musee Royal de l'Afrique Central), Tervuren, Belgium, and the British National Archives in London in 2003. There we found significant new information regarding the 1936–1938 1:100,000 and 1:50,000 maps, the volcanology map used by Fossey, and other historical data for the Virunga region. These data are currently being processed into our Virunga GIS database.

In London, we reviewed text documents and original maps from the British Ministry of Defense, the Colonial Office, and the Foreign Office, all located at the National Archives. Again, high-resolution color scans of 14 maps, dating from between 1906 and 1922, were provided to us on CD-ROM. These data are also currently being processed into our GIS system, and, together with the Belgian historical data, they will provide the basis for our documentation of historic changes in the VCA.

Finally, we plan to expand the scope of our gorilla ranging analyses to include further cross-sectional and longitudinal analyses using environmental and demographic variables contained in our long-term database. These analyses will add to our understanding of gorilla biology and behavior and, most importantly, will provide a sound knowledge base for rational and effective conservation management.

Acknowledgements The initial GIS database development and imagery analyses were conducted at the CRSSA, Rutgers University. Later and current work has been conducted at the University of North Carolina at Chapel Hill, Informatics International, Inc. of Chapel Hill, North Carolina, and the Georgia Tech Research Institute of Atlanta, Georgia. Many research associates and students have contributed their time and expertise to this project over the years. We particularly thank Rich Bochkay (CRSSA, Rutgers University), Paul Beatty (Georgia Institute of Technology), Bob Wiencek (CRSSA, Rutgers University), Dr. Larry Lass (University of Idaho–Moscow), Jen LeClair (Rutgers University), Amy Jacobson (Rutgers University), and Theresa McReynolds (University of North Carolina, Chapel Hill).

We also wish to express our appreciation to the NASA/Cal Tech Jet Propulsion Laboratory SIR-C/X-SAR Program staff for their assistance in acquiring the SIR-C data. Stephan Maas located the historic Belgian 1:50:000 and 100,000 maps for us, and we are grateful to the Belgian Royal Museum of Central Africa in Tervuren and the British National Archives in London for their assistance in providing archival maps of the region. We also thank Dr. Annette Lanjouw of the International Gorilla Conservation Program for the contribution of the Virunga toponym data. Grateful appreciation is extended to the staff, donors, and supporters of the Dian Fossey Gorilla Fund International. Significant funding for this research was provided by the U.S. Agency for International Development and the Georgia

Research Alliance. Additional funding was provided by the MacArthur Foundation, the Daniel K. Thorne Foundation, the National Geographic Society, and ESSI. In-kind donations of equipment and software have been provided by Trimble GPS, ERDAS, and the iPIX Corporation. We thank the three national parks authorities of Rwanda (ORTPN), DRC (ICCN), and Uganda (UWA) for permitting and facilitating our research in the Virunga region. We also thank Sir Arthur C. Clarke for his continuing inspiration and for his assistance in facilitating aspects of this project. Finally, we wish to express our heartfelt thanks and respect to the expatriate DFGFI staff of the Karisoke Research Center, and particularly to the Center's Rwandan field staff, many of whom lost their lives protecting the mountain gorillas, and whose courageous, tireless efforts assure the gorillas' continued survival. In the end, we acknowledge the mountain gorillas of the Virungas, and their right to exist. Information about our research and products can be viewed on the following websites: www.gorillafund.org, www.informatics.org/gorilla.

References

Fischer, E., and Hinkel, H. (1992). *Natur Ruandas: La Nature du Rwanda*. Rhein-Main Druck, Mainz.

Fossey, D. (1974). Observations on the home range of one group of mountain gorillas (*Gorilla gorilla beringei*). *Animal Behaviour*. 22:568–581.

Fossey, D., and Harcourt, A.H. (1977). Feeding ecology of free-ranging mountain gorilla (*Gorilla gorilla beringei*). In: Clutton-Brock, T. H. (ed.) *Primate Ecology: Studies of Feeding and Ranging Behaviour in Lemurs, Monkeys, and Apes*. Academic Press, New York, pp. 415–447.

Gerald, C.N. (1995). Demography of the Virunga mountain gorilla (*Gorilla gorilla beringei*). MSc. Disseration, Princeton University, Princeton, NJ.

Goren, et al., (1993). *GRASS 4.2 Users Manual*. Army Corps of Engineers Construction Engineering Research Laboratories, Champaign, IL.

Harcourt, A.H. (1995). Population viability estimates: theory and practice for a wild gorilla population. *Conservation Biology*. 9:134-142.

Harcourt, A.H. (1999). Biogeographic relationships of primates on South-East Asian Islands. *Global Ecology and Biogeography*. 8(1):55–61.

Hooge, P.N., and Eichenlaub, B. (2000). Animal movement extension to Arcview. Alaska Science Office, U.S. Geological Survey, Anchorage, AK.

IUCN (2002). *2002 IUCN Redlist of Threatened Species*. IUCN, Gland, Switzerland.

Jernvall, J., and Wright, P.C. (1998). Diversity components of impending primate extinctions. *Proc. Natl. Acad. Sci. USA*. 95:11279–11283.

Jordan, R.L., Huneycutt, B.L., and Werner, M. (1991). The SIR-C/X-SAR Synthetic Aperture Radar System. *Proc. of the IEEE*. 79:827–838.

Kalpers, J., Williamson, E.A., Robbins, M.M., McNeilage, A., Nzamurambaho, A., Lola, N., and Mugiri, G. (2003). Gorillas in the crossfire: population dynamics of the Virunga mountain gorillas over the past three decades. *Oryx*. 37(3):326–337.

Knowles, A.K. (2002). *Past Time, Past Place: GIS for History*. Esri Press, Redlands, CA.

McNeilage, A.J. (1995). Mountain Gorillas in the Virunga Volcanoes: Ecology and Carrying Capacity. Ph.D. Dissertation. School of Biological Sciences. University of Bristol, Bristol, UK.

McNeilage, A.J. (2001). Diet and habitat use of two mountain gorilla groups in contrasting habitats in the Virungas. In: Robbins, Sicotte, P., and Stewart, K. J. (eds.) *Mountain Gorillas: Three Decades of Research at Karisoke*. Cambridge University Press, Cambridge, pp. 265–292.

Plumptre, A.J., and Williamson, E. A. (2001). Conservation-oriented research in the Virunga Region. In: Robbins, M. M., Sicotte, P., and Stewart, K. J. (eds.) *Mountain Gorillas: Three Decades of Research at Karisoke*. Cambridge University Press, Cambridge, pp. 361–389.

Robbins, M.M., Sicotte, P., and Stewart, K.J. (2001). *Mountain Gorillas: Three Decades of Research at Karisoke*. Cambridge University Press, Cambridge.

Scott, J.M., Heglund, P.J., Morrison, M.L., Haufler, J.B., Raphael, M.G., Wall, W.A., and Samson, F.B. (eds.) (2002). *Predicting Species Occurrences: Issues of Accuracy and Scale*. Island Press, Washington, D.C.

Steklis, H.D., and Gerald-Steklis, N. (2001). Status of the Virunga mountain gorilla population. In: Robbins, M.M., Sicotte, P., and Stewart, K.J. (eds.) *Mountain Gorillas: Three Decades of Research at Karisoke*. Cambridge University Press, Cambridge, pp. 391–412.

Steklis, H.D., Gerald-Steklis, N., and Madry, S. (1996/1997). The mountain gorilla: Conserving an endangered primate in conditions of extreme political Instability. *Primate Conservation*. 17:145–151.

Stuhr, F., Jordan, R., and Werner, M. (1995). SIR-C/X-SAR: A multifaceted radar. *Aerospace and Electronic Systems Magazine, IEEE*. 10(10):15–24.

Vedder, A. (1984). Movement patterns of a group of free ranging mountain gorillas (*G.g. beringei*) and their relationship to food availability. *American Journal of Primatology*. 7:73–88.

Vedder, A. (1986). Diet selectivity in one group of mountain gorillas (*G. gorilla beringei*). *Primate Report*. 1986:134.

Walsh, P.D., Abernethy, K.A., Bermejo, M., Beyers, R., Wachter, P.D., Akou, M.E., Huijbregts, B., Mambounga, D.I., Toham, A.K., Kilbourn, A.M., Lahm, S.A., Latour, S., Maisels, F., Mbina, C., Minhindou, Y., Obiang, S.N., Effa, E.N., Starkey, M.P., Telfer, P., Thibault, M., Tutin, C.E.G., White, L.J.T., and Wilkie, D.S. (2003). Catastrophic ape decline in western equatorial Africa. *Nature*. (April 6, 2003), pp. 1–3.

Watts, D.P. (1991). Habitat use strategies of mountain gorillas. *Folia Primatologica*. 56:1–16.

Weber, A.W., and Vedder, A. (1983). Population dynamics of the Virunga gorillas. *Biological Conservation*. 26:341–366.

Weber, B. (1995). Le Parc National des Volcans Biosphere Reserve, Rwanda: the role of development in conservation. *Parks*. 10(3):19–21.

White, F. (1978) The afromontane region. In: Werger, M.J.A. (ed.) *Biogeography and Ecology of Southern Africa*. The Hague, Junk. pp. 463–513.

Chapter 12
Biomaterials in Gorilla Research and Conservation

Cathi Lehn

1. Introduction

Biomaterials are defined as any organic piece or derivative of a plant or animal and are used in many disciplines, including taxonomy and systematics, population genetics, reproductive sciences, nutrition, pathology, endocrinology, education, toxicology and veterinary medicine. Examples of biomaterials collected from animals include tissue, urine, feces, skulls, gametes, hair, and DNA. Biomaterials may be collected either from an animal in the wild or from an individual held in captivity (e.g., more than 300 western lowland gorillas (*Gorilla gorilla gorilla*) are held in Association of Zoos and Aquariums (AZA)–accredited zoological parks (D. Wharton, personal communication)). Biological samples such as hair, feces, or urine may be collected noninvasively by either the field biologist or by an animal keeper in a zoological park. Samples may also be collected by a veterinarian either when an animal is handled during a routine procedure or during a postmortem examination. Standard collection methods for any application will dictate that sterile practices are followed, that labeling is accurate and extensive, and that all necessary import and export permits have been secured. Lastly, biological samples should always be collected in a manner that ensures the safety of the collector and the general welfare of the animal.

The most reliable method for guaranteeing both the collector's and animal's safety is to work closely with a veterinarian or other trained personnel, especially for invasive procedures. Two established field veterinary programs for the gorilla are the Mountain Gorilla Veterinary Project (MGVP, Inc.) in the Virunga Volcanoes of Rwanda, the Democratic Republic of Congo, and Uganda and the Wildlife Conservation Society Field Veterinary Program's *Preventive Health Program for Free-Ranging Lowland Gorillas* conducted in Gabon, Congo, and Central African Republic (MGVP/WCS, Chapter 2, this volume).

This chapter will use gorillas to exemplify the many applications for which biomaterials have been used in research and conservation efforts. Some of the results from these studies have had a direct and applied conservation benefit, whereas other studies may be better classified as basic research or for educational purposes. Space constraints limit how much detail may be provided for each application and how many references may be included. The ultimate aim of this chapter, however, is not to provide an exhaustive list of applications or references but to leave the reader with an appreciation for the immense value of biological samples and the importance of collecting, storing, and utilizing these samples for research and education.

2. Health and Nutrition

The diagnosis and prevention of disease is dependent on the availability of biological samples (MGVP/WCS, Chapter 2, this volume). The basic sample collection protocol used by the MGVP, Inc. when an animal is sedated includes the collection of blood, urine, feces, hair, rectal and wound swabs, and live genetic material (e.g., lymphocytes and epithelial cells (Mudakikwa et al., 2001; Cranfield et al., 2002)). A blood sample provides some of the very basic information needed to assess the health of the animal. Several tests may be run from a blood sample, including a complete blood count, serum biochemistry profiles, and vitamin and mineral level assays (Deem et al., 2001; Baitchman et al., 2006). Serum samples may also be used to measure antibodies in the blood as an indication of current and past exposures to infectious agents (Mudakikwa et al., 2001). Diseases that have been diagnosed from a blood sample in captive gorillas include hypothyroidism and leukemia (Barrie et al., 1999; Lair et al., 1999).

The collection of a blood sample requires an invasive procedure (i.e., the sedation of the gorilla), therefore noninvasively collected samples, such as urine and feces, are also utilized for health screenings. A urinalysis includes an assessment of urine proteins, ketones, organ function, carbohydrate metabolism, acid base balance, urinary tract infections, as well as screenings for bacterial and viral diseases (Sleeman and Mudakikwa, 1998; Cranfield et al., 2002). Hormone levels as a measure of stress may also be examined from a urine sample (Stoinksi et al., 2002). A fecal sample may also provide valuable information on the health status of the gorilla and may be examined for parasites, bacteria, and intestinal flora (Redmond, 1983; Ashford et al., 1996; Sleeman et al., 2000; Graczyk et al., 2002; Lilly et al., 2002; Frey et al., 2006). In addition, stress levels in the individual may be assessed by measuring cortisol and dehydroepiandrosterone (DHEA) levels in feces (Czekala and Robbins, 2001; Wasser et al., 2002; Monfort, 2003; Peel et al., 2005).

In the unfortunate event that a gorilla is found dead, whether in captivity or in the wild, every opportunity should be taken to learn as much as possible from that animal (Karesh and Cook, 1995; Deem et al., 2001; Munson and Karesh, 2002). Veterinarians working with endangered species are faced with the challenge of working with an insufficient amount of information about their animals (Karesh and Cook, 1995; McNamara, 1999). Knowledge gained from a postmortem examination, also called a necropsy, may help the pathologist determine the cause of death of an individual animal and may also provide useful information in the development of a preventive health care program. A necropsy includes a complete examination of all organs, as well as the sampling and fixation of all organs in formalin for histopathologic evaluation. Formalin-fixed tissues are then available for electron microscopy studies and cytologic profiles. Tissues are also frozen for future bacterial culture, viral isolation, toxicologic or nutritional evaluation, and for genetic studies (McNamara, 1999). Outbreaks of the Ebola virus (Leroy *et al.*, 2004; Rouquet *et al.*, 2005; Pourrut *et al.*, 2005) and an anthrax infection (Leendertz *et al.*, 2006) in wild gorillas were determined based on the results of post-mortem examinations of carcasses and the biological samples collected during those necropsies. The skull and skeleton of a gorilla may also provide valuable information concerning the incidence of disease and trauma experienced by a gorilla during the course of its life. In 1990, Lovell published her results on the examination of great ape skulls and skeletons housed at the National Museum of Natural History. In her study, teeth were examined for incidence of tooth decay, abscesses, tooth wear, periodontal disease, and tooth loss, and skeletal parts provided information on bone inflammation, arthritic lesions, nutritional deficiencies, and developmental abnormalities. Rothschild and Ruhli (2005) also studied skeletal material collected from wild gorillas to determine the frequency and character of arthritis.

Fecal and serum samples may be used by the nutritionist and can form the basis of a dietary analysis. The contents of a fecal sample, combined with observational data, help the nutritionist better understand what foods are being eaten, during which times of the year and in the captive diet, the effects of variation (Remis, 1997; Doran *et al.*, 2002; Deblauwe *et al.*, 2003; Yamagiwa *et al.*, 2003, 2005; Remis and Dierenfeld 2004). From a serum sample, the nutritionist measures circulating levels of vitamin metabolites, lipids, and carotenoids, which provide information on the animal's health, nutritional status, and the absorption levels of nutrients (Crissey *et al.*, 1999).

3. Morphology

In the late 1800s and early 1900s, Western expeditions were sent to Africa to bring back gorilla specimens for exhibit and research (Ives and Nassau, 1892; Akeley, 1923; Willoughby, 1950; Kennedy and Whittaker, 1976). Schaller (1963) provides a review of the discovery and study of the mountain gorilla and discusses some of these expeditions in more detail. Unfortunately, many of the animals brought back from these early expeditions did not survive but ended up as specimens of wonder and research in museum collections. In 1929, Coolidge estimated that over 800 gorilla specimens were housed in museums around the world. A more recent survey of museums in the United States and Canada found that over 500 gorilla skulls and close to 200 postcranial skeletons are housed in these collections (Albrecht, 1982). The skulls and skeletons collected as a result of these expeditions and curated in museum collections have provided informative characters for the taxonomist ever since the gorilla was first described to science (Savage and Wyman, 1847). Contemporary taxonomists also utilize skull and skeletal characters in the description of gorilla species and subspecies (Sarmiento *et al.*, 1996; Stumpf *et al.*, 2003). Based on museum specimens of gorillas from the Cross River in western Cameroon, Sarmiento and Oates (2000) discuss the distinctiveness of the Cross River gorilla (*G. g. diehli*). Groves (2003) provides a history of gorilla taxonomy, and in this review summarizes some of the studies completed on the anatomy of the gorilla, including the pioneering works of the Henry Cushier Raven Memorial Volume, *The Anatomy of the Gorilla* (Gregory, 1950). Numerous studies have focused on the descriptive nature of the skull (Schmittbuhl *et al.*, 1996; Sherwood, 1999; Lieberman *et al.*, 2000; Preuschoft *et al.*, 2002) and have outlined features that contribute to our understanding of not only great ape evolution but also give insight into our own evolutionary history. In addition, studies on the gorilla skull and skeleton have contributed further to our understanding of adaptations relating to locomotion (Sarmiento, 1994; Taylor, 1997; Inouye, 2003; Payne *et al.*, 2006), diet (Uchida, 1998; Schwartz, 2000; Godfrey *et al.*, 2001; Taylor, 2003), and factors relating to stress (Manning and Chamberlain, 1994; Guatelli-Steinberg, 2001).

4. Genetics

Genetic variation in the gorilla has been examined using several different measures and technologies, including karyotyping (Hamerton *et al.*, 1961; Seuanez, 1986), ABO blood groups (Wiener *et al.*, 1976; Socha and Moor-Jankowski, 1986), protein electrophoresis (Sarich, 1977), DNA-DNA hybridization (Sibley and Ahlquist, 1987; Caccone and Powell, 1989), DNA sequencing (Garner and Ryder, 1996; Ruvolo, 1997), microsatellites (Clifford *et al.*, 1999; Lukas *et al.*, 2004), fluorescence *in situ* hybridization (Mrasek *et al.*, 2001), and array comparative genomic hybridization (Locke *et al.*, 2003). Tissue samples are the preferred materials for genetic studies (Ryder, 1986), however, technology associated with the polymerase chain reaction has allowed the geneticist to extract DNA from noninvasively collected samples, such as feces and hair, as well as museum skins (Saltonstall *et al.*, 1998; Vigilant *et al.*, 2002; Clifford *et al.*, 2003; Oates *et al.*, 2003).

DNA sequencing results have been used to address questions related to the phylogenetic relationships within the Family Hominidae (Ruvolo *et al.*, 1994; Ruvolo, 1997), as well as to examine intraspecific relationships within the genus *Gorilla* (Ruvolo *et al.*, 1994; Garner and Ryder, 1996; Saltonstall *et al.*, 1998; Clifford *et al.*, 2003; Jensen-Seaman *et al.*, 2003; Oates *et al.*, 2003). While the majority of DNA sequencing studies have focused on genes from the mitochondrial genome (Ruvolo *et al.*, 1994; Garner and Ryder, 1996), recent studies have explored the nuclear regions of the genome, including the Y-chromosome (Burrows and Ryder, 1997; Ruvolo, 1997; Jensen-Seaman *et al.*, 2003). A review by Vigilant and Bradley (2004) provides a summary of the studies addressing genetic variation in gorillas.

Genetic studies additionally may be used: 1) to determine the sex of an animal (Ramsay *et al.*, 2000; Ensminger and Hoffman, 2002); 2) for paternity testing (Field *et al.*, 1998; Bradley *et al.*, 2005); 3) to assess population structure (Bradley *et al.*, 2002; Bradley *et al.*, 2004) and gene flow (Saltonstall *et al.* 1998); and 4) to assist in forensic identification in illegal trade (Garner and Ryder, 1996), as well as in human forensics (Doi *et al.*, 2004; Matsuda *et al.*, 2005). Additionally, many of the health screenings mentioned previously incorporate genetic techniques, especially the use of the polymerase chain reaction (PCR), for example, detection of the Ebola virus (Rouquet *et al.*, 2005). Ruvolo (1997) was interested in resolving the relationships between the apes and humans, however, completion of the Human Genome Project (HGP), and the technologies emerging from the HGP, including the completion of the initial sequence of the chimpanzee genome (Chimpanzee Sequencing and Analysis Consortium, 2005), now allow researchers to investigate these relationships further and offer us a new look at ourselves through our genomes (Kim *et al.*, 2003; Ryder, 2003, 2005). By focusing efforts on genomic comparisons of chimps, gorillas, and humans, a better understanding of the evolutionary basis for human traits and their uniqueness can be obtained, including a better understanding of the genetic mechanisms underlying disease, language, reproductive disorders, the influence of environmental factors on behavior and more (McConkey and Varki, 2000; Hacia, 2001; Fisher and Marcus, 2006; Sikela, 2006).

5. Reproductive Biology

Biological samples play a vital role in the monitoring and evaluation of the reproductive status of an individual and in assisted reproduction. The reproductive cycle of the female gorilla is not easily assessed visually; therefore the researcher must rely on other measures to evaluate cycles and reproductive status. Female hormones (e.g., estrogen and progesterone) can give the researcher this information, including pregnancy diagnosis (Lasley *et al.*, 1980; Mitchell *et al.*, 1982; Roser *et al.*, 1986; Czekala *et al.*, 1988; Loskutoff *et al.*, 1991; Bellem *et al.*, 1995). In the past, hormone levels were measured solely by serum analysis, however, today many of these measures can be assessed from urine and feces (Whitten *et al.*, 1998; Czekala and Robbins, 2001; Shimizu *et al.*, 2003; Atsalis *et al.*, 2004). Hormone analyses have also been used to explain maternal behavior (Mitchell *et al.*, 1985; Bahr *et al.*, 1998; Bahr *et al.*, 2001). While the majority of these hormonal analyses have been completed on captive gorillas, in the last few years measurements have also been taken from habituated animals in the wild (Czekala and Sicotte, 2000;

Czekala and Robbins, 2001). Hormone levels, as measured from urine and feces, have also been used to assess male hormones and fertility (Wildt, 1996; Robbins and Czekala, 1997; Stoinski *et al.*, 2002). Fertility measures in the male can also be assessed from sperm and semen (Seuanez *et al.*, 1977; Gould and Kling, 1982; Seager *et al.*, 1982) or from a testicular biopsy (Foster and Rowley, 1982). Assisted reproduction is also dependent on the collection of biological samples or, more specifically, the collection of gametes (Loskutoff *et al.*, 1991). Gametes have been used for *in vitro* fertilization (Lanzendorf *et al.*, 1992; Dresser *et al.*, 1996; Pope *et al.*, 1997), for intracytoplasmic sperm injection of oocytes (Kurz *et al.*, 1996), as well as for artificial insemination (Douglass and Gould, 1981; Tribe *et al.*, 1989). The latest technology used in assisted reproduction efforts for the gorilla is the predetermination of an offspring's sex using sperm sorting. This technology will assist the captive manager by producing female embryos from genetically valuable parents and transferring them to less genetically valuable females proven to be good parents (O'Brien *et al.*, 2002, 2005).

6. Population Biology

Direct observation of gorillas in the forest is rare and therefore indirect census techniques are often used to determine population size and composition. Field biologists can use hair and fecal samples to verify species and to determine group composition of animals and use dung size to determine age (Schaller, 1963; Blom *et al.*, 2001; McNeilage *et al.*, 2001). Additionally, by utilizing scanning electron microscopy the researcher can determine if a hair sample found in a nest belongs to a chimp or a gorilla (Furuichi *et al.* 1997).

7. Education

An excellent example of the historical use of gorilla biomaterials for educational purposes can be viewed in the Akeley Hall of African Mammals in the American Museum of Natural History in New York City. In this hall, the visitor experiences the extraordinary taxidermy style of Carl Akeley. His realistic mounts and landscapes revolutionized the art of taxidermy and the museum diorama (Lucas, 1927). One diorama in the hall depicts a gorilla family set at Lake Kivu, Zaire (now the Democratic Republic of Congo). The mountain gorilla diorama preserves for the museum visitor an exact reproduction of the landscape and animals observed by Carl Akeley in the early 1900s while on a collecting trip for the museum (Akeley, 1923). This scene has inspired many visitors over the years at the museum, including students, researchers, and conservationists (Bodry-Sanders, 1998). The museum's dioramas display perhaps the ultimate application of gorilla biomaterials for educational use, although the more traditional use of gorilla biomaterials in the classroom is the use of the gorilla skull as a means of teaching.

8. Banking and Biomaterials

The above discussion has focused on numerous applications for gorilla biomaterials. It must be emphasized, however, that these studies were possible only because biological samples were collected, stored properly, and made available for research or education. Future studies will also depend on the use of properly collected and banked samples. It is crucial, therefore, that these resources are collected now (Munson, 2002). Ryder et al., (2000, 2005) and others (Oldham and Geschwind, 2006) have stressed the importance of coordinated efforts for the collecting and banking of samples, especially in the case of endangered species when there may be a limited amount of time to acquire biological resources. Reproductive scientists have for many years made a plea for a coordinated national and/or international effort for the banking of genetic resources for research and conservation (Wildt et al., 1997; Holt et al., 2003). It is also critical that several different types of samples are collected because of unknown research possibilities in the future (Rideout, 2002; Ryder, 2002). In addition, for some applications it is desired that sampling is completed from the same individual over an extended period of time or that sampling from the same locality is done multiple times (e.g., disease monitoring (Deem et al., 2001; Munson and Karesh, 2002)). Technological advances ensure that new techniques, analyses, and storage parameters will be used in the future, therefore it is advised to be aware of current protocols and to work closely with researchers in order to collect and store samples properly (e.g., Whittier et al., 2004; Smith and Morin, 2005; Waits and Paetkau, 2005).

The curation of skins, skulls, and skeletons has for centuries been the responsibility of natural history museums (Mehrhoff, 1996); however, the curation of tissues and gametes is a relatively new discipline and relies on a set of specific curatorial methods (Dessauer et al., 1996; Sheldon and Dittman, 1997; Prendini et al., 2002). Comprehensive, as well as taxon-specific, tissue collections have been established worldwide (Dessauer et al., 1996; Prendini et al., 2002). A recent initiative funded by the National Science Foundation, the Integrated Primate Biomaterials and Information Resource, was established to specifically bank nonhuman primate samples (Stone, 2003). In addition, the MGVP, Inc. has begun to bank mountain gorilla samples at a facility in Denver, Colorado (Cranfield et al., 2001).

9. Summary and Conclusions

This review has introduced the reader to just some of the numerous applications for gorilla biomaterials. Biological samples may be collected noninvasively by either the field biologist or the animal keeper, or alternatively, they may be collected by a veterinarian during a medical procedure. The research conducted using these samples may be directly applied to conservation efforts or may be used for unforeseen applications in the future. It is hoped that this chapter has given the reader a better appreciation for the multitude of applications for biological samples and recognition of the extreme urgency for collecting and archiving these samples for use now and in the future.

Acknowledgments. The author thanks Tara Stoinski for her invitation to contribute to this volume. I also thank Mike Cranfield, Randy Junge, Billy Karesh, Bruce Latimer, Naida Loskutoff, Lisa Starr, and two anonymous reviewers for their valuable comments.

References

Akeley, C.E. (1923). *In Brightest Africa.* Doubleday, Page and Company, Garden City, NY.

Albrecht, G.H. (1982). Collections of nonhuman primate skeletal materials in the United States and Canada. *Am. J. Phys. Anthrop.* 57:77–97.

Ashford, R.W., Lawson, H., Butynski, T.M., and Reid, G.D.F. (1996). Patterns of intestinal parasitism in the mountain gorilla *Gorilla gorilla* in the Bwindi-Impenetrable Forest, Uganda. *J. Zool.* 239(3):507–514.

Atsalis, S., Margulis, S.W., Bellem, A., and Wielebnowski, N. (2004). Sexual behavior and hormonal estrus cycles in captive aged lowland gorillas (*Gorilla gorilla*). *Am. J. Primatol.* 62:123–132.

Bahr, N.I., Pryce, C.R., Dobeli, M., and Martin, R.D. (1998). Evidence from urinary cortisol that maternal behavior is related to stress in gorillas. *Physiol. Behav.* 64(4):429–437.

Bahr, N.I., Martin, R.D., and Pryce, C.R. (2001). Peripartum sex steroid profiles and endocrine correlates of postpartum maternal behavior in captive gorillas (*Gorilla gorilla gorilla*). *Horm. Behav.* 40(4):533–541.

Baitchman, E.J., Calle, P.P., Clippinger, T.L., Deem, S.L., James, S.B., Raphael, B.L., and Cook, R.A. (2006). Preliminary evaluation of blood lipid profiles in captive western lowland gorillas (*Gorilla gorilla gorilla*). *J. Zoo Wildl. Med.* 37:126–129.

Barrie, M.T., Backues, K.A., Grunow, J., and Nitschke, R. (1999). Acute lymphocyte leukemia in a six-month-old western lowland gorilla (*Gorilla gorilla gorilla*). *J. Zoo Wildl. Med.* 30:268–272.

Bellem, A.C., Monfort, S.L., and Goodrowe, K.L. (1995). Monitoring reproductive development, menstrual cyclicity and pregnancy in the lowland gorilla (*Gorilla gorilla*) by enzyme immunoassay. *J. Zoo Wildl. Med.* 26:24–31.

Blom, A., Almasi, A., Heitkonig, I.M.A., Kpanou, J.B., and Prins, H.H.T. (2001). A survey of the apes in the Dzanga-Ndoki National Park, Central African Republic: A comparison between the census and survey methods of estimating the gorilla (*Gorilla gorilla gorilla*) and chimpanzee (*Pan troglodytes*) nest group density. *Afr. J. Ecol.* 39(1):98–105.

Bodry-Sanders, P. (1998). *African Obsession: The Life and Legacy of Carl Akeley*, revised 2nd edition. Batax Museum Publishing, Jacksonville.

Bradley, B.J., Doran, D., Robbins, M.M., Williamson, E., Boesch, C., and Vigilant, L. (2002). Comparative analyses of genetic social structure in wild gorillas (*Gorilla gorilla*) using DNA from feces and hair. *Am. J. Phys. Anthr. Suppl.* 34:47–48.

Bradley, B.J., Doran-Sheehy, D.M., Lukas, D., Boesch, C., and Vigilant, L. (2004). Dispersed male networks in western gorillas. *Current Biol.* 14:510–513.

Bradley, B.J., Robbins, M.M., Williamson, E.A., Steklis, H.D., Steklis, N.G., Eckhardt, N., Boesch, C., and Vigilant, L. (2005). Mountain gorilla tug-of-war:

Silverbacks have limited control over reproduction in multimale groups. *Proc. Nat. Acad. Sci.* 102:9418–9423.

Burrows, W., and Ryder, O.A. (1997). Y-chromosome variation in great apes. *Nature* 385:125–126.

Caccone, A., and Powell, J.R. (1989). DNA divergence among hominoids. *Evolution* 43(5):925–942.

Chimpanzee Sequencing and Analysis Consortium. (2005). Initial sequence of the chimpanzee genome and comparison with the human genome. *Nature* 437:69–87.

Clifford, S.L., Jeffrey, K., Bruford, M.W., and Wickings, E.J. (1999). Identification of polymorphic microsatellite loci in the gorilla (*Gorilla gorilla gorilla*) using human primers: Application to noninvasively collected hair samples. *Mol. Ecol.* 8: 1551–1561.

Clifford, S.L., Abernethy, K.A., White, L.J.T., Tutin, C.E.G., Bruford, M.W., and Wickings, E.J. (2003). Genetic studies of western gorillas. In: Taylor, A.B. and Goldsmith, M.L. (eds.), *Gorilla Biology: A Multidisciplinary Perspective.* Cambridge University Press, Cambridge, pp. 269–292.

Coolidge, Jr., H.J. (1929). *Revision of the Genus* Gorilla. Memoirs of the Museum of Comparative Zoology at Harvard College. Vol. L. No. 4.

Cranfield, M., Gaffikin, L., and Cameron, K. (2001). Conservation medicine as it applies to the mountain gorilla (*Gorilla gorilla beringei*). *The Apes: Challenges for 21st Century Conference Proceedings*, May 10–13, 2001. Brookfield, IL, pp. 238–240.

Cranfield, M., Gaffikin, L., Sleeman, J., and Rooney, M. (2002). The mountain gorilla and conservation medicine. In: Aguirre, A.A., Ostfeld, R.S., Tabor, G.M., House, C., and Pearl, M.C. (eds.), *Conservation Medicine: Ecological Health in Practice.* Oxford University Press, Oxford, pp. 282–298.

Crissey, S.D., Barr, J.E., Slifka, K.A., Bowen, P.E., Stacewicz-Sapuntzakis, M., Langman, C., Ward, A., and Ange, K. (1999). Serum concentrations of lipids, vitiamins A and E, vitamin D metabolites, and carotenoids in nine primate species at four zoos. *Zoo Biol.* 18(6):551–564.

Czekala, N.M., Roser, J.F., Mortensen, R.B., Reichard, T., and Lasley, B.L. (1988). Urinary hormone analysis as a diagnostic tool to evaluate the ovarian function of female gorillas (*Gorilla gorilla*). *J. Reprod. Fertil.* 82(1):255–261.

Czekala, N., and Sicotte, P. (2000). Reproductive monitoring of free-ranging female mountain gorillas by urinary hormone analysis. *Am. J. Primatol.* 51(3):209–215.

Czekala, N., and Robbins, M.M. (2001). Assessment of reproduction and stress through hormone analysis in gorillas. In: Robbins, M.M., Sicotte, P., and Stewart, K.J. (eds.), *Mountain Gorillas: Three Decades of Research at Karisoke.* Cambridge University Press, Cambridge, pp. 317–340.

Deblauwe, I., Dupain, J., Nguenang, G.M., Werdenich, D., and Van Elsacker, L. (2003). Insectivory by *Gorilla gorilla gorilla* in southeast Cameroon. *Int. J. Primatol.* 24:493–502.

Deem, S.L., Karesh, W.B., and Weisman, W. (2001). Putting theory into practice: Wildlife health in conservation. *Conserv. Biol.* 15(5):1224–1233.

Dessauer, H.C., Cole, C.J., and Hafner, M.S. (1996). Collection and storage of tissues. In: Hillis, D.M., Moritz, C., and Mable, B.K. (eds.), *Molecular Systematics*, Second Edition. Sinauer Associates, Inc., Sunderland, pp. 29–47.

Doi, Y., Yamamoto, Y., Inagaki, S., Shigeta, Y., Miyaishi, S., and Ishizu, H. (2004). A new method for ABO genotyping using a multiplex single-base primer extension reaction and its application to forensic casework samples. *Leg. Med.* 6:213–223.

Doran, D.M., McNeilage, A., Greer, D., Bocian, C., Mehlman, P., and Shah, N. (2002). Western lowland gorilla diet and resource availability: New evidence, cross-site comparisons, and reflections on indirect sampling methods. *Am. J. Primatol.* 58(3):91–116.

Douglass, E., and Gould, K. (1981). Artificial insemination in the gorilla. *Proc. Amer. Assoc. Zoo Vet.*, pp. 128–130.

Dresser, B.L., Pope, C.E., Chin, N., Liu, J., Loskutoff, N.M., Behnke, E., Brown, C., McRae, M., Sinoway, C., Campbell, M., Cameron, K., Evans, R., Owens, O., Johnson, C., and Cedars, M. (1996). Successful in vitro fertilization, embryo transfer and pregnancy in a western lowland gorilla. *Theriogenology* 45:248.

Ensminger, A., and Hoffman, S.M.G. (2002). Sex identification assay useful in great apes is not diagnostic in a range of other primate species. *Am. J. Primatol.* 56:129–134.

Field, D., Chemnick, L., Robbins, M.M., Garner, K., and Ryder, O. (1998). Paternity determination in captive lowland gorillas and orangutans and wild mountain gorillas by microsatellite analysis. *Primates* 39(2):199–209.

Fisher, S.E., and Marcus, G.F. (2006). The eloquent ape: Genes, brains and the evolution of language. *Nature Rev. Genet.* 7:9–20.

Foster, J.W., and Rowley, M.J. (1982). Testicular biopsy in the study of gorilla infertility. *Am. J. Primatol. Suppl.* 1:121–125.

Frey, J.C., Rothman, J.M., Pell, A.N., Nizeyi, J.B., Cranfield, M.R., and Angert, E.R. (2006). Fecal bacterial diversity in a wild gorilla. *Appl. Environ. Microbiol.* 72:3788–3792.

Furuichi, T., Inagaki, H., and Angoue-Ovono, S. (1997). Population density of chimpanzees and gorillas in the Petit Loango Reserve, Gabon: Employing a new method to distinguish between nests of the two species. *Int. J. Primatol.* 18(6):1029–1046.

Garner K.J., and Ryder, O.A. (1996). Mitochondrial DNA diversity in gorillas. *Mol. Phylogenet. Evol.* 6:39–48.

Godfrey, L.R., Samonds, K.E., Jungers, W.L., and Sutherland, M.R. (2001). Teeth, brains, and primate life histories. *Am. J. Phys. Anthropol.* 114(3):192–214.

Gould, K.G., and Kling, O.R. (1982). Fertility in the male gorilla (*Gorilla gorilla*): Relationship to semen parameters and serum hormones. *Am. J. Primatol.* 2:311–316.

Graczyk, T.K., Bosco-Nizeyi, J., Ssebide, B., Thompson, A., Read, C., and Cranfield, M.R. (2002). Anthropozoonotic *Giardia duodenalis* genotype (assemblage) A infections in habitats of free-ranging human-habituated gorillas, Uganda. *J. Parasit.* 88:905–909.

Gregory, W.K. (ed.). (1950). *The Anatomy of the Gorilla*. Columbia University Press, New York.

Groves, C.P. (2003). A history of gorilla taxonomy. In: Taylor, A. B. and Goldsmith, M. L. (eds.), *Gorilla Biology: A Multidisciplinary Perspective*. Cambridge University Press, Cambridge, pp. 15–34.

Guatelli-Steinberg, D. (2001). What can developmental defects of enamel reveal about physiological stress in nonhuman primates? *Evol. Anthropol.* 10:138–151.

Hacia, J.G. (2001). Genome of the apes. *Trends Genet.* 17:637–645.

Hamerton, J.L., Fraccaro, M., De Carli, L., Nuzzo, F., Klinger, H.P., Hulliger, L., Taylor, A., and Lang, E.M. (1961). Somatic chromosomes of the gorilla. *Nature* 192:225.

Holt, W.V., Abaigar, T., Watson, P.F., and Wildt, D.E. (2003). Genetic resource banks for species conservation. In: Holt, W.V., Pickard, A.R., Rodger, J.C. and Wildt, D.E. (eds.), *Reproductive Science and Integrated Conservation*. Cambridge University Press, Cambridge, pp. 267–280.

Inouye, S.E. (2003). Intraspecific and ontogenetic variation in the forelimb morphology of the *Gorilla*. In: Taylor, A.B., and Goldsmith, M. L. (eds.), *Gorilla Biology: A Multidisciplinary Perspective*. Cambridge University Press, Cambridge, pp. 194–235.

Ives, J.E., and Nassau, R.H. (1892). Collecting gorilla brains. *Science* 19(482): 240–241.

Jensen-Seaman, M.I., Dienard, A.S., and Kidd, K.K. (2003). Mitochondrial and nuclear DNA estimates of divergence between western and eastern gorillas. In: Taylor, A.B., and Goldsmith, M.L. (eds.), *Gorilla Biology: A Multidisciplinary Perspective*. Cambridge University Press, Cambridge, pp. 247–268.

Karesh, W.B., and Cook, R.A. (1995). Applications of veterinary medicine to *in situ* conservation efforts. *Oryx* 29(4):244–252.

Kennedy, K.A.R., and Whittaker, J.C. (1976). The ape in Stateroom 10. *Nat. Hist.* 85:48–53.

Kim, C.-G., Fujiyama, A., and Saitou, N. (2003). Construction of a gorilla fosmid library and its PCR screening system. *Genomics* 82:571–574.

Kurz, S.G., Barnes, A.M., Ramey, J.W., Brown, C., Loskutoff, N.M., Simmons, L.G., Armstrong, D.L., and De Jong, C.J. (1996) Semen characteristics of a western lowland gorilla determined by manual and computer-assisted motion analysis. *Biol. Reprod.* 54, Suppl. 1:301.

Lair, S., Crawshaw, G.J., Mehren, K.G., and Perrone, M.A. (1999). Diagnosis of hypothyroidism in a western lowland gorilla (*Gorilla gorilla gorilla*) using human thyroid-stimulating hormone assay. *J. Zoo Wildl. Med.* 30:537–540.

Lanzendorf, S.E., Holmgren, W.J., Schaffer, N., Hatasaka, H., Wentz, A.C., and Jeyendran, R.S. (1992). In vitro fertilization and gamete micromanipulation in the lowland gorilla. *J. Assist. Reprod. Genet.* 9:358–364.

Lasley, B.L., Hodges, J.K., and Czekala, N.M. (1980). Monitoring the female reproductive cycle of great apes and other primate species by determination of oestrogen and LH in small volumes of urine. *J. Reprod. Fertil. Suppl.* 28:121–129.

Leendertz, F.H., Yumlu, S., Pauli, G., Boesch, C., Couacy-Hymann, E., Vigilant, L., Junglen, S., Schenk, S., and Ellerbrok, H. (2006). A new *Bacillus anthracis* found in wild chimpanzees and a gorilla from West and Central Africa. *PLoS Pathog.* 2:e8.

Leroy, E.M., Rouquet, P., Formenty, P., Souquière, S., Kilbourne, A., Froment, J.-M., Bermejo, M., Smit, S., Karesh, W., Swanepoel, R., Zaki, S.R., and Rollin, P.E. (2004). Multiple Ebola virus transmission events and rapid decline of Central African wildlife. *Science* 303:387–390.

Lieberman, D.E., Ross, C.F., and Ravosa, M.J. (2000). The primate cranial base: Ontogeny, function, and integration. *Am. J. Phys. Anthropol.* 113(S31):117–169.

Lilly, A.A., Mehlman, P.T., and Doran, D. (2002). Intestinal parasites in gorillas, chimpanzees, and humans at Mondika Research site, Dzanga-Ndoki National Park, Central African Republic. *Int. J. Primatol.* 23(3):555–573.

Locke, D.P., Segraves, R., Carbone, L, Archidiacono, N., Albertson, D.G., Pinkel, D., and Eichler, E.E. (2003). Large-scale variation among human and great ape genomes determined by array comparative genomic hybridization. *Genome Res.* 13(3):347–357.

Loskutoff, N.M., Kraemer, D.C., Raphael, B.L., Huntress, S.L., and Wildt, D.E. (1991). Advances in reproduction in captive female great apes: An emphasis on the value of biotechniques. *Am. J. Primatol.* 24:151–166.

Lovell, N.C. (1990). *Patterns of Injury and Illness in Great Apes: A Skeletal Analysis*. Smithsonian Institution Press, Washington and London.

Lucas, F.A. (1927). Akeley as a taxidermist: A chapter in the history of museum methods. *Natural History* 27:142–152.

Lukas, D., Bradley, B.J., Nsubuga, A.M., Doran-Sheehy, D., Robbins, M.M., and Vigilant, L. (2004). Major histocompatibility complex and microsatellite variation in two populations of wild gorillas. *Mol. Ecol.* 13:3389–3402.

Manning, J.T., and Chamberlain, A.T. (1994). Fluctuating asymmetry in gorilla canines: A sensitive indicator of environmental stress. *Proc. R. Soc. Lond., Ser. B: Biol. Sci.* 255:189–193.

Matsuda, H., Seo, Y., Kakizaki, E., Kozawa, S., Muraoka, E., and Yukawa, N. (2005). Identification of DNA of human origin based on amplification of human-specific mitochondrial cytochrome *b* region. *Forensic Sci. Int.* 152:109–114.

McConkey, E.H., and Varki, A. (2000). A primate genome project deserves high priority. *Science* 289:1295.

McNamara, T. (1999). The role of pathology in zoo animal medicine. In: Fowler, M.E., and Miller, R.E. (eds.), *Zoo and Wild Animal Medicine: Current Therapy.* W.B. Saunders Co., Philadelphia, pp. 3–7.

McNeilage, A., Plumptre, A.J., Brock-Doyle, A., and Vedder, A. (2001). Bwindi Impenetrable National Park, Uganda: Gorilla census 1997. *Oryx* 35:39–47.

Meehan, T., and Lowenstine, L. (1994). Causes of mortality in captive lowland gorillas: A survey of the SSP population. *Proceedings of the ARAV and AAZV Annual Meeting,* p. 216.

Mehrhoff, L.J. (1996). Museums, research collections, and the biodiversity challenge. In: Reaka-Kudla, M.L., Wilson, D.E. and Wilson E.O. (eds.), *Biodiversity II: Understanding and Protecting Our Biological Resources.* Joseph Henry Press, Washington, D.C., pp. 447–465.

Mitchell, W.R., Presley, S., Czekala, N.M., and Lasley, B.L. (1982). Urinary immunoreactive estrogen and pregnanediol-3-glucuronide during the normal menstrual cycle of the female lowland gorilla (*Gorilla gorilla*). *Am. J. Primatol.* 2:167–175.

Mitchell, W.R., Lindburg, D.G., Shideler, S.E., Presley, S., and Lasley, B.L. (1985). Sexual behavior and urinary ovarian hormone concentrations during the lowland gorilla menstrual cycle. *Int. J. Primatol.* 6:161–172.

Monfort, S.L. (2003). Non-invasive endocrine measures of reproduction and stress in wild populations. In: Holt, W.V., Pickard, A.R., Rodger, J.C. and Wildt, D.E. (eds.), *Reproductive Science and Integrated Conservation.* Cambridge University Press, Cambridge, pp. 147–165.

Mrasek, K., Heller, A., Rubtsoz, N., Trifonov, V., Starke, H., Rocchi, M., Claussen, U., and Liehr, T. (2001). Reconstruction of the female *Gorilla gorilla* karyotype using 25-color FISH and multicolor banding (MCB). *Cytogenet. Cell Genet.* 93:242–248.

Mudakikwa, A.B., Cranfield, M.R., Sleeman, J.M., and Eilenberger, U. (2001). Clinical medicine, preventative health care and research on mountain gorillas in the Virunga Volcanoes region. In: Robbins, M. M., Sicotte, P., and Stewart, K. J. (eds.), *Mountain Gorillas: Three Decades of Research at Karisoke.* Cambridge University Press, Cambridge, pp. 341–360.

Munson, L. (2002). The living dead: Keeping wildlife alive through scientific use of biomaterials. In: Baer, C. K. (ed.), *Proceedings of the American Association of Zoo Veterinarians, October 5-10, 2002.* Milwaukee, Wisconsin, pp. 269–271.

Munson, L., and Karesh, W.B. (2002). Disease monitoring for the conservation of terrestrial animals. In: Aguirre, A.A., Ostfeld, R.S., Tabor, G.M., House, C., and Pearl, M.C. (eds.), *Conservation Medicine: Ecological Health in Practice.* Oxford University Press, Oxford, pp. 95–103.

O'Brien, J.K., Crichton, E.G., Evans, K.M., Schenk, J.L., Stojanov, T., Evans, G., Maxwell, W.M.C., and Loskutoff, N.M. (2002). Sex ratio modification using sperm sorting and assisted reproductive technology—a population management strategy.

Proc. of the Second International Symposium on Assisted Reproductive Technology (ART) for the Conservation and Genetic Management of Wildlife, pp. 224–231.

O'Brien, J.K., Stojanov, T., Crichton, E.G., Evans, K.M., Leigh, D., Maxwell, W.M., Evans, G., and Loskutoff, N.M. (2005). Flow cytometric sorting of fresh and frozen-thawed spermatozoa in the western lowland gorilla (*Gorilla gorilla gorilla*). *Am. J. Primatol.* 66:297–315.

Oates, J.F., McFarland, K.L., Groves, J.L., Bergl, R.A., Linder, J.M., and Disotell, T.R. (2003). The Cross River gorilla: Natural history and status of a neglected and critically endangered subspecies. In: Taylor, A.B. and Goldsmith, M.L. (eds.), *Gorilla Biology: A Multidisciplinary Perspective*. Cambridge University Press, Cambridge, pp. 472–497.

Oldham, M.C., and Geschwind, D.H. (2006). Grasping human transcriptome evolution: What does it all mean? *Heredity* 96:339–340.

Payne, R.C., Crompton, R.H., Isler, K. Savage, R., Vereecke, E.E., Gunther, M.M., Thorpe, S.K., and D'Aout, K. (2006). Morphological analysis of the hindlimb in apes and humans. I. Muscle architecture. *J. Anat.* 208:709–724.

Peel, A.J., Vogelnest, L., Finnigan, M., Grossfeldt, L., and O'Brien, J.K. (2005). Non-invasive fecal hormone analysis and behavioral observations for monitoring stress responses in captive western lowland gorillas (*Gorilla gorilla gorilla*). *Zoo Biol.* 24:431–445.

Pope, C.E., Dresser, B.L., Chin, N.W., Liu, J.H, Loskutoff, N.M., Behnke, E.J., Brown, C., McRae, M.A., Sinoway, C.E., Campbell, M.K., Cameron, K.N., Owens, O.M., Johnson, C. A., Evans, R.R., and Cedars, M.I. (1997). Birth of a western lowland gorilla (*Gorilla gorilla gorilla*) following *in vitro* fertilization and embryo transfer. *Am. J. Primatol.* 41:247–260.

Pourrut, X., Kumulungui, B., Wittmann, T., Moussavou, G., Délicat, A., Yaba, P., Nkoghe, D., Gonzalez, J.-P., and Leroy, E.M. (2005). The natural history of Ebola virus in Africa. *Microb. Infect.* 7:1005–1014.

Prendini, L., Hanner, R., and DeSalle, R. (2002). Obtaining, storing and archiving specimens for molecular genetic research. In: DeSalle, R., Giribet, G., and Wheeler, W. (eds.), *Techniques in Molecular Systematics and Evolution*. Birkhauser, Basel, pp. 176–248.

Preuschoft, H., Witte, H., and Witzel, U. (2002). Pneumatized spaces, sinuses and spongy bones in the skulls of primates. *Anthropol. Anz.* 60(10):67–79.

Ramsay, P.A., Boardman, W., MacDonald, B., Roberts, C., and Fraser, I.S. (2000). Chorionic villus sampling for sex determination in a western lowland gorilla (*Gorilla gorilla gorilla*). *J. Zoo Wildl. Med.* 31 (4):532–538.

Redmond, I. (1983). Summary of parasitological research, November 1976 to April 1978. In: Fossey, D. (ed.), *Gorillas in the Mist*. Houghton Mifflin, Boston, pp. 271–278.

Remis, M.J. (1997). Western lowland gorillas (*Gorilla gorilla gorilla*) as seasonal frugivores: Use of variable resources. *Am. J. Primatol.* 43:87–109.

Remis, M.J., and E.S. Dierenfeld. (2004). Digesta passage, digestibility and behavior in captive gorillas under two dietary regimes. *Int. J. Primatol.* 25:825–845.

Rideout, B.A. (2002). Creating and maintaining a postmortem biomaterials archive: Why you should do it and what's in it for you. In: Baer, C.K. (ed.), *Proceedings of the American Association of Zoo Veterinarians, October 5–10, 2002*. Milwaukee Wisconsin, pp. 272–274.

Robbins, M.M., and Czekala, N.M. (1997). A preliminary investigation of urinary testosterone and cortisol levels in wild male mountain gorillas. *Am. J. Primatol.* 43(1):51–64.

Roser, J., Czekala, N.M., Mortensen, R., and Lasley, B.L. (1986). Daily urinary hormone assays as a diagnostic tool to evaluate infertility in gorillas. *Biol. Reprod.* 34(Suppl. 1):131.

Rothschild, B.M., and Ruhli, F.J. (2005). Comparison of arthritis characteristics in lowland *Gorilla gorilla* and mountain *Gorilla beringei. Am. J. Primatol.* 66:205–218.

Rouquet, P., Froment, J.-M., Bermejo, M., Kilbourn, A., Karesh, W., Reed, P., Kumulungui, B., Yaba, P., Délicat, A., Rollin, P.E., and Leroy, E.M. (2005). Wild animal mortality monitoring and human Ebola outbreaks, Gabon and Republic of Congo, 2001–2003. *Emerg. Infect. Diseases* 11:283–290.

Ruvolo, M. (1997). Molecular phylogeny of the Hominoids: Inferences from multiple independent DNA sequence data sets. *Mol. Biol. Evol.* 14(3):248–265.

Ruvolo, M., Pan, D., Zehr, S., Goldberg, T., Disotell, T. R., and von Dornum, M. (1994). Gene trees and hominoid phylogeny. *Proc. Natl. Acad. Sci. USA* 91:8900–8904.

Ryder, O.A. (1986). Appendix B: The collection of samples for genetic analysis: Principles, protocols, and pragmatism. In: Benirschke, K. (ed.), *Primates: The Road to Self-Sustaining Populations.* Springer-Verlag, New York, pp. 1031–1036.

Ryder, O.A., McLaren, A., Brenner, S., Zhang, Y.-P., and Benirschke, K. (2000). DNA banks for endangered animal species. *Science* 288:275–277.

Ryder, O.A. (2002). Evaluating the importance of biomaterials banking: Converging interests and diversifying opportunities for conservation efforts. In: Baer, C.K. (ed.), *Proceedings of the American Association of Zoo Veterinarians, October 5–10, 2002.* Milwaukee Wisconsin, pp. 268.

Ryder, O.A. (2003). An introductory perspective: Gorilla systematics, taxonomy, and conservation in the era of genomics. In: Taylor, A. B. and Goldsmith, M. L. (eds.), *Gorilla Biology: A Multidisciplinary Perspective.* Cambridge University Press, Cambridge, pp. 239–246.

Ryder, O.A. (2005). Conservation genomics: Applying whole genome studies to species conservation efforts. *Cytogenet. Genome Res.* 108:6–15.

Ryder, O.A., McLaren, A., Brenner, S., Zhang, Y.-P., and Benirschke, K. (2000). DNA banks for endangered species. *Science* 288:275–277.

Saltonstall, K., Amato, G., and Powell, J. (1998). Mitochondrial DNA variability in Grauer's gorillas of Kahuzi-Biega National Park. *J. Hered.* 89:129–135.

Sarich, V.M. (1977). Rates, sample sizes, and the neutrality hypothesis for electrophoresis in evolutionary studies. *Nature* 265:24–28.

Sarmiento, E.E. (1994). Terrestrial traits in the hands and feet of gorillas. *Am. Mus. Novit.* 3091:1–56.

Sarmiento, E.E., Butynski, T.M., and Kalina, J. (1996). Gorillas of the Bwindi-Impenetrable Forest and the Virunga Volcanoes: Taxonomic implications of morphological and ecological differences. *Am. J. Primatol.* 40(1):1–21.

Sarmiento, E.E., and Oates, J.F. (2000). The Cross River gorillas: A distinct subspecies, *Gorilla gorilla diehli* Matschie 1904. *Am. Mus. Novit.* 3304:1–55.

Savage, T.S., and Wyman, J. (1847). Notice of the external characters and habits of *Troglodytes gorilla*, a new species of orang from the Gaboon River; Osteology of the same. *Boston Journal of Natural History* 5:417–442.

Schaller, G.B. (1963). *The Mountain Gorilla: Ecology and Behavior.* University of Chicago Press, Chicago.

Schmittbuhl, M., Le Minor, J.M., and Schaaf, A. (1996). Relative position and extent of the nasal and orbital openings in *Gorilla, Pan* and the human species from the study of their areas and centres of area. *Folia Primatol.* 67(4):182–192.

Schwartz, G.T. (2000). Taxonomic and functional aspects of the patterning of enamel thickness distribution in extant large-bodied hominoids. *Am. J. Phys. Anthropol.* 111(2):221–244.

Seager, S.W.J., Wildt, D.E., Schaffer, N., and Platz, C.C. (1982). Semen collection and evaluation in *Gorilla gorilla gorilla. Am. J. Primatol. Suppl.* 1:13.

Seuanez, H.N. (1986). Chromosomal and molecular characterization of the primates: Its relevance in the sustaining of primate populations. In: Benirschke, K. (ed.), *Primates: The Road to Self-Sustaining Populations.* Springer-Verlag, New York, pp. 887–910.

Seuanez, H.N., Carothers, A.D., Martin, D.E., and Short, R.V. (1977). Morphological abnormalities in spermatozoa of man and great apes. *Nature* 270:345–347.

Sheldon, F.M., and Dittman, D.L. (1997). The value of vertebrate tissue collections in applied and basic science. In: Hoagland, K.E., and Rossman, A.Y. (eds.), *Global Genetic Resources: Access, Ownership, and Intellectual Property Rights.* Association of Systematics Collections, Washington, D.C., pp. 151–164.

Sherwood, R.J. (1999). Pneumatic processes in the temporal bone of chimpanzee (*Pan troglodytes*) and gorilla (*Gorilla gorilla*). *J. Morphol.* 241(2):127–137.

Shimizu, K., Udono, T. Tanaka, C., Narushima, E., Yoshihara, M., Takeda, M., Tanahashi, A., van Elsackar, L., Hayashi, M., and Takenaka, O. (2003). Comparative study of urinary reproductive hormones in great apes. *Primates* 44:183–190.

Sibley, C.G., and Ahlquist, J.E. (1987). DNA hybridization of evidence of hominoid phylogeny: Results from an expanded data set. *J. Mol. Evol.* 26:99–121.

Sikela, J.M. (2006). The jewels of our genome: The search for the genomic changes underlying the evolutionarily unique capacities of the human brain. *PLoS Genet.* 2(5):e80.

Sleeman, J.M., and Mudakikwa, A.B. (1998). Analysis of urine from free-ranging mountain gorillas (*Gorilla gorilla beringei*) for normal physiological values. *J. Zoo Wildl. Med.* 29(4):432–434.

Sleeman, J.M., Meader, L.L., Mudakikwa, A.B., Foster, J.W., and Patton, S. (2000). Gastrointestinal parasites of mountain gorillas (*Gorilla gorilla beringei*) in the Parc National Des Volcans, Rwanda. *J. Zoo Wildl. Med.* 31(3):322–328.

Smith, S., and Morin, P.A. (2005). Optimal storage conditions for highly dilute DNA samples: A role for trehalose as a preserving agent. *J. Forensic Sci.* 50:1101–1108.

Socha, W.W., and Moor-Jankowski, J. (1986). Blood groups of apes and monkeys. In: Benirschke, K. (ed.), *Primates: The Road to Self-Sustaining Populations.* Springer-Verlag, New York, pp. 921–932.

Stoinski, T.S., Czekala, N., Lukas, K.E., and Maple, T.L. (2002). Urinary androgen and corticoid levels in captive, male western lowland gorillas (*Gorilla g. gorilla*): Age- and social group-related differences. *Am. J. Primatol.* 56(2):73–87.

Stone, A. (2003). IPBIR update for AAPA. *Physical Anthropology* 4(1):2.

Stumpf, R.M., Polk, J.D., Oates, J.F., Jungers, W.L., Heesy, C.P., Groves, C.P., and Fleagle, J.G. (2003). Patterns of diversity in gorilla cranial morphology. In: Taylor, A.B., and Goldsmith, M.L. (eds.), *Gorilla Biology: A Multidisciplinary Perspective.* Cambridge University Press, Cambridge, pp. 35–61.

Taylor, A.B. (1997). Scapula form and biomechanics in gorillas. *J. Hum. Evol.* 33(5):529–553.

Taylor, A.B. (2003). Ontogeny and function of the masticatory complex in *Gorilla*: Functional, evolutionary, and taxonomic implications. In: Taylor, A.B., and Goldsmith, M.L. (eds.), *Gorilla Biology: A Multidisciplinary Perspective.* Cambridge University Press, Cambridge, pp. 132–193.

Tribe, A., Butler, R., Butler, C., McBain, J., Martin, M., Galloway, D., and Moriarty, K. (1989) Artificial insemination of gorillas at Melbourne Zoo. *Proc. Symp. Fertility in the Great Apes*, Atlanta, Georgia, pp. 45–46.

Uchida, A. (1998). Variation in tooth morphology of *Gorilla gorilla. J. Hum. Evol.* 34(1):55–70.

Vigilant, L., Van Neer, W., Siedel, H., and Hofreiter, M. (2002). Gorillas then and now: Genetic analysis of museum specimens as a way to examine temporal changes in the distribution and diversity of gorilla populations [Abstract]. In: *Caring for primates: Abstracts of the XIXth Congress of the International Primatological Society*; Beijing: Mammalogical Society of China. pp. 47–48.

Vigilant, L., and Bradley, B.J. (2004). Genetic variation in gorillas. *Am. J. Primatol.* 64:161–172.

Waits, L.P., and Paetkau, D. (2005). Noninvasive genetic sampling tools for wildlife biologists: A review of applications and recommendations for accurate data collection. *J. Wildl. Manage.* 69:1419–1433.

Wasser, S.K., Hunt, K.E., and Clarke, C.M. (2002). Assessing stress and population genetics through noninvasive means. In: Aguirre, A.A., Ostfeld, R.S., Tabor, G.M., House, C., and Pearl, M.C. (eds.), *Conservation Medicine: Ecological Health in Practice*. Oxford University Press, Oxford, pp. 130–144.

Whitten, P.L., Brockman, D.K., and Stavisky, R.C. (1998). Recent advances in noninvasive techniques to monitor hormone-behavior interactions. *Am. J. Primatol.* 107:1–23.

Whittier, C.A., Horne, W., Slenning, B., Loomis, M., and Stoskopf, M.K. (2004). Comparison of storage methods for reverse-transcriptase PCR amplification of rotavirus RNA from gorilla (*Gorilla g. gorilla*) fecal samples. *J. Virol. Methods* 116:11–17.

Wiener, A.S., Socha, W.W., Arons, E.B., Mortelmans, J., and Moor-Jankowski, J. (1976). Blood groups of gorillas: Further observations. *J. Med. Primatol.* 5:317–320.

Wildt, D.E. (1996). Male reproduction: Assessment, management, and control of fertility. In: Kleiman, D.G., Allen, M.E., Thompson, K.V., and Lumpkin, S. (eds.), *Wild Mammals in Captivity: Principles and Techniques*. University of Chicago Press, Chicago, pp. 429–450.

Wildt, D.E., Rall, W.F., Critser, J.K., Monfort, S.L., and Seal., U.S. (1997). Genome resource banks: Living collections for biodiversity conservation. *Bioscience* 47(10):689–698.

Willoughby, D.P. (1950). The gorilla—largest living primate. *Scientific Monthly* 70(1):48–57.

Yamagiwa, J., Basabose, K., Kaleme, K., and Yumoto, T. (2003). Within-group feeding competition and sociological factors influencing social organization of gorillas in the Kahuzi-Biega National Park, Democratic Republic of Congo. In: Taylor, A.B., and Goldsmith, M.L. (eds.), *Gorilla Biology: A Multidisciplinary Perspective*. Cambridge University Press, Cambridge, pp. 328–357.

Yamagiwa, J., Basabose, A. K., Kaleme, K., and Yumoto, T. (2005). Diet of Grauer's gorillas in the montane forest of Kahuzi, Democratic Republic of Congo. *Int. J. Primatol.* 26:1345–1373.

Section 4
Approaches—Building Regional and International Alliances

Chapter 13
Transboundary Conservation in the Virunga-Bwindi Region

Annette Lanjouw

1. Introduction

The Virunga Volcano massif and Bwindi Impenetrable Forest are two forest blocks found in the Albertine Rift, the western branch of the Great Rift Valley. This area used to form an extensive forest massif, which has slowly been eroded by human use, encroachment, and accelerated deforestation during the 20th century and has resulted in a number of fragmented islands of forest separated by large expanses of agricultural and pastoral land.

The two forest blocks make up the last remaining habitat of the mountain gorilla (*Gorilla beringei beringei*). Based on the 2006 census undertaken in the Bwindi Impenetrable National Park, and the census undertaken in the Virunga Massif in November 2003, the total number of mountain gorillas is now estimated at approximately 720 (for the purposes of this chapter the term "mountain gorilla" will be used to refer to both the Bwindi and Virunga populations). These two forest blocks are rich in species diversity, with many endemics as well as rare and threatened species. The expanse of high altitude forest plays an important water-catchment function and ensures the stability of soils on the cultivated slopes at lower altitude. As a consequence, these forests are not only important in their role of providing habitat for wildlife, but also for the maintenance of the ecological processes necessary for the agricultural livelihoods of the people in this region. Most international conservation organizations rate the mountain forests of the Albertine Rift in the highest priority for conservation in Africa (Hamilton, 1996).

The forest habitat of the mountain gorillas is composed of two separate ecological units. The Bwindi Impenetrable Forest is located primarily in Uganda, with a small portion crossing the border into DR Congo

(Sarambwe Forest). The Virunga forest massif is bisected by the international boundaries between Rwanda, DR Congo, and Uganda, and is composed of three contiguous but individually managed national parks. Since the ecological processes within each of these units are continuous, effective management and conservation requires collaboration among the countries sharing them. An activity or event on one side of the border will generally have an impact across the entire unit. The threats to the forest and its wildlife are primarily of human origin and relate to the high human population density in the region, the subsistence livelihoods and heavy reliance on natural resources by a population living in poverty. In addition, the conflict in the region, weak governance structures and political/economic instability have contributed to the unsustainable use of resources by people affected by and involved in conflict.

Transboundary Conservation and Development Initiatives have recently been defined (Braak *et al.*, 2004) as a process of cooperation across boundaries where at least one of the primary objectives includes the protection and maintenance of biodiversity and ecological processes, and whereby the achievement of the biodiversity objectives depends on cooperation across jurisdictional boundaries, involving a range of stakeholders. In the Virunga-Bwindi region, the work toward implementing a strategy for transboundary conservation and development began in earnest in the early 1990s, through the work of the International Gorilla Conservation Programme. Prior to that date, a traditional approach of national-level conservation and ecosystem management was applied, focusing only on each portion of the ecosystem under the sovereign rule of the respective countries.

2. Transboundary Conservation and Development: Phases and Emphasis

The strategy adopted by the protected area authorities in the three countries focused on a bottom-up approach, building on practical collaboration at the field level and moving the process of cooperation over time along increasingly sophisticated and formal structures. This strategy was adopted partly as a result of the existing political climate in the region, which limited communication between the countries. It was also a result of the explicit recognition of the need to have, from the onset, the active involvement and buy-in from the people essential to transboundary collaboration: local people and field-based protected area staff. Building on the demonstrated benefits and impacts of increasing collaboration, the strategy could then progressively involve higher levels of decision-makers. Transboundary conservation and development in the region is being led by the authorities of Rwanda (Office Rwandais de Tourisme et des Parcs Nationaux-ORTPN), DR Congo (Institut Congolais pour la Conservation de la Nature-ICCN), and Uganda (Uganda Wildlife Authority-UWA), and is facilitated by the International Gorilla Conservation Programme.[1]

Effective conservation of the shared ecosystem was the primary goal for transboundary conservation in each of the three countries. Many other objectives can also be included, but the principal one was conservation. The habitat and wildlife have a key role to play in the economic development of the region, through the generation of income (tourism and wildlife-based enterprise) as well as through the conservation of soil and watershed. In order to achieve this goal, the countries must

1. Understand the threats to the ecosystem and use this information to guide management
2. Have the capacity to deal with the threats to protect and effectively manage the parks
3. Ensure that local people benefit from the forest and that decision-makers understand the value of the forest

A phased approach was adopted to implement the strategy for regional collaboration, along a continuum moving from no collaboration to the creation of a transboundary protected area where all investments and benefits are shared.

Phase I on this continuum focuses on the agreement to harmonize and coordinate management of the parks, as well as the development of field-based mechanisms for collaboration. Examples of activities undertaken during this phase include the implementation of joint law-enforcement patrols by rangers of two or more park management authorities along border areas; adopting similar protocols to collect data on the use of forest resources and pooling the data for regional analysis; joint training courses for park staff from more than one country; development of a regional vision and strategy for the shared habitat (adopted in 2001 by the three countries); quarterly meetings by park staff to discuss issues of regional concern and sharing information.

Phase II focuses on the formal adoption of the regional strategy and development of a regional management plan, with resources allocated by each of the countries to the implementation of activities harmonized at a regional level (monitoring, law-enforcement, training, communication, planning, data sharing, etc). In each country, staffs have the responsibility to ensure that the regional agreements and processes are adhered to and that activities are implemented according to the plan.

---

[1] The International Gorilla Conservation Programme (IGCP) is a program of the African Wildlife Foundation (AWF), Fauna and Flora International (FFI), and WorldWide Fund for Nature (WWF), established in 1991, based on these organizations' previous collaboration in the Mountain Gorilla Project.

Phase III includes the formal agreement to manage the shared ecosystem as one unit, with a single regional management plan and secretariat. The national park management plans will form specific parts of the regional management plan. It is not an *a priori* condition that the ecosystem be managed as a single national park with a single budget and shared benefits. It is, however, understood that the contiguous parks will form part of a unit that the three countries manage in a fully collaborative manner. International recognition of the status of the Virunga forest block as a shared management unit with World Heritage Status will be sought from the UNESCO World Heritage Centre. Phase II differs from Phase III in that the former seeks to formalize approaches that are regionally harmonized, whereas the latter formalizes the status of the site as one regional unit managed in a collaborative manner. Phase III involves the signing of a formal agreement among the three governments, establishing a Transboundary Protected Area for Conservation and Development (adopted in 2006). It is understood, however, that the phases are steps along a continuum and that the divisions between one step and another are not defined in absolute terms.

Informal collaboration in Phase I can achieve all the benefits that more formal structures can achieve. Yet the formalization of field-based coordination and collaboration is necessary in order to ensure that the principles are institutionalized and not dependent on individuals who know and trust one another. In order to provide both the structure and principles for sustained collaboration over time and through changing political and economic circumstances, the processes and activities involved in regional collaboration must be included in strategic and operational planning, and time and other resources must be allocated to these activities. The formal agreements, however, are dependent on a minimum level of political entente among the official governments of the three countries, and this has been a major constraint in the region for the past 10 years.

Although the transborder protected areas have yet to be formed officially, the work of the last decade—collaborating with the protected area authorities, strengthening their ability to effectively manage the protected areas, and demonstrating the potential economic as well as ecological value of the forest—has increased the importance attributed to environmental issues. Due to the emphasis on informal, field-based mechanisms for collaboration, the political tensions in the region have not impeded regional collaboration throughout the past 10 years of conflict, and this collaboration has strengthened the impact of environmental activities.

2.1. Mechanisms Established for Transboundary Collaboration

Three primary mechanisms were used to implement the conservation goals highlighted above. These mechanisms include communication, monitoring and management, and economic development.

2.1.1. Regional-Level Communication, Planning and Cooperation in Collaborative Activities

One of the key tools for collaboration is communication. For this reason the transboundary program emphasizes open communication between the three protected area authorities. Specific mechanisms have been developed to enable a regular exchange of information and joint planning:

- *Regional Meetings*—These quarterly meetings bring together key protected area personnel of the three countries as well as technical partners (Lanjouw et al., 2001). The regional meetings allow the exchange of information relevant to the conservation of these ecosystems and stimulate discussions on specific issues. In order to maintain neutrality, each country hosts a meeting on a rotational basis.
- *Wardens Coordination Meetings*—One of the many products of the regional meetings was the development of more regular meetings between the wardens of the four protected areas to once again tackle technical issues but in a more focused environment. Several joint patrols and cross visits have been implemented as a result of these meetings.
- *Joint Patrols*—With the support of IGCP, the protected area authorities (PAA) regularly conduct joint surveillance and antipoaching patrols. During the patrols, Park staffs come together to patrol border areas to share information and logistics (Lanjouw et al., 2001). They work as a team, and the patrols tend to be successful due to increased patrolling of border areas and exchange of relevant information.
- *Cross Visits*—Because of the differing political and economic situation between the three countries, IGCP encourages the PAA staff to visit sites in the neighboring countries to gain a better understanding of the different challenges faced by the respective PAA towards the protection and conservation of the mountain gorilla habitat.
- *Gorilla Census*—Censuses of the gorilla populations in Bwindi and the Virungas have involved staff from the parks authorities in the three countries and many conservation organizations working together (Lanjouw et al., 2001). Not only did this result in training the park staff of all three countries but it also strengthened the regional links between them.

2.1.2. Regional-Level Ecological Monitoring and Management

The foundation for effective management and conservation of the forest is a strong understanding of the threats to the forest and the needs of key species within that forest. These threats change over time and, for this reason, a three-pronged information gathering and monitoring program was developed to inform park management. This Regional Information System (RIS) was established to be used by the three PAAs responsible for the management of the Virunga-Bwindi forests.

The first component of RIS, ranger-based monitoring (RBM), was initiated in 1996. The objective of the program is regular monitoring of the forest by park rangers, in order to understand the illegal human use of the habitat (poaching, woodcutting, etc.), ecological processes in the forest, and distribution and habitat-use of specific key species (including the mountain gorilla). The monitoring feeds directly into the day-to-day management of the park and enables surveillance and specific interventions to be based on solid data. This can include where to send patrols, based on activities of poachers, availability of seasonal resources and presence of snares. It can also include the movements of key species, such as the gorillas and their use of the habitat. The RBM has produced effective field maps for the park staff and patrolling rangers, using topographic features and toponyms. At present, the data are being analyzed in each park, as well as at the headquarters of the PAAs. A centralized, regional database is also being developed so that the data will be available for the entire ecosystem, thus allowing park staff to manage the forest as one ecological unit.

The second RIS component involves data collection on socioeconomic factors outside the forest, such as number of people living near the forest, their distance from the forest, their livelihood activities, and their needs and dependence on resources from the forest. These data are analyzed in conjunction with the illegal activities in the forest to gain a better under-standing of the threats to the forest and its wildlife.

The third RIS component is remote sensing of the gorilla habitat to detect changes in vegetation cover and land-use over time. This is being conducted in conjunction with the European Space Agency and World Heritage Center of UNESCO. Through satellite imagery, the changes in forest cover and effects of the human population can be measured, and understood in conjunction with the socioeconomic and ecological data.

A final approach in addition to the RIS is joint surveillance and antipoaching patrols conducted by the PAAs with the support of IGCP. In the joint patrols, the staffs of contiguous parks come together to patrol the border areas to share information and logistics and to work as a team. These border areas are often very vulnerable and insecure and recently have involved the military of all three countries, thus bringing together not only park staff, but also military staff from across the borders. The 1998 census of the gorillas at Bwindi Impenetrable National Park involved not only staff from the Uganda Wildlife Authority, but also staff of the ORTPN from Rwanda and the ICCN from DR Congo. Again, the objective was training of park staff in all three countries as well as strengthening the regional links among them.

2.1.3. Economic Mechanisms

Northwestern Rwanda, eastern DR Congo, and southwestern Uganda have a large proportion of the population living below the poverty line, with insufficient land to meet their most basic needs (Waller, 1996). Very few alternatives exist to subsistence agriculture on steep slopes and plots that are too small to feed the average family. Numerous efforts have been made, and consultant hours spent, searching for alternatives for the local people in this region. Tourism, and more specifically, nature-tourism, offers one of the few viable options. Although a fragile industry, easily affected by political, economic, and social changes, tourism nonetheless poses real economic potential for the region.

The risk of tourism, however, to both the mountain gorillas and their habitat is also considerable (MGVP/WCS, Chapter 2, and Litchfield, Chapter 4, this volume). The potential of transmission of diseases from humans to gorillas, thus possibly infecting the entire population, poses one of the greatest threats to gorilla conservation (Homsy, 1999). Transmission of diseases is not only a potential risk between tourists or researchers and gorillas, however. It is equally, if not more likely between gorillas and poachers, local farmers, harvesters of natural resources, park staff, military and rebels. Efforts to sensitize some of these groups are underway by conservation organizations in the region (IGCP, Institute of Tropical Forest Conservation, Dian Fossey Gorilla Fund International, and Mountain Gorilla Veterinary Project). Public health and conservation workshops have been organized in villages near the forest to teach people some of the basic principles of disease transmission, hygiene and waste disposal, thereby benefiting communities as well as potentially reducing the risks of disease transmission to wildlife. Health education programs have been provided to park staff and local communities, and medical support has been made available for certain highly contagious diseases (e.g., HIV AIDS). The impact of these programs will have to be monitored over time, to determine whether behavioral changes have occurred and the risks of disease transmission are reduced.

The park authorities have worked together to harmonize tourism by establishing common rules to manage and control tourism in all three countries. These rules focus on reducing the risks of disease transmission, overexploitation of the gorillas for tourism, and reducing the stress to the gorillas. Having the same rules at each tourism site will strengthen collaboration and reduce competition among the three countries and will ensure that visitors to multiple sites receive consistent messages regarding their behavior and visit. Common approaches are also being applied with respect to interpretation and development of joint messages for conservation, handling procedures, and training for tourism staff. These activities are forming a fledgling regional tourism approach that will ensure consistent communication between sites once peace and stability returns to the region.

To spread the economic benefits of tourism to the parishes (the smallest administrative unit applied in Uganda and Rwanda), the countries have worked with their conservation and development partners to develop tourism-linked enterprises as well as more generally conservation-based enterprises for the local communities. A Regional Enterprise Forum was established, bringing communities from all sides of the border together to share experiences and expertise in enterprise development, as well as to actively involve them in transboundary collaboration. Enterprise activities include agriculturally based and tourism-related activities (tourism facilities, marketing of crafts and nature-based products such as honey and candles, and cultural activities). Tourism accommodation facilities, owned and managed by local communities, are now available in both Rwanda and Uganda, in partnership with the private sector. These partnerships built capacity within the local communities for management and tourism operations, and enabled local communities to access capital investments that would otherwise not have been available to them. Although the scale of these interventions has been small, due to the limited number of permits available for gorilla tourism in each park, it has made direct economic benefits to local communities from gorilla tourism possible.

3. Achievements and Effects

The Virunga-Bwindi region is still in the midst of acute political turmoil and conflict. This is leading to severe pressures on the environment through the unsustainable exploitation of natural resources and breakdown of social and economic structures that had been protecting the environment. Despite this, the transboundary conservation program has made considerable progress. This includes:

1. Bilateral and trilateral meetings between the PAAs of the three countries and the four parks
2. Communication network and system for regular information exchange between the three countries (regional meetings, warden's coordination meetings, etc.)
3. Regular joint law-enforcement and monitoring patrols between field-based park staff of DRC, Uganda and Rwanda
4. Improved understanding of the habitat, priorities for conservation and threats, through the Regional Information System. This includes data on human use of the forest, key species in the forest, socioeconomic indicators outside the forest, and habitat change through remote sensing data
5. Capacity building of PAA staff
6. Improved communication and relations between the three countries
7. Increase in flow of benefits to local communities and involvement of local people in enterprises linked to conservation.

4. Lessons Learned

Looking at the transboundary work that was implemented in the Virunga-Bwindi region, and placing it in the political and social context of the region, a number of lessons can be identified. Most of the transboundary natural resource management lessons cannot be examined in isolation from the context of conflict of the region, however. A great deal of overlap therefore exists between lessons learned on the potential and importance of focusing on conservation during conflict (Cairns, 1997; Lanjouw, 2000), and the potential and experience in transboundary natural resource management (Kalpers and Lanjouw, 1997; Muruthi et al., 2000).

4.1. Top Down Versus Bottom Up

One of the greatest difficulties with transboundary conservation and development initiatives across the globe is that they tend to be imposed by governments or leaders onto the communities and people that need to cooperate. This has resulted in what could become a reversal in the trend towards building on community governance of conservation and development initiatives, by bringing wildlife conservation back to an activity imposed by centralized government on local people, without taking into consideration their needs and practices. The challenge in transboundary initiatives is involving local stakeholders from the onset, in the design, management and implementation of transboundary efforts and ensuring that the benefits of collaboration flow to the local people. The "bottom-up" approach applied in the Virunga-Bwindi region has avoided many of the governance tensions between an "authoritarian/protectionist" approach versus a more locally empowered and decentralized approach. By involving local stakeholders, many of the negative impacts frequently associated with transboundary initiatives, such as increasing human-wildlife conflict, have been avoided. The Regional Enterprise Forum, involving conservation enterprise activities on all sides of the border, has served to bring communities together and to involve them in transboundary collaboration.

4.2. Transboundary Conservation as a Continuum of Strategies

Effective conservation involves the abatement of threats to natural resources, ecosystems or species. When those threats come from more than one side of a border, it is necessary to focus on threat abatement at a regional level. Given the sovereignty of nations, this requires coordination and, where possible, collaboration on conservation activities. The stronger the ability and willingness to coordinate and collaborate, the more effective the conservation will be.

At one end of the continuum, efforts can be made to make conservation approaches in each country harmonious, or nonconflicting. At the other end is full transfrontier management of one shared ecosystem, or a formal Transboundary Protected Area. When strategies move along the continuum toward increased collaboration, the habitat can be managed more effectively. The more people, institutions, and sectors are involved, however, the more difficult and complicated conservation becomes. For this reason, it is not always possible, or even desirable, to establish full regional management of an area as one shared unit. Such political level involvement can delay, or even impede effective collaboration on the ground. Effective transboundary natural resource management must be argued as the combination of strategies along the continuum that has the optimum net gain in terms of positive conservation outcomes, relative to costs. It needs to be flexible over time and evolve based on needs and opportunities. The development of human and institutional capacity is a critical emphasis. Collaboration across borders only happens among people, either as individuals or as members of institutions. To collaborate effectively, a basic level of trust and understanding is required. In addition, the institutions need to be strong enough to be able to coordinate their activities with others. To be able to accomplish this, it is critical to build organizational capacity and to develop a clear understanding of the issues involved. Once mechanisms for effective coordination have been developed, and institutionalized, collaboration becomes routine.

There has been much discussion about the inefficacy of "paper parks," which have formally designated protection but no effective management or protection on the ground. Unless these areas are effectively managed on the ground, the formal designation of a "park" has little effect on conservation and natural resource management. Working out the complex mechanisms, both institutional and personal, that make collaboration work on the ground is the fundamental basis of effective conservation. Once established and implemented by all parties, formalization of these mechanisms and relationships is often a much simpler process. By involving the many stakeholders on the ground, and ensuring that their needs are being met, transboundary processes can become sustainable.

Transboundary collaboration is a process rather than a goal. The formal designation of a transboundary park is not what will make collaboration take place—it is the process of working together, of communicating and coordinating activities, developing joint plans and implementing joint or coordinated activities. The objectives attained through this process are building a framework for collaboration, involving people from all three countries in this process, and making sure that objectives are perceived as shared.

4.3. Measuring the Impact of Transboundary Collaboration

A large range of objectives have been identified for transboundary conservation and development initiatives, ranging from conservation to building peace and reducing poverty. Although many of these may be applicable, it is important to have explicit objectives identified in order to monitor the impact of the initiative over time. It is also important to differentiate between national-level objectives and transboundary objectives. Not all the objectives can be compatible for all activities. For example, certain measures taken for conservation purposes (e.g., limiting the use of forest resources by local users, in order to reduce stress to vulnerable populations in the ecosystem) can go against the development objectives, where it is desirable to have all people access resources. In the Virunga-Bwindi case, the objective for transboundary collaboration was primarily for conservation of the forest habitat and the endangered populations of mountain gorillas. The impact of the transboundary initiative on those objectives can be clearly measured. In addition, secondary objectives that include poverty alleviation at a local scale and contributing to the process of building peace and reducing conflict can also be measured, along specific criteria. The effort of involving local communities in conservation decision-making, empowering them in negotiating private sector partnerships and developing conservation-related enterprises were primarily national in nature, rather than transboundary. Efforts were made to learn from each other, build a regional network, and harmonize approaches so that communities can operate on an equal basis.

References

Braack, L. et al., (2004). In: *Beyond Boundaries: Enabling Effective Transboundary Conservation*, Sandwith, T., and Besançon, C. (eds). IUCN.

Cairns, E. (1997). *A Safer Future: Reducing the Human Cost of War*. Oxfam Publications, London.

Hamilton, L.S. (1996). The role of protected areas in sustainable mountain development. *Parks*, 6:2–13.

Homsy, J. (1999). Ape tourism and human diseases: how close should we get? A critical review of rules and regulations governing park management and tourism for the wild mountain gorilla, *Gorilla gorilla beringei*. Consultancy for the International Gorilla Conservation Programme.

Kalpers, J., and Lanjouw, A. (1997). Potential for the Creation of a Peace Park in the Virunga Volcano Region. *PARKS: The International Journal for Protected Area Managers* 7(3):25–35.

Lanjouw, A. (2000). Building partnerships in the face of political and armed crisis. New Haven, Yale University, School of Forestry and Environmental Studies. New perspectives on conservation in periods of armed conflict. Conference Proceeding.

Lanjouw, A., Kayitare, A., Rainer, H., Rutagarama, E., Sivha, M., Asuma, S., and Kalpers, J. (2001). Beyond Boundaries: Transboundary Natural Resource Management for Mountain Gorillas in the Virunga-Bwindi Region. Biodiversity Support Program, Washington, D.C. pp. 78.

Muruthi, P., Soorae, P., Moss, C., Stanley-Price, M., and Lanjouw, A. (2000). Conservation of large mammals in Africa. What lessons and challenges for the future? In: Entwistle, A., and Dunstone, N. (eds.). *Priorities for the Conservation of Mammalian Diversity: Has the Panda Had Its Day?* Cambridge University Press, Cambridge, pp. 207–219.

Waller, D. (1996). *Rwanda: Which Way Now? An Oxfam Country Profile*, Oxfam, Oxford, U.K.

Chapter 14
The Great Ape World Heritage Species Project

Richard W. Wrangham, Gali Hagel, Mark Leighton,
Andrew J. Marshall, Paul Waldau, and Toshisada Nishida

1. Introduction

The mission of the Great Ape World Heritage Species Project is to offer a new way to help avert the extinction crisis that currently faces chimpanzees, bonobos, gorillas, and orangutans, and in so doing to assist the plight of these apes in captivity also.

We believe that a higher international profile for the great apes is necessary if they are to survive in the wild. Our goal is therefore to launch a collaboration that will lead to designating the great apes as World Heritage Species. This designation of World Heritage Species would denote a new internationally protected category of species. The essential notion of World Heritage Species status is that any species so named would be recognized to be of outstanding universal value, and to need special help if they are to be conserved in the wild. Outstanding universal value is the operational criterion for nominations to the World Heritage Convention, so designation of World Heritage Species might be through a protocol to this convention. The great apes would be the first set of species to be so named. Others would be expected to follow.

We consider that the designation of great apes as World Heritage Species would advance their conservation by accelerating international cooperation in three main ways, signified by *attention, resources,* and *mechanism.*

Attention means elevating awareness of the value and plight of great apes, particularly among political leaders.

Resources means increasing the resources needed to help the great apes, especially by tapping into the worldwide interest in great ape welfare as a result of their unique relationship with humans.

Mechanism means creating a new international mechanism for organizing great ape conservation in the wild, given that no such mechanism currently exists.

The Great Ape World Heritage Species Project (GAWHSP) was initiated in January 2001 with the appointment by the International Primatological Society of an Ad-hoc Committee for the World Heritage Status for the Great

Apes (Anon, 2001). Toshisada Nishida proposed the project, and was appointed as the first chair of GAWHSP, a position that he continues to hold. Richard Wrangham has acted as co-chair since August 2002.

Since August 2002, GAWHSP has been an independent international initiative, with activists united through email and occasional meetings in Japan, the United States, Europe, and Africa. Key participants and supporters have been the International Primatological Society, the Chimpanzee Collaboratory (initiated and funded by the Glaser Progress Foundation), the Primate Society of Japan, the Wild Chimpanzee Foundation, SAGA, Japan (Support for African/Asian Great Apes), the Great Ape Action Group and the Great Ape World Heritage Species Project, Inc. This chapter summarizes the rationale for GAWHSP, its development over its first two years, and its prospects for promoting great ape conservation. [Note that three years have elapsed since this chapter was accepted for publication in this volume, and significant developments have occurred to further efforts for collaborative international great ape conservation. These chiefly concern the evolution of GRASP, the Great Ape Survival Project Partnership established under a joint UNEP and UNESCO Secretariat (www.unep.org/grasp). We provide an addendum at the end of this chapter to update readers of relevant developments while preserving the historical time frame of this chapter.]

2. The Severity of the Problem

Currently, six species of nonhuman great ape are recognized: Sumatran and Bornean orangutans, eastern and western gorillas, chimpanzees, and bonobos. Predictions of great ape extinctions began at least as early as 1867 (Darwin, 1871). Pessimistic forecasts have subsequently been common because of the great apes' slow reproductive rates, need for large areas, and competition with humans over habitat. Reliable data on the severity of the crisis are elusive, however, because great ape population densities are difficult to measure. Estimates therefore come from indirect data such as the predicted rates of forest loss, calculations of losses from hunting, and occasional detailed counting of nests in a few key areas. Frequent conclusions from such methods are that without dramatic changes to current conservation strategy, global extinctions of great ape species will start during the present century (e.g., Rijksen and Meijaard, 1999; Nishida et al., 2001; van Schaik et al., 2001). The Sumatran orangutan will probably go first. Recent survey work suggests that there are currently only 7,500 orangutans remaining on Sumatra, and that by 2010 they will become the first ape species to be functionally extinct in the wild (Wich et al., 2003, Singleton et al., 2004). Some estimates suggest that chimpanzees in central and eastern Africa are the only great ape that is likely to survive in the wild to 2100, and even then in much diminished numbers (Nishida et al., 2001).

The problem is acute because almost all great ape populations need large expanses of fruit-rich forest. These habitats are in steep decline throughout the tropics as a result both of conversion to agriculture and of logging. The effects of logging on ape populations vary with the intensity of timber extraction. Light to moderate selective logging need not completely destroy ape habitat, and most evidence suggests that ape populations can be maintained at somewhat reduced densities in degraded habitats (e.g., Rijksen and Meijaard, 1999; Felton et al., 2003). However, as apes have long lifespans and slow reproductive rates, the long-term effects of habitat degradation on individual fitness, and therefore ultimately population viability, are difficult to assess. Nevertheless, it is reasonable to assume that habitat degradation will lower female fecundity and lead to additional time-delayed but deterministic population declines ("extinction debt" sensu Tilman, 1994). Furthermore, many logging operations are accompanied by collateral damage that endangers ape populations even more gravely than does the timber extraction itself. For example, unsustainable levels of hunting and elevated transmission rates of epidemic diseases associated with logging operations will likely result in the local extinction of several ape populations in Africa (Rose, 1998; Wilkie and Carpenter, 1999; Peterson, 2003).

If the crisis itself is not surprising, it has nevertheless emerged into the consciousness of the primatological community with surprising suddenness during the 1990s. Until that time, particular populations such as the Virunga gorillas were famously under threat and were the subject of major conservation efforts. The change during the last decade is that over most of their ranges, it has now become clear that the majority (rather than a select minority) of great ape populations are rapidly losing numbers and habitat (Beck et al., 2001).

In spite of the newly appreciated scale of the problem, attempts to solve it have followed traditional paths. Thus, they have been directed largely toward particular populations or areas that happen to be of interest to specific supporters or donors (e.g., the Virunga gorillas, Tanjung Puting orangutans, or National Parks and Reserves such as Taï, Mahale, Korup, and others). These local efforts have had important successes. For example, the Virunga gorilla population has risen in number steadily since the 1970s and continues to flourish despite occasional episodes of disease and poaching (Robbins et al., 2001).

More often, however, they have failed. Even some of the best-known great ape populations have suffered heavily. Logging has advanced rapidly in the key orangutan habitats of Tanjung Puting and Gunung Palung in Borneo, despite strong protests (C. Knott, personal communication). There has been severe population loss of gorillas in Kahuzi-Biega, Democratic Republic of Congo (DRC) (J. Yamagiwa, personal communication). In Gombe, Tanzania, only one community of chimpanzees (the research and tourism community of Kasekela) appears viable (A. Pusey, personal communication). Poaching has begun in the longest-studied bonobo community, at Wamba in DRC (T. Kano, personal communication).

It might be argued that some such reversals are bound to happen, given that there are many populations of great apes. But the emerging picture does not support such a comforting view.

Instead, we must reluctantly conclude that the current strategy is a failure. Throughout their range in the wild, great ape populations are plummeting (Nishida *et al.*, 2001). Unless something drastic reverses the trend, they are doomed to frequent national extinctions, which for some subspecies will likely become worldwide during the 21st century.

3. The High Value of Apes

The four species of great ape are unique among animals in their human-like characteristics, including their emotional lives, mental abilities, and genetic make-up. This phenomenon is readily recognized by untrained people who spend time with great apes in the wild or in captivity. At a scientific level, advances in genetics, comparative psychology, and ethology mean that with every decade this close proximity of humans to the great apes has become more vivid. As a result, the great apes are widely thought of as a kind of bridge between humans and the rest of the animal world.

The special concern that people feel for the great apes is particularly prominent among people who have had the opportunity for contact with individual apes. Such contact comes about not only through sanctuaries, nature tourism, and zoos but also through films, books, and magazines. Education through such means has created large numbers of people interested in seeing great apes treated in humanitarian ways.

The great apes thus have particularly high value for a wide range of people. But, so far, conservationists have done little to harness this widespread popular interest. As a result, the strong empathy that exists in many parts of the world for great apes has done little to reduce the threats to their continued survival in the wild.

This means that in an effort to ensure great ape survival, there are important opportunities to tap the energies and commitment of large numbers of passionate, educated people ranging from zoo administrators to academics, across the professions, to individuals involved in local animal shelters and the zoo-going public, and more. Many of these potential supporters have important political and economic power.

To harness these sources of support, the great apes need to be given both a substantially higher international profile and a mechanism for taking advantage of it.

4. The Benefits of a Higher Profile for the Great Apes

The first major benefit of designating the great apes as World Heritage Species is that it would allow the passions of those who care about the great apes to be represented forcefully to key political and cultural leaders. Such leaders include powerful opinion-makers in both the non-range states and the range states. The active support of such leaders is critical if conservation programs are going to work, and can be obtained if advocates present their case with sufficient strength, clarity and unity.

But currently, the sad fact is that the importance of the great apes has not reached many of the key decision-makers.

A few of the non-range states, such as New Zealand and the United States, have legislated support for the great apes. The United States and the European Community have provided important funds in support of great ape conservation. But the great apes have not become an object of widespread international concern. Since there are 160+ non-range states, and they control most of the world's wealth, the importance of bringing the plight of the great apes to their attention is clear.

Meanwhile, in range states, the substantial efforts of conservationists on the ground are often met with little support from political leaders. The leadership problems include tolerance of illegal activities, tacit support for viewing great apes as meat, and a lack of resolve in planning conservation.

The inattention to the problems of the great apes is easily understood. Many of these countries are faced with massive problems of war, poverty, hunger, ecological unpredictability, and corruption. And although there are some places where great apes are cherished by traditional cultural values (such as bonobos protected by the people of Wamba (Kano, 1992)), there are many others with cultural values that treat the apes as unimportant. These problems contribute to explaining why the great apes are in such a precarious state. But they do not mean that efforts to save the apes are hopeless. Instead, they mean that a particularly strong initiative on ape conservation is required as soon as possible.

Accordingly, we believe that a profile-raising legal mechanism that publicizes both the high value placed on the great apes by many people throughout the world and the prospect of great ape extinctions holds vital promise for the development of effective conservation strategies. The world's leaders need to agree that it is time to make the great apes a priority.

5. The Benefits of an International Treaty

The current and future efforts of so many on behalf of the great apes can only benefit from the designation of the great apes as World Heritage Species, because it is from that designation that critical legal ramifications will flow.

First, why a treaty? And, we might add, why yet another treaty? A treaty because it is a mechanism that recognizes both the sovereignty of the range states and the need for support of those range states by the international community. A treaty because a collective effort among nations, completely voluntary in nature but with the force of international law and the availability of agreed remedies for those nations that decide to participate, can be very effective at implementing change. A treaty so that the range states that choose to participate can take the lead in the conservation of their own natural resources, in a unified cross-boundary effort. A treaty to create a vehicle for the financial, technical, and scientific support of the range states by non-range states as needed and requested. Finally, a treaty because that is an effective mechanism by which the range states that choose to do so can evaluate their internal laws and ensure their consistency with a voluntary international standard.

But we already have the Convention on International Trade in Endangered Species (CITES), and the Convention on Biodiversity, and half a dozen other treaties, so why another one? The short answer is that none of the existing treaties we have identified accomplish the same purpose as this one: to create a global protection strategy specifically for the great apes. CITES, for example, addresses cross-border trade in endangered species, and the Biodiversity Convention addresses across-the-board conservation measures for all species.

We envision that the treaty will consist of two documents: a Declaration for the Protection of the Great Apes, and a Convention. The Declaration will set forth the philosophical, moral, and scientific basis for the Convention, which will contain the substantive provisions of the treaty.

In the Declaration, signatory countries will acknowledge the close genetic relationship of great apes to humans; their exceptional intelligence, social interaction, and capacity for symbolic thought and cultural sophistication; their inherent dignity and worth; and that all these factors together entitle the great apes to the new special status of World Heritage Species, which in turn will entitle them to the protection of all signatory range states and indeed of the entire participating international community.

In essence, parties to the treaty will commit to protect the great apes from injury, imprisonment, destruction, and removal from their habitat (other than to protect them from further destruction). Specific measures will include a prohibition against activities likely to cause physical injury or death to great apes. Each signatory range state will agree to closely monitor the population, health, and well-being of the great apes, and to create educational programs designed to increase awareness of the value of and threats to the great apes. And, the internal laws and enforcement practices in each signatory nation will implement the obligations assumed in the treaty.

Non-range states, in turn, will commit among other things to providing scientific support as well as financial assistance when appropriate, and to ensuring that their own activities will not injure great apes or their habitats located in other nations.

At this time, we are eager for the Declaration to be signed by 2005. The Convention will follow, with a target date for signing of 2010. Though the date may seem remote, the process of achieving a treaty may be almost as important as the treaty itself if it promotes sufficient awareness of the problem and thereby contributes to initiatives.

6. The Need for an International Institution Overseeing Great Ape Conservation

The increased attention and resources promised by a higher international profile are valuable only if they are complemented with an appropriate institution for conserving and protecting the great apes. We believe that a major problem with the current system is that there is no such mechanism.

As noted above, no existing treaty aims specifically at protecting the great apes. Equally importantly, there has been no international institution responsible for the great apes (but see Addendum). There have not even been any conservation organizations dedicated to the conservation of great apes as a taxon (as opposed to advocating on behalf of either individual species, such as chimpanzees, or of animals as a whole). Nor, indeed, has there been any conservation meeting at which governments, conservation NGOs, and scientists have met to generate a series of regional plans, let alone a global plan. At no level has there been any significant attempt to organize an international conservation strategy for the great apes. Thus, conservation efforts have tended to be fragmented, and as a result, have not necessarily been well placed. They are often limited in scope and poorly informed by key general principles.

For example, basic tenets of conservation biology and population ecology clearly suggest that the best way to prevent species extinction is through the protection of a network of independent, viable populations (MacArthur and Wilson, 1967; Soulé, 1987; Primack, 1993; Pimm and Raven, 2000). It should therefore be a priority to identify those large populations of each species that might survive over the long term. Resources should be especially devoted to protecting these populations and the habitats that support them. Long-term survival of species and their geographic variability would be best insured if the risk of local extinction, from whatever cause, is spread among many of these populations.

In practice, however, few if any great ape conservation efforts have achieved the broad perspective necessary to address these major challenges (Whitten et al., 2001). Instead, disproportionate effort has been put into ape populations that are so small or threatened that their long-term survival has both a low probability and a low global significance. Substantial funding has also been diverted into projects that integrate conservation and development in ways that neglect the attainment or assessment of actual conservation outcomes (Oates, 1999; Terborgh, 1999; Wells et al., 1999; Whitten et al., 2001). Meanwhile, some of the limited number of viable populations of each species have failed to capture the attention of conservationists. As a result, we have missed opportunities to reduce the rate of ape population declines.

We find it hard to imagine that the great apes will survive without the systematic adoption of a larger-scale integrated perspective that solves these problems. A major missing component required for the conservation of great ape species is their management as meta-populations, without regard to national boundaries or affiliation with particular research teams or conservation organizations. Conservation efforts must take into account the need to protect several large habitat blocks that contain populations with the greatest chances of long-term viability. There is therefore the need for a formally recognized and scientifically respected international body to make decisions about the allocation of resources. This body would also help in many other ways, such as developing monitoring systems of ape populations, assessing the efficacy of various conservation efforts, coordinating the management of systems of protected areas that span several countries, and addressing the political problems of trans-national collaboration.

Accordingly, we view the first practical benefit of World Heritage Species status as the establishment of an international institution dedicated to the protection of the great apes. We conceive of some form of "International Great Ape Commission," which would bring governments, scientists, and NGOs together into a common forum for recognizing the global concern about the great apes, and for planning, implementing, and monitoring an appropriate conservation strategy. A recent collaborative effort to address the conservation crisis related to the spread of the Ebola virus through many separate gorilla populations in Central West Africa is encouraging in this respect. Researchers, policy-makers, and conservation professionals have stepped back from the concerns of their specific areas to seek a broad solution. The creation of a formal international institution would provide the mechanism and have the authority to address such crises quickly and efficiently.

Such a commission could in theory develop out of existing institutions. For example, it is possible that it might evolve out of the recently instituted Great Apes Survival Project (GRASP) of the United Nations Environmental Programme (UNEP) and the United Nations Educational, Scientific and Cultural Organization (UNESCO), or the Primate Specialist Groups of the International Union for the Conservation of Nature (IUCN). We would welcome such a development. A key aim of the Great Ape World Heritage Species Project, therefore, is to promote the establishment of some such mechanism for uniting and accelerating current efforts.

As this chapter goes to press, GAWHSP is working with UNESCO and UNEP to plan a "summit meeting" on the great apes. We hope that one outcome of this meeting will be such a commission.

7. The Value Problem in Conservation

The GAWHSP proposal is that the great apes be formally recognized at the global level as having outstanding universal value for all mankind.

But for at least two reasons this proposal is problematic for many conservationists. First, it challenges the conventional wisdom that all nonhuman species should be treated equally. As Hargrove (1989) wrote, the predominant quest in environmental ethics has been for a nonanthropocentric philosophy of conservation. This tends to lead to the view that all life has equal inherent value (e.g., Naess, 1986).

Accordingly, the priorities for conservationists are to save as many forms of life as possible, which they do by directing resources to those species that are most endangered (Hargrove, 1989; Harcourt, 2000), and to areas containing particularly high biodiversity (Myers, 1988; Mittermeier et al., 1998; Olson and Dinerstein, 2000). (Depending on how these guidelines were interpreted, some species of great apes would not be given special attention. For example, there are many species more immediately threatened than the great apes; and the first 24 "biodiversity hotspots" identified by Mittermeier et al. (1998) did not include gorilla or bonobo habitat.)

The GAWHSP argument that the great apes should be given a special conservation status has therefore sometimes been seen as a threat to this conventional conservationist philosophy. For example, the concern has been expressed by some people that efforts to save more threatened species of primates, such as some gibbons and monkeys, would be undermined if the great apes become "World Heritage Species."

However, although endangerment and biodiversity are key criteria for setting conservation priorities, they need not be the only ones. In practice, different species are valued for many different reasons, including economic, spiritual, scientific, educational, and strategic reasons, as well as their uniqueness (Hunter, 1996; Kellert, 1997). Particular species or taxa often tend to be singled out for special attention, including those that are more closely related to humans. For example, the U.S. Fish and Wildlife Service (1981–1983) assumed that greater phylogenetic proximity to humans represented greater value to humans (e.g., mammals outranked birds) (Norton, 1987). The general public clearly feels the same way.

A focus on the great apes because of the empathy that humans feel for them, therefore, fits public sentiment and can be used to the advantage of other species, including the small apes (gibbons and siamang) and other primates. We suggest that new ways of raising public awareness will bring new economic, political, and activist resources to the problem. Furthermore, because the great apes can act as umbrella species (having large home ranges that encompass many other species), flagship species (having broad and intensely personal appeal), and indicator species (being particularly sensitive to threats to their habitats), they have strong strategic value.

In fact, there is much overlap in conservation priorities of great apes versus other tropical plant and animal life. Because all great apes live at relatively low population densities, large areas need to be protected for each population. These large areas of habitat are the optimal umbrella for the conservation of all habitat and species diversity.

As noted, we also propose that the great apes be merely the first World Heritage Species. We would expect other species to follow, if they would benefit from a global support system with a new international mechanism for integrating their conservation.

The second (and closely related) difficulty that GAWHSP introduces for traditional conservationist philosophy is that it aims to unite environmental ethics with support for individual interests of nonhuman species. Advocates of the latter are in conflict with the "sustainable use" paradigm of conservation. However, no great ape conservation group advocates harvest or killing of great apes, and this is illegal in every great ape range state. Thus, GAWHSP aims to enlist the passions of advocates for the individual interests of great apes in the mission of conserving these species in the wild. But, unlike those interested in individual welfare, conservationists tend to play down the importance of individuals, personalities, and emotional lives in the species that they try to save. Indeed, they often regard animal welfare as in conflict with environmentalism, particularly because an interest in welfare tends to be associated with an animal rights philosophy more concerned with human-like species than with biodiversity (Hargrove, 1989).

The GAWHSP philosophy, by contrast, is that biodiversity is an important criterion of value, but it is not the only one. For strategic reasons, we think it unwise to advocate for animal rights since the rights question involves legal and philosophical issues that are unlikely to be viewed in similar ways worldwide. Nevertheless, we view the relationship between conservationists and advocates of great ape welfare as a coalition with potentially much greater power than has to date been achieved. We expect this increased power to come partly by galvanizing widespread support from animal-welfare groups, a sector that has to date been co-opted relatively little in the conservation movement.

In sum, we suggest that the singling out of great apes for special attention is justified by popular interest, and that, rather than jeopardizing the conservation of other species, it will significantly help other species.

8. The Development of GAWHSP, January 2001 to January 2003

The project's first task has been to make the scientific case that the great apes need stronger protection, in order to find out whether international agencies would support efforts to obtain a higher profile for the great apes. This phase, organized by the Ad-hoc Committee for the World Heritage Status for the Great Apes, culminated at a meeting in Paris in October 2001. Various UNESCO officers (concerned with the Convention on World Heritage Sites) agreed that improved international legislation to protect great apes is desirable and practicable, and encouraged the Ad-Hoc Committee to explore ways of achieving World Heritage Species status. We were also advised not to seek modification to the 1972 Convention on World Heritage Sites, because, in practice, UN Conventions are very rarely modified.

From October 2001 to August 2002, the Ad-Hoc Committee worked with the International Committee of the Chimpanzee Collaboratory to begin the drafting of potential legislative instruments. It also approached various organizations, individuals, and governments in an attempt to gauge interest and support for the concept of World Heritage Species status for the great apes. This led to public and private expressions of support for GAWHSP by representatives from Uganda, the Democratic Republic of the Congo, and Indonesia. Key luminaries have written letters directly in support of GAWHSP, including Jane Goodall and Edward O. Wilson.

In August 2002, GAWHSP was discussed at the International Primatological Congress in Beijing. The achievements and goals of the IPS Ad-hoc Committee for World Heritage Species for the Great Apes were reviewed first in a two-hour workshop and subsequently by the International Primatological Society (IPS) General Assembly. The General Assembly voted to approve the effort to seek World Heritage Species status for the Great Apes. The Assembly also proposed that the Ad-hoc Committee evolve into an independent body which would continue its work by attempting to develop a Convention on World Heritage Species, with the great apes as the first such species. This proposal was accepted.

Since then, the IPS Ad-Hoc Committee has therefore officially transformed itself into the Steering Committee for GAWHSP. It is this body that continues to interact with UNESCO and other organizations to develop an International Declaration, followed by a Convention, as proposed by the IPS Ad-Hoc Committee. For continuing news on these endeavors, see www.4greatapes.com.

9. Addendum

This chapter was written in 2003. As it goes to press (May 2006), we wish to note several positive developments over the last three years. UNEP and UNESCO's Great Ape Survival Project GRASP) has undergone institutional revisions that address some of these issues, and GAWHSP has been a strong supporter for GRASP's increased effectiveness. At GRASP's inaugural Council Meeting in September 2005, the Kinshasa Declaration was unanimously approved and now has been signed by nearly all government and NGO partners, with others intending to do so. The Declaration includes much of the sentiment and commitments we had hoped might be in a declaration establishing great apes as world heritage species. Further, in late 2004 GRASP incorporated a Scientific Commission, and its initial objective has been to focus GRASP actions on the identification and protection of those great ape wild populations that will preserve the genetic, ecological and cultural diversity of the great apes. This commitment is explicitly stated in the Kinshasa Declaration.

The World Conservation Union's (IUCN) Primate Specialist Group established a Section on Great Apes in 2004 that has begun addressing a number of international collaborative issues to improve great ape conservation. Chief among these have been regional workshops to develop conservation action plans for specific great ape species and subspecies. Other taxon-specific, but transnational workshops have helped identify priority populations for conservation attention and funding. We expect the IUCN/PSG's Section on Great Apes and GRASP's Scientific Commission to fill mutually supportive and complementary roles. So the crisis in great ape conservation is now benefiting from international scientific collaboration and advice.

However, these positive developments have not diminished the need for vastly improved political commitment and funding, both of which would be advanced by pursuing a formal status of great apes as World Heritage Species. As we pursue this objective, it is critical to tie these elements together so this status confers tangible and sustained benefits for the protection of great ape wild populations and individuals.

Acknowledgments. This chapter represents the energies, ideas, and activity of many individuals and organizations, including Christophe Boesch, Debby Cox, Sally Coxe, Doug Cress, Jim Else, Takeshi Furuichi, Michele Goldsmith, Jane Goodall, Chie Hashimoto, Holly Hazard, Jan van Hooff, Gilbert Isabirye-Basuta, Jamie Jones, Sonya Kahlenberg, Cheryl Knott, Sarah Luick, Tetsuro Matsuzawa, Martin Muller, Dale Peterson, Herman Pontzer, Ian Redmond, Vernon Reynolds, Tony Rose, Norm Rosen, Anne Russon, John Scherlis, Craig Stanford, Janette Wallis, David Watts, Steven Wise, and Juichi Yamagiwa. Particular thanks go to David Burmon and Kayo Burmon for their development of Great Ape World Heritage Species Project, Inc., and to the Glaser Progress Foundation for funding the Chimpanzee Collaboratory. GAWHSP also appreciates the support of the Alexander Abraham Foundation, the Carr Foundation, and the Shared Earth Foundation. TN thanks the Global Environment Research Fund from the Ministry of the Environment, Japan.

References

Anonymous (2001). The great ape declaration preventing the extinction of the great apes by awarding them World Heritage Status. *Pan Africa News* 8:2–17.

Beck, B.B., Stoinski, T.S., Hutchins, M., Maple, T.L., Norton, B., Rowan, A., Stevens, E.F., and Arluke, A. (eds.). (2001). *Great Apes and Humans: The Ethics of Coexistence*. Smithsonian Institute, Washington, DC.

Darwin, C. (1871). *The Descent of Man and Selection in Relation to Sex*. John Murray, London.

Felton, A.M., Engstrom, L.M., Felton, A., and Knott, C.D. (2003). Orangutan population density, forest structure and fruit availability in hand-logged and unlogged peat swamp forest in West Kalimantan, Indonesia. *Biological Conservation* 114:91–101.

Harcourt, A.H. (2000). Coincidence and mismatch of biodiversity hotspots: a global survey for the order, primates. *Biological Conservation* 93:163–175.

Hargrove, E.C. (1989). An overview of conservation and human values: are conservation goals merely cultural attitudes? In: Western, D., and Pearl, M.C. (eds), *Conservation for the Twenty-First Century*. Oxford University Press, New York, pp. 227–231.

Hunter, M.L.J. (1996). *Fundamentals of Conservation Biology*. Blackwell Science, Cambridge, MA.

Kano, T. (1992). *The Last Ape: Pygmy Chimpanzee Behavior and Ecology*. Stanford University Press, Stanford, CA.

Kellert, S.R. (1997). *Kinship to Mastery: Biophilia in Human Evolution and Development*. Island Press, Washington, DC.

MacArthur, R.H., and Wilson, E.O. (1967). *Island Biogeography*. Princeton University Press, Princeton, NJ.

Mittermeier, R.A., Myers, N., Thomsen, J.B., da Fonseca, G.A.B., and Olivieri, S. (1998). Biodiversity hotspots and major tropical wilderness areas: approaches to setting conservation priorities. *Conservation Biology* 12:516–520.

Myers, N. (1988). Threatened biotas: 'hotspots' in tropical forests. *Environmentalist* 8:187–208.

Naess, A. (1986). Intrinsic value: will the defenders of nature please rise? In: Soulé, M.E. (ed.), *Conservation Biology: The Science of Scarcity and Diversity*. Sinauer, Sunderland, MA, pp. 504–516.

Nishida, T., Wrangham, R.W., Jones, J.H., Marshall, A., and Wakibara, J. (2001). Do chimpanzees survive the 21st century? In: Brookfield Zoo (ed.), *The Apes: Challenges for the 21st Century*. Chicago Zoological Society, Brookfield, IL, pp. 43–51.

Norton, B.G. (1987). *Why preserve natural variety?* Princeton University Press, Princeton, NJ.

Oates, J.F. (1999). *Myth and Reality in the Rain Forest: How Conservation Strategies are Failing in West Africa*. University of California Press, Berkeley, CA.

Olson, D.M., and Dinerstein, E. (2000). The Global 2000: A representation approach to conserving the earth's most biologically valuable ecoregions. *Conservation Biology* 12:502–515.

Peterson, D. (2003). *Eating Apes*. University of California Press, Berkeley, CA.

Pimm, S.L., and Raven, P. (2000). Extinction by numbers. *Nature* 403:843–845.

Primack, R.B. (1993). *Essentials of Conservation Biology*. Sinauer Associates, Sunderland, MA.

Rijksen, H., and Meijaard, E. (1999). *Our Vanishing Relative: The Status of Wild Orang-utans at the Close of the Twentieth Century*. Tropenbos Publications, Wageningen, the Netherlands.

Robbins, M., Sicotte, P., and Stewart, K.J. (eds.) (2001). *Mountain Gorillas: Three Decades of Research at Karisoke*. Cambridge University Press, New York.

Rose, A. (1998). Growing commerce in bushmeat destroys great apes and threatens humanity. *African Primates* 3:6–12.

Singleton, I.S. Wich, S., Husson, S., Stephens, S., Utami Atmoko, S., Leighton, M., Rosen, N., Traylor-Holzer, K., Lacy, R., and Nyers, O. (eds.). 2004 Orangutan Populations and Habitat Viability Assessment: Final Report. IUCN/SSC Conservation Breeding Specialist Group, Apple Valley, MN. http://www.cbsg.org/reports/exec_sum/Orangutan PHVA04_LowRes.pdf

Soulé, M.E., (ed.). (1987). *Viable Populations for Conservation*. Cambridge University Press, Cambridge.

Terborgh, J. (1999). *Requiem for Nature*. Island Press, Washington, DC.

Tilman, D., May, R.M., Lehman, C.L., and Nowak, M.A. (1994). Habitat destruction and the extinction debt. *Nature* 371:65–66.

van Schaik, C.P., Monk, K.A., and Robertson, J.M.Y. (2001). Dramatic decline in orang-utan numbers in the Leuser ecosystem, northern Sumatra. *Oryx* 35:14–25.

Wells, M.S., Guggenheim, S., Kahn, A., Wardojo, W., and Jepson, P. (1999). *Investing in Biodiversity: A Review of Indonesia's Integrated Conservation and Development Projects*. The World Bank, Washington DC.

Whitten, T., Holmes, D., and MacKinnon, K. (2001). Conservation biology: a displacement behavior for academia? *Conservation Biology* 15:1–3.

Wich, S.A., Singleton, I., Utami-Atmoko, S.S., Geurts, M.L., Rijksen, H.D., and van Schaik, C.P. (2003). The status of the Sumatran orang-utan *Pongo abelii*: an update. *Oryx* 37:49–54.

Wilkie, D.S., and Carpenter, J.F. (1999). Bushmeat hunting in the Congo Basin: An assessment of impacts and options for mitigation. *Biodiversity and Conservation* 8:927–955.

Chapter 15
Conservation Through Scientific Collaboration: Case Study—Western-gorilla.org

Emma J. Stokes

1. Introduction

It is widely recognized that well-focused research by conservation biologists and scientists in related fields can yield important information and planning guidelines for conservation and wildlife managers (e.g., Soulé and Kohm, 1989; Groves *et al.*, 2002; Salfasky *et al.*, 2002).

As the human population continues to increase, the political and physical landscape of the natural world is in constant flux, and increasing demands are placed on the Earth's natural resources. As a result, priorities for research must similarly adapt. The opportunities to study large, undisturbed ecosystems and predominately intact animal populations are rapidly diminishing, and whole populations are disappearing before their basic biology and functional relationships with other species are even documented. As such, conservation biologists are under increasing pressure to focus their efforts on research that can deliver important and useful results quickly (e.g., Soulé and Orians, 2001). Time constraints, coupled with both political, and practical difficulties of conducting research in many of those habitats most threatened, demands a larger-scale collaboration across the conservation and research community so that informed conservation and management decisions can be made quickly and efficiently (Western, 2003). Conservation funds are limited, and the financial cost of assembling research expertise must be outweighed by the potential benefits to the species or situation in question. Given the appropriate conditions, a unified approach to conservation with significant backing of the research community holds considerable potential for the implementation of an effective conservation strategy.

This chapter illustrates how the conservation potential of scientific collaboration might be harnessed using a case study on western gorillas (*Gorilla gorilla* sp.).

Western-gorilla.org was set up as an informal collaboration of western gorilla researchers and conservationists with a considerable combined field experience across the Central Africa region. The collaboration followed a workshop on western gorilla conservation biology held at the Max Planck

Institute of Evolutionary Anthropology at Leipzig in May 2002; Western-gorilla.org was subsequently set up as a web-based information network and bulletin board for funding and media agencies. Few examples exist of scientific collaboration toward species conservation on a regional scale (Rainer *et al.*, 2003), and this case study provides a useful model for future collaborative efforts. This chapter summarizes the workshop objectives and achievements and discusses the potential of such a collaboration in implementing an effective conservation strategy for western gorillas.

2. Case-Study—Western-gorilla.org

2.1. Current Status of Western Gorilla Populations

Central Africa is one of the few remaining places in the world with intact areas of primary rainforest harboring viable populations of gorillas and other large mammals (Harcourt, 1995). However, the threats currently facing western gorilla populations are numerous and steadily increasing.

Western gorillas occur throughout the lowland forests of central West Africa (Figure 15.1) occupying Gabon, Republic of Congo, Cameroon, Equatorial Guinea, and Central African Republic, with outlier populations in the Cross

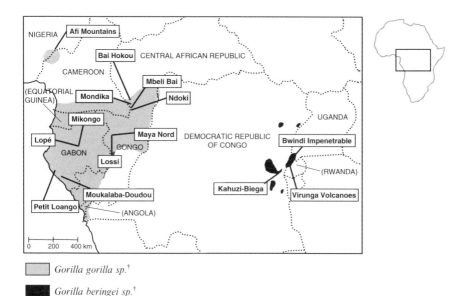

▨ *Gorilla gorilla sp.*[†]

▧ *Gorilla beringei sp.*[†]

FIGURE 15.1. Current distribution of gorillas, highlighting respective ranges of western and eastern species and the location of participating research sites. [†]Taxonomy given according to Morell, 1994; Ruvolo et al., 1994, but see Doran and McNeilage, 1998, for a historical account of the debate.

River region on the border of Nigeria and Cameroon, and in Cabinda, Angola, and southwestern Democratic Republic of Congo (Mehlman, Chapter 1, this volume). The history of western gorilla populations is now being revealed by genetic studies (Gardener and Ryder, 1996; Clifford *et al.*, 2002, but see Thalmann, 2004) and it is clear that some large rivers have been barriers to gene flow, and that some outlier populations have been separated for a long time from the core population. Human disturbance throughout western gorilla range has resulted in remaining populations becoming increasingly fragmented, an extreme case of which is the Cross River gorillas on the border of Nigeria and Cameroon (Oates *et al.*, 1999; Sarmiento and Oates, 2000). This subspecies is now thought to exist as a critically endangered population of only 200 individuals that is further fragmented into at least nine isolated habitat blocks (Groves *et al.*, 2005). Any remaining western gorilla populations in Cabinda, Angola, and Democratic Republic of Congo are likely to be similarly small, fragmented, and critically endangered.

The commercial bushmeat trade likely represents the single most significant threat to western gorilla populations (Robinson, 1999; Robinson and Bennett, 1999; Wilkie and Carpenter, 1999; Mehlman, Chapter 1, this volume). Gorillas are particularly vulnerable to large-scale hunting due to slow rates of reproduction and certain aspects of their social behavior. As the human population, particularly in urban areas, continues to grow, the demand for bushmeat increases (Chardonnet *et al.*, 1995; Barnes, 2002). At the same time, logging activities are spreading across Central Africa, increasing local population density and pressure on the surrounding resources, and opening up previously inaccessible areas that hasten the passage of bushmeat from the remote forests to the urban markets (Wilkie and Carpenter, 1999). In addition, populations of gorillas are facing local extinction through the Ebola virus (e.g., northern Congo and northeastern Gabon (Walsh *et al.*, 2003), and human encroachment and development (e.g., Cameroon/Nigeria; Bassey and Oates, 2001). Given the current trend toward gorilla-based ecotourism in Central Africa, further concerns arise through human-gorilla disease transmission (Woodford *et al.*, 2002), particularly for small and fragmented populations. Despite several individually successful conservation projects in place across Central Africa, the latest predictions suggest that western gorilla populations are declining (Walsh *et al.*, 2003).

2.2. Current Status of Western Gorilla Field Research

In spite of the fact that western gorillas account for more than 90% of all gorillas in the wild (Figure 15.1), our knowledge of their basic biology has fallen considerably behind that of the well-studied mountain gorillas in the Virunga Volcanoes. A combination of dense forest habitat and hunting pressures has severely restricted efforts at habituating and observing western gorillas in the wild, and civil unrest across much of Central Africa has rendered many areas inaccessible to researchers.

Much of the research to date has, therefore, focused on diet and ranging behavior, which can be studied using indirect methods such as feeding trails and fecal analysis. However, these studies have illustrated that broad ecological differences exist between mountain gorillas and western gorillas (e.g., Sabater-Pi, 1977; Tutin and Fernandez, 1983; Rogers, 1987; Rogers, 1988; Williamson et al., 1990; Tutin et al., 1991; Malenky et al., 1994; Goldsmith, 1996; Kuroda et al., 1996; Fay, 1997; Remis, 1997; Doran et al., 2002). Increased frugivory combined with comparatively greater home ranges make gorillas extremely important seed dispersers (Voysey et al., 1999b, 1999b), and feeding on leaves and stems exerts selective pressures on shoot regeneration (Watts, 1987). Western gorillas are therefore important umbrella species for a variety of other animals and plants that share their habitat, and play an important role in maintaining general forest structure, forest species composition, and overall resilience to disturbance, disease, and climate change.

Our knowledge of the social organization and behavior of western gorillas is still in its infancy: problems associated with visibility and habituation largely preclude long-term studies based upon direct observation of individuals. The exceptions to this are studies based at large swampy clearings or "bais," which provide an unobstructed view of gorillas. These studies have already begun to redress the imbalance in our knowledge of gorilla social organization and demography (Magliocca et al., 1999; Parnell, 2002; Stokes et al., 2003), and hold considerable potential for more detailed behavioral studies.

Given the rapid increase in threats to western gorillas across their geographical range, it was considered time to take stock of the current level of knowledge of western gorillas and its implications for conservation. The goal of the workshop, therefore, was to assemble expertise on western lowland gorilla behavior, genetics, and conservation biology in order to set conservation priorities based on the most current and complete data available.

2.3. Participating Research Sites

The workshop in Leipzig was a culmination of over 65 years of research on western gorillas from 11 different sites across Central Africa (Tables 15.1 and 15.2). Sites were selected based upon their proven or potential contribution to our knowledge of western gorillas, and representatives from each site were invited to present the most recent results from the field. The sites covered a total of four countries and a range of different habitat types (Table 15.1), including montane forest (Afi Mountains), savanna-forest mosaic (Lopé, Moukalaba-Doudou), coastal forests (Petit Loango), *terra firma* closed-canopy forest (e.g., Ndoki), and swamp forest with natural clearings (Maya Nord, Mbeli Bai). Each site is exposed to human disturbance to varying degrees (Table 15.1): from relatively intact ecosystems with no hunting or logging activities (Ndoki), to previously logged areas (Bai Hokou, Mikongo, Lossi), to sites close to human habitation and subject to high hunting levels (Afi Mountains).

TABLE 15.1. Description of study area for all western gorilla research sites.

Study site	Maya Nord	Lossi	Ndoki	Mbeli Bai	Bai Hokou	Mondika	Mikongo	Petit Loango	Moukalaba-Doudou	Lopé	Afi mountains
Protected area	Odzala-Kokoua National Park	Lossi Faunal Reserve	Nouabalé-Ndoki National Park	Nouabalé-Ndoki National Park	Dzanga-Ndoki National Park	Dzanga-Ndoki National Park (75%)	Lopé-Okanda National Park	Loango National Park	Moukalab-aDoudo National Park	Lopé-Okanda National Park	Afi Mountains Wildlife Sanctuary
Region	Cuvette Ouest	Cuvette Ouest	Sangha	Sangha	Sangha-Mbaere	Sangha-Mbaere	Ogoué-Ivindo	Ogoué Maritime	Nyanga	Ogoué-Ivindo	Cross River State
Country	Republic of Congo	R. Congo	R. Congo	R. Congo	Central African Republic	CAR/R. Congo	Gabon	Gabon	Gabon	Gabon	Nigeria
Study area	0.2 km²	50 km²	20 km²	0.13 km²	40 km²	50 km²	50 km²	20 km²	40 km²	100 km²	35 km²
Altitude	300–600 m	300–450 m	330–600 m	330 m	400–480 m	<400 m	300 m	0 m	100–700 m	100–700 m	200–1,300 m
Mean annual rainfall	1,500 mm	1,509 mm	1,430 mm	1,376 mm (1,265–693)	1,813 mm (1,544–083)	1,430 mm (1,084–779)	—	2,376 mm	—	1,500 mm (± 213)	3,528 mm (3,422–652)
Vegetation: [a]											
SAV								X	X	X	
GF								X	X	X	X
OUF		X	X		X	X	X	X	X	X	X
BCF					X	X	X	X	X	X	X
LF		X			X		X	X	X	X	
MONO			X		X	X	X				
MAN		X	X		X	X	X				
SF		X	X	X	X	X		X	X	X	X
BAI	X			X							

300

Other mammals:

Chimpanzees	Not in bai	X	X	Not in bai	X	X	X	X	X	X	X
Elephants	X	X	X	X	X	X	X	X	X	X	X
Buffalo	X		X	X	X	X	X	X	X	X	X
Red river hog	X	X	X	X	X	X	X	X	X	X	X
Giant forest hog	X	X	X	X	X	X					
Primate sp. no.	Not in bai	8	9	1 in bai	8	8	6	8	8	6	4
Nearest village	50 km	28 km	30 km	35 km	25 km	25 km	10 km	<10 km	4 km	12 km	<4 km
Human influence:											
Logging	No	No	No	No	Past	No	Past	Past	Past	Past	No
Hunting	No	Occasional	Rare	No	Occasional	Rare	Present	No	Occasional	No	Frequent
Agriculture[b]	50 km	>5 km	25 km	30 km	>25 km	25 km	>5 km	>5 km	<5 km	>5 km	1 km
Tourism	Occasional	Occasional	No	Occasional	Occasional	No	Occasional	Occasional	No	Frequent	No

[a] Vegetation types: SAV, savanna; GF, gallery forest; OUF, open understorey forest (primary); BCF; broken canopy forest (light gaps); LF, logged forest (secondary); MONO, monodominant *Gilbertiodendron* forest; MAN, marantaceae forest; SF, swamp forest; BAI, bais (natural forest clearing).
[b] Distance to nearest field; X, presence.

TABLE 15.2. Nature and focus of research program at western gorilla study sites.

Study site	Maya Nord	Lossi	Ndoki	Mbeli Bai	Bai Hokou	Mondika	Mikongo	Petit Loango	Moukalaba-Doudou	Lopé	Afi mountains
Study period	10/96– 6/97, 8–9/99, 2–4/00	1994–96, 1998, 2000–2002	1989–92, 1994–99	1995–present	1986–89, 1992–94, 1998–present	1995–present	4/00–12/00	8/97–12/98, 4-6/99	1999–present	1983–present	1996, 11/97–present
Research presence	Shortterm	Permanent	None	Permanent	Permanent	Permanent	Shortterm	None	Permanent since 2001	Permanent	Permanent
Research projects:											
Diet		X			X	X	X		X	X	X
Ranging		X	X		X	X	X		X	X	X
Feeding behavior	X			X							
Social organization	X	X		X		X			X		
Population structure	X			X							
Demography		X		X						X	
Social behavior		X		X	X						X
Habitat utilization				X		X				X	
Habitat quantification			X					X		X	X
Nutritional ecology	X									X	

	Excellent in bai	Good	None	Excellent in bai	Fair	Good	Fair	None	Fair	Fair	None
Population genetics				X						X	X
Health				X					X		
Censusing	X							X			
Habituation	Excellent in bai	Good	None	Excellent in bai	Fair	Good	Fair	None	Fair	Fair	None
Methods: [a]											
Indirect		X	X	X	X	X	X	X	X	X	X
Nest		X	X	X	X	X	X			X	X
Direct	X	X	X	X	X			X	X	X	
Focal	X	X		X							

[a] Methodology: Indirect, includes feeding trail and fecal analysis; Nest, nest to nest follows; Direct, direct observation; Focal, focal subject sampling; X, presence.

The nature of research conducted at each site fell broadly into three categories, corresponding to the degree of gorilla habituation (Table 15.2): nest to nest follows and analysis of home-range patterns and diet (gorillas habituated); direct observation and long-term monitoring of population dynamics (gorillas habituated to fixed presence of observers in bais); and habitat quantification, utilization, and gorilla density estimates (no habituation). Collaboration between sites already exists in the form of population genetic analysis (coordinated through separate programs run by the Max Planck Institute of Evolutionary Anthropology, Leipzig and by Cardiff University, UK and CIRMF, Gabon), and gorilla health monitoring [coordinated by the Wildlife Conservation Society (WCS) gorilla health program]. In order to place the most current data set in a comparative context, representatives from mountain gorilla and eastern lowland gorilla research sites, namely, Kahuzi-Biega, Bwindi Impenetrable Forest, and the Virunga Volcanoes (Figure 15.1) were also present. Participants were asked to compare their data both across and within taxa in order to explore ecological, social, demographic, human, and site-specific factors that might influence behavioral ecology.

Many of the western gorilla research sites are already part of a protected area management program and had an applied research focus (e.g., Mbeli Bai: WCS-Nouabalé-Ndoki Project, Bai Hokou: WWF-Dzanga-Sangha Project, Maya Nord: ECOFAC Odzala, Afi Mountains: WCS, Fauna Flora International, Pandrillus, Cross River State Forestry Commission). Furthermore, some sites had established conservation education programs (Mbeli Bai), national capacity building programs (Ndoki, Lopé), and ecotourism projects (Bai Hokou, Mikongo, Maya Nord, Mbeli Bai). For the majority of sites, the permanent presence of researchers provided an effective deterrent to poachers. As a result, hunters within the study area did not specifically target gorillas. These factors, in particular, lent a considerable experience of NGO and local government involvement to the workshop, which would be important in focusing conservation strategies to their greatest effect.

2.4. Objectives of the Workshop

The workshop was structured according to three broad questions as follows:

- *What do we need to know about western gorillas to plan an effective conservation strategy?*
- *What type of information is available?*
- *What recommendations can be made based upon current information?*

2.4.1. What Do We Need to Know in Order to Protect Western Gorillas?

Figure 15.2 gives an overview of the framework of the workshop. Research topics specific to conservation issues were addressed, and the functional relationships between resource availability, population size and structure, demography, and

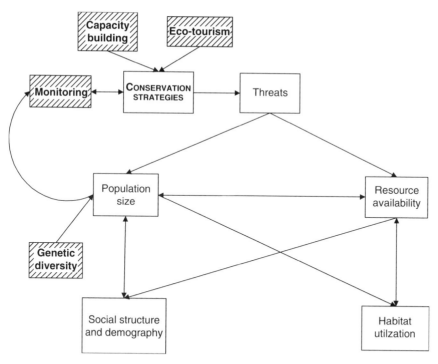

FIGURE 15.2. Overview of research topics and their conservation significance. Hatched boxes represent cross-cutting issues related to conservation strategy (in caps). Open boxes in lowercase denote research topics presented at the workshop and arrows between boxes denote inter-relationships between one research topic and another.

human disturbance, in particular, were investigated. The field expertise of the Leipzig group allowed four additional crosscutting issues to be fleshed out in the domains of monitoring, genetics, ecotourism, and capacity building. Monitoring was considered as an evaluative tool for management decisions, and the role of genetics in setting priority populations, as well as an alternative tool to censusing, was discussed. The relative economic advantages of ecotourism were weighed against the risks associated with habituation and disease transmission. Capacity building was discussed in terms of the capacity required and the identification of trainees and trainers. For all of these issues, cost-efficiency is paramount. The efficacy of current conservation strategies was considered in the light of new information on western gorilla behavioral ecology and threats.

2.3.2. What Type of Information Is Available?

2.3.2.1. Resource availability

Vast quantities of data on vegetation type can now be collected by remote-sensing techniques using satellite imagery. These techniques are only just beginning to be explored but are a cost-effective way of quantifying habitat at large spatial scales. Field data from a number of different research sites provide useful "ground-truthing" data for image analysis.

2.3.2.2. Habitat utilization

Given the breadth of habitat types covered by the participating research sites, a considerable database on diet and ranging already exists. Keystone food items and rare, yet important, resources (e.g., swamps) were identified, and the temporal and spatial use of these resources is currently being investigated.

2.3.2.3. Population structure and demography

Although several sites are able to provide accurate data on social organization (Lossi, Maya Nord, Mbeli Bai), there are few data on important demographic variables such as interbirth intervals and generation time, currently only available from one research site (Mbeli Bai). Long-term study sites, such as Mbeli Bai, were considered important in monitoring trends in population growth and structure.

2.3.2.4. Population size

There is a considerable lack of reliable data on gorilla densities and distributions. Seemingly straightforward questions have been largely hampered by cost and methodological issues (for a review, see Plumptre, 2000). Study sites that have already made considerable progress with habituation were encouraged to focus research efforts toward reducing the many sources of error inherent in current censusing methods (e.g., nest decay rate), and investigating alternative censusing techniques (e.g., dung defecation rate and rate of dung decay).

2.3.2.5. Threats

Qualitative data exist on the threats currently faced by western gorillas, although quantification of such data, particularly on a spatial scale, is largely absent. The impact of threats on gorilla populations is variable according to the nature of the threat posed, and quantifiable data are largely precluded by the lack of accurate information on gorilla density and distribution (see above). Long-term monitoring of population size and structure was considered an important tool in evaluating threat-levels.

2.3.3. What Recommendations Can Be Made Based upon Current Information?

The killing of gorillas for bushmeat is illegal in all range states, and yet gorillas are still being hunted across their geographic range. It was agreed, therefore, that western gorilla conservation efforts must, in the short-term, focus on improved law enforcement at all levels if we are to see a decline in the commercial trade of gorillas for food.

The eight recommendations listed below are the result of a series of workgroups based upon current data sets and expert opinion, and they apply to the entire western gorilla range. Capacity building and monitoring, in particular, should be considered as linking several different activities simultaneously. The recommendations are presented in order of urgency of implementation and can be considered as the backbone of the conservation strategy, as follows:

1. Effectively enforce existing national laws and international conventions that protect gorillas.
2. Create a network of ecologically representative protected areas across the geographical range of western gorillas.
3. Create sustainable funding mechanisms such as Trust Funds to ensure stable and sufficient revenues for management and research within protected areas.
4. Obtain precise estimates of the numbers of western gorillas remaining and put in place a system to monitor future population trends.
5. Reduce negative impact of selective logging on western gorillas.
6. Minimize detrimental effects of economic development on western gorillas.
7. Establish independent evaluation of conservation and research activities.
8. Form a network linking all efforts to conserve western gorillas.

2.4. Next Steps—Implementing an Effective Conservation Strategy

The real test now facing the follow-up to the workshop is in distilling the list of recommendations into specific fundable activities, which can be presented to donors and implemented in a timely fashion. In order to achieve this, an action plan must be developed, which works toward the ultimate objective of conserving ecologically functioning populations of gorillas throughout their geographical range.

The plan needs to consider the threats to this objective the proposed interventions with which to address these threats, and the relationships that exist between them. Moreover, this needs to be tightly linked to a monitoring and evaluation plan with two major components: performance monitoring, in order to evaluate whether our interventions are being implemented in a timely manner, and impact monitoring, so as to evaluate the impact of conservation management decisions on the ape populations that we are trying to conserve.

For each of these components, we need to clearly state our targets, specify a realistic timeframe with which to achieve these targets, and describe how they will be measured. To monitor, in turn, the effect of interventions on our overall objective, that is the status of gorilla populations, we need a system that is sufficiently sensitive to detect trends in gorilla population density and distribution. Finally, we need to specify who will implement the activities, who will train the necessary field personnel, and who will provide the funds. Most importantly, the involvement of range state protected area and wildlife managers and scientists must be encouraged at the outset if a region-wide action plan stands any chance of long-term success.

In addressing these issues, western-gorilla.org is best considered as a facilitating network or independent steering committee that aims to work through field-based conservation NGOs, international and national governments, institutions, and donors. In reality, however the transition from sound conservation science-based recommendations to conservation policy is far from straightforward (Meffe, 1998, 1999; Galusky, 2000; Letnic, 2000) and well-intentioned collaborative efforts amongst governmental and nongovernmental conservation groups are frequently fraught with institutional and political wrangling. Given the finite source of conservation money available, these efforts are more often dominated by the responsibilities of individual organizations to their respective donors. Lobbying donors to fund large-scale collaborative efforts will help to encourage and promote a cooperative and coordinated conservation approach among international and range state institutions and governments.

At the same time, collaborations must be seen to offer more than an individual organization working to its own agenda, be it through fundraising or the provision of resources, technical advice, or trained personnel. In this sense, the considerable body of technical expertise offered by western-gorilla.org holds great potential, and lends a considerable authority to both lobbying potential donors and informing conservation policy and decision making. Furthermore, an independent network is more likely to be successful in encouraging collaborative efforts that would otherwise be left to institutional and political will. However, this, in turn, has its own potential pitfalls. Independent networks risk becoming overwhelmed by administration and bureaucracy, and losing touch with the situation in the field.

Here, western-gorilla.org has a head start, consisting mostly of field based scientists and conservationists. However, dedicated people are still required to run the network, and, in spite of the benefits afforded by its relatively independent status, the network lacks institutional support and funds. As such, an independent funding source must be found, and the follow-up has until now largely depended upon the voluntary efforts of a few individuals, albeit successful, in maintaining the momentum generated in the aftermath of the workshop. Finally, once a logistical and administrative system is put into place, clear guidelines need to be put in place for collaborating partners, to ensure transparency and accountability at all levels.

3. Summary

The workshop at Leipzig fulfils a number of criteria that qualify its usefulness as a case study and potential model for species conservation through scientific collaboration. Firstly, there was a recognized increase in the level of threats to western gorilla populations and predicted declines in gorilla populations across Central Africa. Both factors highlighted the inefficiency of current conservation strategies and identified a need for a timely change in conservation focus. Secondly, a growing number of western gorilla researchers with considerable field-based experience and expertise suggested the need to assess the rapidly increasing database on western gorilla behavioral ecology and conservation biology and provide a vital first-hand source with which to address the problem. Thirdly, the significance of such a meeting had far wider implications for Central African forest biodiversity and conservation in general, and, finally, there was a sense of urgency to the current situation with the real possibility of populations becoming extinct within two generations if current policies are not rethought.

The aim of this workshop was to synthesize available information on western gorillas and to refocus current gorilla conservation strategy. To this end, the workshop succeeded in providing a list of recommendations with which to align conservation efforts. In addition, it succeeded in establishing a network of individuals and organizations with which to link these efforts. Furthermore, the workshop brought the potential crisis faced by western gorillas to the attention of the general public, through considerable press coverage that coincided with the release of the recommendations and publication of the major findings (Bermejo, 2004; Cipolletta, 2004; Doran-Sheehy and Boesch, 2004; Doran-Sheehy et al., 2004; Robbins et al., 2004; Rogers et al., 2004; Stokes, 2004; Vigilant and Bradley, 2004). However, the workshop now requires considerable follow-up in order to convert these findings into a clear set of actions that will ultimately result in the long-term survival of western gorillas. Western-gorilla.org is in a potentially strong position to build consensus and develop an effective action plan. However, in order to do so, it needs to consider, firstly, the human and financial resources required to co-ordinate the considerable network that has been built up, and, secondly, how to maintain the spirit of collaboration through to the implementation phase.

The situation currently faced by western gorilla conservation is critical but not hopeless. Individual conservation projects that have already independently adopted a similar strategy to that proposed here have met with considerable success, which holds realistic potential for the adoption and implementation of such a strategy on a regional scale.

4. Postscript

Six months subsequent to the Leipzig workshop, the Lossi Gorilla Sanctuary in the Republic of Congo suffered a catastrophic decline in its gorilla population. Between October 2002 and January 2004, it is estimated that approximately 5000 gorillas died in the Lossi study area as a result of the Ebola virus (Bermejo et al., 2006). Three years on and Ebola now represents one of the most potent threats to western gorilla populations in Central Africa. It currently threatens some of the largest remaining strongholds of apes on the planet, with population declines of up to 90% now estimated to have occurred in some areas of Gabon and Congo (Huijbregts *et al.*, 2003; Walsh *et al.*, 2003; Tutin *et al.*, 2005; Bermejo *et al.*, 2006; Caillaud *et al.*, 2006). For western gorillas, the combination of Ebola and commercial hunting has resulted in escalating rates of population decline of up to 50% over the last decade (Walsh *et al.*, 2003; Walsh, 2006). The situation for western gorillas is rapidly deteriorating toward crisis point, with the need for innovative and large-scale collaboration now paramount. Faced with these rapidly shifting conservation goalposts, the scientific and conservation community has had to respond quickly and to this end a number of significant collaborative processes have been developed since this manuscript was originally prepared. Of particular note are the creation of interdisciplinary networks to develop innovative science-based solutions to the issue of great ape health and Ebola, i.e. GAHMU[1] (Great Ape Health Monitoring Unit) and GRAET[2] (Great Ape Ebola Taskforce) as well as the bolstering of ongoing efforts to shape great ape conservation practice and policy, i.e., the new Section for Great Apes (SGA) of the IUCN Primate Specialist Group and the recently published Regional Action Plan for the conservation of chimpanzees and gorillas in West Central Africa (Tutin *et al.*, 2005). The 2002 Leipzig workshop was instrumental in the creation of the Regional Action Plan and succeeded in developing a regional conservation strategy grounded in conservation science and based on a consensus approach. In this way, the process has stayed true to the original values of the western-gorilla.org network. As threats to western gorillas continue to escalate in West Equatorial Africa, the pressure is now on politicians, decision-makers, and donors to ensure the effective implementation of this strategy across western gorilla range.

For information on the current status of western gorillas and useful links to current conservation initiatives, please visit www.western-gorilla.org. This website continues to be maintained on a voluntary basis by Mark Gately and Emma Stokes, and welcomes suggestions and contributions.

Acknowledgments. I thank in particular the organizers of the workshop, Caroline Tutin, Christophe Boesch, and Diane Doran. I also thank Peter

---

[1] GAHMU is a network of researchers from different disciples concerned about diseases of great apes. It was created following the "1st Conference on: Diseases—the third major threat for wild Great Apes?" held in May 2004 at the Max Planck Institute for Evolutionary Anthropology in Leipzig

[2] GRAET is a network of great ape scientists, molecular biologists, virologists, and veterinarians with the aim of developing innovative research partnerships to evaluate and implement effective Ebola control measures. It was created following an interdisciplinary meeting held in Washington, DC, in March 2005.

Walsh for his significant contribution during and after the workshop and to Mark Gately for developing and updating the website. The success of the workshop was in large part due to its participants: Kate Abernethy, Magdalena Bermejo, Allard Blom, Brenda Bradley, Chloe Cippoletta, Stephen Clifford, Jacqui Groves, Daniel Idiata Manbounga, Marie Laure Klein, Stephanie Latour, Alistair McNeilage, Florence Magliocca, Kelly McFarland, Tomo Nishihara, John Oates, Richard Parnell, Melissa Remis, Martha Robbins, Emma Stokes, Linda Vigilant, Liz Williamson, and Juichi Yamagiwa, with guests Nicola Anthony, Richard Bergl, Angelique Todd, and Jean Wickings. The workshop was hosted by the Max Planck Institute of Evolutionary Anthropology and funded by a grant from the USFWS Great Ape Conservation Fund. Finally, thank you to Tara Stoinski for representing the goals of western-gorilla.org at the AZA meeting in September 2002 and for inviting this manuscript.

References

Barnes, R. (2002). The bushmeat boom and bust in Central Africa. *Oryx* 36:236–242.

Bassey, A.E., and Oates, J. (eds.) (2001). *Proceedings of the International Work shop and Conference on the Conservation of the Cross River gorillas.* Calabar, Nigeria.

Bermejo, M. (2004). Home range use and intergroup encounters in western gorillas (*Gorilla g. gorilla*) at Lossi forest, North Congo. *American Journal of Primatology* 64:223–232.

Bermejo, M., Rodríguez-Teijeiro, J.D., Illera, G., Barroso, A., Vilà, C., and Walsh, P.D. (2006). "Ebola Outbreak Killed 5000 Gorillas." *Science* 314:1564.

Caillaud, D., Levréro, F., Cristescu, R., Gatti, S., Dewas, M., Douadi, M., Gautier-Hion, A., Raymond, M., and Ménard, N. (2006). "Gorilla susceptibility to Ebola virus: The cost of sociality." *Current Biology.*

Chardonnet, P., Fritz, H., Zorzi, N., and Feron, E. (1995). Current importance of traditional hunting and major contrasts in wild meat consumption in sub-Saharan Africa. In: Bissonette, J. and Krausman, P. (eds.), *Intergrating People and Wildlife for a Sustainable Future.* Bethesda, MD, The Wildlife Society, pp. 304–307.

Cipolletta, C. (2004). Effects of group dynamics and diet on the ranging patterns of a western gorilla group (*Gorilla gorilla gorilla*) at Bai Hokou, Central African Republic. *American Journal of Primatology* 64:193–205.

Clifford, S., Abernethy, K., White, L., Tutin, C., Bruford, M., and Wickings, E. (2002). Genetic studies of western gorillas. In: Taylor, A. and Goldsmith, M. (eds.), *Gorillas in the 21st Century,* Cambridge, Cambridge University Press, pp. 269–292.

Doran, D.M., and McNeilage, A. (1998). Gorilla ecology and behaviour. *Evolutionary Anthropology* 6:120–131.

Doran, D., McNeilage, A., Greer, D., Bocian, C., Mehlman, P., and Shah, N. (2002). "Western lowland gorilla diet and resource availability: new evidence, cross-site comparisons, and reflections on indirect sampling methods." *American Journal of Primatology* 58:91–116.

Doran-Sheehy, D., and Boesch, C. (2004). Behavioral ecology of western gorillas: new insights from the field. *American Journal of Primatology* 64:139–143.

Doran-Sheehy, D.M., Greer, D., Mongo, P., and Schwindt, D. (2004). Impact of Ecological and social factors on ranging in western gorillas. *American Journal of Primatology* 64:207–222.

Fay, J.M. (1997). *The Ecology, Social Organization, Populations, Habitat and History of the Western Lowland Gorilla (Gorilla gorilla gorilla* Savage and Wyman 1847). PhD Thesis, Washington University, Saint Louis, pp. 416.

Galusky, J. (2000). The Promise of conservation biology: the professional and political challenges of an explicitly normative science. *Organization and Environment* 13(2):226–232.

Gardener, K.J., and Ryder, O.A. (1996). Mitochondrial DNA diversity in gorillas. *Molecular Phylogenetics and Evolution* 6:39–48.

Goldsmith, M.L. (1996). *Ecological Influences on the Ranging and Grouping Behavior of Western Lowland Gorillas at Bai Hokou, Central African Republic.* PhD Thesis, State University of New York at Stony Brook, New York, pp. 403.

Groves, C., Jensen, D., Valutis, L., Redford, K., Shaffer, M., Scott, J., Baumgartner, J., Higgins, J., Beck, M., and Anderson, M. (2002). Planning for biodiversity conservation: putting conservation science into practice. *BioScience* 52(6):499–512.

Harcourt, A. (1995). Population viability estimates: Theory and practice for a wild gorilla population. *Conservation Biology* 9:134–142.

Huijbregts, B., DeWachter, P., Sosthene, L., Obiang, N., and Akou, M. (2003). Ebola and the decline of gorilla *(Gorilla gorilla)* and chimpanzee *(Pan troglodytes)* populations in Minkebe Forest, north-eastern Gabon. *Oryx* 37(4):437–443.

Kuroda, S., Nishihara, T., Suzuki, S., and Oko, R. (1996). Sympatric chimpanzees and gorillas in the Ndoki Forest, Congo. In: McGrew, W, Marchant, L and Nishida, T. (eds.) *Great Ape Societies*, Cambridge University Press Cambridge, pp. 71–81.

Letnic, M. (2000). The politics of science and conservation: why not anonymous publication? *Biodiversity and Conservation* 9(5):707–709.

Magliocca, F., Querouil, S., and Gautier-Hion, A. (1999). Population structure and group composition of western lowland gorillas in North-Western Republic of Congo. *American Journal of Primatology* 48:1–14.

Malenky, R., Kuroda, S., Vineberg, E., and Wrangham, R. (1994). The significance of terrestrial herbaceous foods for bonobos, chimpanzees and gorillas. In: Wrangham, R, McGrew, W, De Waal, F., and Heltne, P. (eds.) *Chimpanzee Cultures*, Harvard University Press, Cambridge, MA, pp. 59–75.

Meffe, G. (1998). Conservation science and the policy process. *Conservation Biology* 12(4):741–742.

Meffe, G. (1999). Conservation science and public policy: Only the beginning. *Conservation Biology* 13(3):463–464.

Morell, V. (1994). Will primate genetics split one gorilla into two? *Science* 265:1661.

Oates, J., McFarland, K., Stumpf, R., Fleagle, J., and Disotell, T. (1999). The Cross River gorillas: a distinct subspecies, *Gorilla gorilla diehli. American Journal of Physical Anthropology* 28:213–214.

Parnell, R.J. (2002). Group size and structure in Western Lowland Gorillas *(Gorilla gorilla gorilla)* at Mbeli Bai, Republic of Congo. *American Journal of Primatology* 56:193–206.

Plumptre, A. (2000). Monitoring mammal populations with line transect techniques in African forests. *Journal of Applied Ecology* 37:356–368.

Rainer, H., Asuma, S., Gray, M., Kalpers, J., Kayitare, A., Rutagarama, E., Sivha, M., and Lanjouw, A. (2003). Regional conservation in the Virunga-Bwindi region: The impact of Transfrontalier collaboration through the experiences of the International Gorilla Conservation Programme. *Journal of Sustainable Forestry* 17(1/2):183–198.

Remis, M.J. (1997). Western lowland gorillas (*Gorilla gorilla gorilla*) as seasonal frugivores: Use of variable resources. *American Journal of Primatology* 43:87–109.

Robbins, M., Bermejo, M., Cipolletta, C., Magliocca, F., Parnell, R., and Stokes, E. (2004). Social structure and life-history patterns in western gorillas (*Gorilla gorilla gorilla*). *American Journal of Primatology* 64:145–159.

Robinson, J., and Bennett, E. (eds.) (1999). *Hunting for Sustainability in Tropical Forests*. Columbia University Press, New York.

Robinson, J.G., Redford, K.H., and Bennett, E.L. (1999). Wildlife harvest in logged tropical forests. *Science* 284:595–596.

Rogers, E., and Williamson, E. (1987). Density of herbaceous plants eaten by gorillas in Gabon: Some preliminary data. *Biotropica* 19(3):278–281.

Rogers, M., Abernethy, K., Bermejo, M., Cipolletta, C., Doran, D., McFarland, K., Nishihara, T., Remis, M., and Tutin, C. (2004). Western gorilla diet: A synthesis from six sites. *American Journal of Primatology* 64:173–192.

Rogers, M.E., Williamson, E.A., Tutin, C.E.G., and Fernandez, M. (1988). Effects of the dry season on gorilla diet in Gabon. *Primate Report* 19:29–34.

Ruvolo, M., Pan, D., Zehr, S., Goldberg, T., Disotell, T., and von Dornum, M. (1994). Gene trees and hominoid phylogeny. *Proceedings of the National Academy of Science, USA* 91:8900–8904.

Sabater-Pi, J. (1977). Contribution to the study of the alimentation of lowland gorillas in the natural state in Rio Muni, Republic of Equatorial Guinea (West Africa). *Primates* 18:183–204.

Salfasky, N., Margoluis, R., Redford, K., and Robinson, J. (2002). Improving the practice of conservation: A conceptual framework and research agenda for conservation science. *Conservation Biology* 16(6):1469–1479.

Sarmiento, E., and Oates, J. (2000). The Cross River gorillas: a distinct subspecies, *Gorilla gorilla diehli* Matschie 1904. *American Museum Noviates*: 3304.

Soulé, M., and Kohm, K. (eds.) (1989). *Research Priorities for Conservation Biology*. Island Press, Washington DC, pp. 96.

Soulé, M., and Orians, G. (eds.) (2001). *Conservation Biology: Research Priorities for the Next Decade*. Island Press, Washington, DC, pp. 307.

Stokes, E. (2004). Within-group social relationships among females and adult males in wild western lowland gorillas (*Gorilla gorilla gorilla*). *American Journal of Primatology* 64:223–246.

Stokes, E.J., Parnell, R.J., and Olejniczak, C. (2003). Female dispersal and reproductive success in wild western lowland gorillas (*Gorilla gorilla gorilla*). *Behavioural Ecology and Sociobiology* 54:329–339.

Sunderland-Groves, J., Oates, J., and Bergl, R. (2005). The Cross River gorilla (*Gorilla gorilla diehli*). In: Caldecott, J., and Miles, L. (eds.) *World Atlas of Great Apes and Their Conservation*. Berkeley, University of California Press, pp. 109.

Thalmann, O.H., Hebler, J., Poinar, H.N., Pääbo, S., and Vigilant, L. (2004). Unreliable mtDNA data due to nuclear insertions: a cautionary tale from analysis of humans and other great apes. *Molecular Ecology* 13:321–335.

Tutin, C., Stokes, E., Boesch, C., Morgan, D., Sanz, C., Reed, T., Blom, A., Walsh, P., Blake, S., and Kormos, R. (2005). *Regional Action Plan for the Conservation of Chimpanzees and Gorillas in Western Equatorial Africa*. Conservation International, Washington, DC.

Tutin, C.E.G., and Fernandez, M. (1983). Composition of the diet of chimpanzees and comparisons with that of sympatric lowland gorillas in the Lope Reserve, Gabon. *International Journal of Primatology* 30:195–211.

Tutin, C.E.G., Fernandez, M., Rogers, M.E., Williamson, E.A., and McGrew, W.C. (1991). Foraging profiles of sympatric lowland gorillas and chimpanzees in the Lope Reserve, Gabon. In: Whiten, A., and Widdowson, E.M. (eds.) *Foraging Strategies and Natural Diet of Monkeys, Apes and Humans,* Clarendon Press, Oxford, pp. 179–186.

Vigilant, L., and Bradley, B. (2004). Genetic variation in gorillas. *American Journal of Primatology* 64:161–172.

Voysey, B.C., McDonald, K.E., Rogers, M.E., Tutin, C.E.G., and Parnell, R.J. (1999a). Gorillas and seed dispersal in the Lope Reserve, Gabon. I: Gorilla acquisition by trees. *Journal of Tropical Ecology* 15:23–58.

Voysey, B.C., McDonald, K.E., Rogers, M.E., Tutin, C.E.G., and Parnell, R.J. (1999b). Gorillas and seed dispersal in the Lope Reserve, Gabon. II: Survival and growth of seedlings. *Journal of Tropical Ecology* 15:39–60.

Walsh, P. (2006). Ebola and commercial hunting: dim prospects for African apes. In: Peres, C.A., and Laurance, W. Chicago, *Emerging Threats to Tropical Forests.* University of Chicago Press: 175–197.

Walsh, P., Abernethy, K., Bermejo, M., Beyers, R., Wachter, P.D., Akou, M.E., Huijbregts, B., Idiata, D.M., Toham, A.K., Kilbourn, A., Lahm, S., Latour, S., Maisels, F., Mbina, C., Mihindou, Y., Obiang, S.N., Effa, E.N., Starkey, M., Telfer, P., Thibault, M., Tutin, C., White, L., and Wilkie, D. (2003). Commercial hunting, Ebola and catastrophic ape decline in Western Equatorial Africa. *Nature* 422:611–614.

Watts, D. (1987). Effects of mountain gorilla foraging activities on the productivity of their food plant species. *African Journal of Ecology* 25:155–163.

Western, D. (2003). Conservation science in Africa and the role of international collaboration. *Conservation Biology* 17(1):11–19.

Wilkie, D., and Carpenter, J. (1999). Bushmeat hunting in the Congo Basin: An assessment of impacts and options for mitigation. *Biodiversity and Conservation* 8:927–955.

Williamson, E.A., Tutin, C.E., Rogers, M.E., and Fernandez, M. (1990). Composition of the diet of lowland gorillas at Lope in Gabon. *American Journal of Primatology* 21:265–277.

Woodford, M., Butynski, T., and Karesh, W. (2002). Habituating the great apes: the disease risks. *Oryx* 36(2):153–160.

Chapter 16
Zoos and Conservation: Moving Beyond a Piecemeal Approach

Tara S. Stoinski, Kristen E. Lukas, and Michael Hutchins

1. The Contribution of AZA Institutions to Gorilla Conservation

With approximately 360 western lowland gorillas (*Gorilla g. gorilla*) in 52 North American zoos, institutions accredited by the Association of Zoos and Aquariums (AZA) have both an opportunity and responsibility to make meaningful contributions to the *in situ* conservation of gorillas. Numerous chapters within this volume have illustrated the extreme threats facing wild populations of gorillas and the urgent need for new and expanded approaches to their conservation. We believe zoos have an important role to play in gorilla conservation but to date have not sufficiently met this challenge. Many institutions have identified opportunities for supporting research and conservation of gorillas in the wild, but the approach has been largely piecemeal in that zoos have not coordinated efforts, integrated information, or pooled resources to maximize returns on investments. A cooperative, collaborative approach is needed to ensure long-term support of initiatives that have been evaluated and prioritized by experts in gorilla conservation. Our goal in this chapter thus is to describe the current status of zoo support of gorilla conservation and examine how it can be reconfigured to ensure that zoos make a substantial contribution to gorilla conservation initiatives.

To characterize the current status of zoo support for gorilla conservation, information on ways AZA institutions contribute to *in situ* conservation of eastern (*G. beringei* species) and western (*G. gorilla* species) populations of gorillas was compiled through a search for the term "*gorilla(s)*" in the *AZA Annual Report on Conservation and Science* (ARCS) for 1997–2005; a survey sent to the Ape Taxon Advisory Group (TAG) listserv; a survey sent to the gorilla keepers listserv; the Bushmeat Crisis Task Force Projects Database; and through personal communication with zoo personnel. Only projects with direct links to *in situ* conservation were included in the database (i.e., research projects on captive gorillas were not included). Additionally, only projects that listed gorillas as a specific focus were included; thus, we are underestimating the overall contribution of zoos to *in situ* gorilla conservation, as projects that contributed to the conservation of gorilla habitat but were not focused on gorillas are not represented.

It is important to mention difficulties associated with tracking zoo contributions to conservation. Although in theory a tracking system exists (ARCS), only about half of the projects included in our database were listed in ARCS; the other half were found through surveys and personal contacts. For zoos to understand their collective contribution to conservation, it is imperative that they improve their reporting of conservation activities. The ARCS reporting process has been greatly simplified and streamlined in the last few years, and we would encourage all institutions to submit yearly reports.

1.1. Contributions to Eastern Gorilla Conservation

Fifty percent of AZA-accredited institutions housing gorillas supported *in situ* eastern gorilla conservation projects during the nine-year period (Table 16.1). A total of 37 individual projects were identified; 30 (82%) were focused on mountain gorillas and 7 (18%) involved eastern lowland gorillas. Zoo support for eastern gorilla conservation was largely related to strengthening infrastructure, assisting local communities, addressing veterinary issues, conducting basic research, and providing financial support to established conservation organizations (Table 16.1).

Zoos offered infrastructure support by providing supplies to parks and park staff in the form of financial donations, uniforms, and technology; by building staff capacity through training; and by supporting on-site research facilities. Zoo support also addressed community development and education needs, and community-conservation initiatives. For example, education programs in and around the Volcanoes National Park in Rwanda and Bwindi Impenetrable Forest National Park in Uganda received funds, as did an ecosystem health program designed to improve the health of humans living near gorilla populations by identifying and treating intestinal parasites in Rwanda and the Democratic Republic of the Congo. Finally, a community-initiated conservation project in the Democratic Republic of Congo, the Tayna Gorilla Reserve, aimed at conserving eastern lowland gorillas, also received support.

Support for veterinary health primarily consisted of financial and technical aid for The Mountain Gorilla Veterinary Project, which provides health care and monitoring of mountain gorilla populations and conducts health-related research. Basic research projects aimed at gathering information on eastern gorilla populations, including studies of demography, distribution, biology, and socioecology, were also supported. Finally, some AZA institutions provided general financial support for conservation organizations working with eastern populations of gorillas. Institutions that received support include The Dian Fossey Gorilla Fund International, Partners in Conservation, and the Mountain Gorilla Veterinary Project.

TABLE 16.1. AZA institutions' support of conservation of Eastern gorillas.

Species	Project/Location	Project Type	Country**	Zoos Providing Support
G.b.beringei	Bwindi Impenetrable Forest NP	Community: Education	UGA	Columbus Zoo
G.b.beringei	Bwindi Impenetrable Forest NP	Infrastructure Support	UGA	Woodland Park Zoo
G.b.beringei	Bwindi Impenetrable Forest NP	Research: Ecological	UGA	Zoo Atlanta
G.b.beringei	Dian Fossey Gorilla Fund International	Various	RWA	Cleveland Metroparks Zoo Calgary Zoo Columbus Zoo Denver Zoo Disney's Animal Kingdom Lincoln Park Zoo Little Rock Zoo Los Angeles Zoo Milwaukee County zoo Saint Louis Zoo Santa Barbara Zoo Sedgwick County Zoo Tulsa Zoo* Woodland Park Zoo Zoo Atlanta
G.b.beringei	"DNA Fingerprinting for Positive Gorilla Identification"	Research: Genetic	RWA	Calgary Zoo San Diego Zoo Woodland Park Zoo
G.b.beringei	"Hormones to Indicate the Well Being of Free-Living Mountain Gorillas"	Research: Welfare	UGA	Smithsonian National Zoological Park
G.b.beringei	"Monitoring Mountain Gorilla Populations"	Research: Demographic	RWA	Wildlife Conservation Society
G.b.beringei	Mountain Gorilla PHVA	Research: Behavior/Ecology	RWA/UGA	Columbus Zoo Denver Zoo
G.b.beringei	Mountain Gorilla Veterinary Project	Veterinary Care	RWA/UGA	Baltimore Zoo Cincinnati Zoo

(*Continued*)

TABLE 16.1. AZA institutions' support of conservation of Eastern gorillas—Cont'd.

Species	Project/Location	Project Type	Country**	Zoos Providing Support
G.b.beringei	Partners in Conservation	Community: Development	RWA/ UGA	Columbus Zoo Sacramento Zoo* Columbus Zoo Memphis Zoo
G.b.beringei	"Screening of mountain gorilla for possible causes of diarrhea"	Research: Veterinary	RWA/ UGA	Saint Louis Zoo
G.b.beringei	Virunga NP	Infrastructure	DRC	Disney's Animal Kingdom
G.b.graueri	Tayna Gorilla Reserve	Research: Demography, census	DRC	Louisville Zoo Oklahoma City Zoological Park
G.b.graueri	Supplies for staff at Kahuzi-Biega NP	Infrastructure	DRC	Houston Zoo Wildlife Conservation Society
G.b.graueri	Census at Kahuzi-Biega NP	Research: Census	DRC	Columbus Zoo Dallas Zoo Wildlife Conservation Society

* Institution does not house gorillas
**Country: Democratic Republic of Congo (DRC), Rwanda (RWA), Uganda (UGA)
Note: Not all individual projects are listed.

1.2. Contributions to Western Gorilla Conservation

Only 30% of AZA institutions housing gorillas supported *in situ* conservation of western gorillas (Table 16.2). Approximately 25 individual projects were identified, 5 (20%) of which supported Cross River gorilla projects, with the remaining 20 (80%) supporting western lowland gorilla conservation. Zoos reported providing assistance to projects that included basic research, park management, education, sanctuaries, fund-raising, and a bushmeat project (Table 16.2).

Basic research projects at five sites (Mbeli Bai, Afi Mountain, Goualougo Triangle, Mongambe and Lope National Park) received support for studies focused on distribution, behavior, genetics, nutrition, and ecology. Applied studies examining the commercial logging trade and bushmeat issue also received funding. Direct financial support for park management was provided to the Counkouati-Douli National Park, Nouabale-Ndoki National Park, and Lac Tele Reserve. Additionally, capacity building of park staff was supported through a regional workshop on gorilla research and conservation that involved protected area managers, conservation organizations, and researchers from Cameroon and Nigeria. Community-oriented programs were sponsored, although perhaps slightly less than in eastern Africa where human and gorilla populations are in much closer proximity. Monitoring and treatment of human intestinal parasites in the Central African Republic received zoo support, as did *in situ* educational initiatives. For example, educational materials were developed for the Mbeli Bai education project "Club Ebobo." Additionally, zoos supported sanctuaries, which assist in conservation through activities such as educating local populations about indigenous wildlife and enforcing wildlife laws by providing space for confiscated animals (Farmer and Courage, Chapter 3, this volume). Mefou Sanctuary and Limbe Wildlife Center, both in Cameroon, and Project Protection des Gorillas in Republic of Congo have all received support from zoos. Finally, we feel it is important to mention that the Bushmeat Crisis Task Force has received support from over 25 zoos. Although this collaborative organization includes much more than gorillas in its conservation objectives, its focus on the primary cause of western gorilla decline means that funds provided will hopefully contribute to their conservation (Eves *et al.*, Chapter 17, this volume).

1.3. Are Zoos Making a Difference?

Combining the data presented for both species, we find that slightly more than half (58%) of AZA institutions housing gorillas were involved in or supported *in situ* conservation for the genus during the nine-year period. Approximately 60 projects were supported: 7% focused on the *diehli* subspecies, 10% on the *graueri* subspecies, 37% on the *gorilla* subspecies, and 47% on the *beringei* subspecies. Exactly half (50%) of the supported projects involved basic and applied research on gorillas. Eighteen percent supported park infrastructure, 15% supported veterinary work, 8% supported community

TABLE 16.2. AZA institutions' support of conservation of Western gorillas.

Species	Project/Location	Project Type	Country**	Zoos Providing Support
G.g.diehli	Afi Mountain Wildlife Sanctuary/ Cross River Gorilla Project	Research: Behavior/Ecology	NIG	Columbus Zoo Disney's Animal Kingdom Lincoln Park Zoo Pittsburgh Zoo Wildlife Conservation Society
G.g.diehli	Bamenda Highlands	Research: Demography	CAM	Wildlife Conservation Society
G.g.diehli	Boshi Forest	Research: Behavior/Ecology	NIG	Lincoln Park Zoo
G.g.diehli	Regional Workshop	Park Management	CAM/ NIG	Wildlife Conservation Society
G.g.gorilla	Bai Hokou	Research: Ecology	CAR	Cleveland Metroparks Zoo
G.g.gorilla	Counkouati-Douli National Park and Lac Tele Reserve	Park Management	ROC	Wildlife Conservation Society
G.g.gorilla	"Digestive Strategies in Lowland Gorillas"	Research: Nutrition	CAR	Columbus Zoo
G.g.gorilla	"Effective Population Sizes for Western Lowland Gorillas"			
G.g.gorilla	"Forensic Techniques for Identifying Gorillas"	Research: Genetic	GAB	Wildlife Conservation Society
G.g.gorilla	"Intestinal Parasite Loads in Indigenous People Living Near Lowland Gorillas"	Research: Genetic	GAB	Wildlife Conservation Society
		Research: Human Influence	CAR	Cleveland Metroparks Zoo
G.g.gorilla	Goualougo Triangle	Research: Ecology	DOC	Brevard Zoo* Columbus Zoo
G.g.gorilla	Limbe Wildlife Center	Sanctuary	CAM	Columbus Zoo Little Rock Zoo
G.g.gorilla	Lope National Park	Research: Behavior/Ecology	GAB	Wildlife Conservation Society
G.g.gorilla	Mbeli Bai	Education	ROC	Brevard Zoo* Columbus Zoo Wildlife Conservation Society
G.g.gorilla	Mbeli Bai	Research: Behavior/Ecology	ROC	Brevard Zoo* Brookfield Zoo Burnet Park Zoo* Busch Gardens Cincinnati Zoo

Taxon	Project	Type	Country	Institution(s)
G.g.gorilla	Mefou Sanctuary	Sanctuary	CAM	Columbus Zoo, Little Rock Zoo, Pittsburgh Zoo, Toronto Zoo, Wildlife Conservation Society, Woodland Park Zoo
G.g.gorilla	Mongambe Research Camp	Research: Behavior/Ecology	CAR	Cleveland Metroparks Zoo, Denver Zoo, Zoo New England
G.g.gorilla	"National Trends in *G. g .gorilla* populations"	Research: Demography	GAB	Wildlife Conservation Society
G.g.gorilla	Nouabale-Ndoki	Education	ROC	Woodland Park Zoo
G.g.gorilla	Nouabale-Ndoki	Park Management	ROC	Wildlife Conservation Society
G.g.gorilla	Nouabale-Ndoki	Research: Bushmeat	ROC	Lincoln Park Zoo
G.g.gorilla	"Phylogeography of the genus Gorilla"	Research: Genetic	GAB	Wildlife Conservation Society
G.g.gorilla	Pan African Sanctuary Alliance	Sanctuary	VAR	Brevard Zoo*, Cleveland Metroparks Zoo, Oakland Zoo*
G.g.gorilla	"Preventative Health Program for Lowland Gorillas"	Research: Veterinary	ROC/ CAR	Wildlife Conservation Society
G.g.gorilla	Project Protection des Gorillas	Sanctuary	ROC	Columbus Zoo
G.g.gorilla	"Quantifying Threats to Gorilla Populations"	Research: Human Influence	ROC/ GAB/ CAM	Wildlife Conservation Society
G.g.gorilla	"Strategic Action Plan for West and Central African Great Apes"	Action plan	VAR	Cleveland Metroparks Zoo
G.g.gorilla	Wildlife Management and Reduction of Bushmeat Trade	Park Management	ROC	Wildlife Conservation Society

* Institution does not house gorillas
** Country: Cameroon (CAM), Central African Republic (CAR), Gabon (GAB), Nigeria (NIG), Republic of Congo (ROC), Various (VAR)
Note: Not all individual projects are listed.

development projects, 5% supported sanctuaries, and 2% supported community conservation work. In terms of the level of financial support given to these projects, we have data for 32 projects supported in 2000–2005, as financial data were not provided in ARCS before this period, and not all projects included financial information in their reports. The total amount provided to these projects equaled $212,863.

Further analyses of the database highlight several critical aspects of zoo support for the *in situ* conservation of gorillas. First, support is short term; very few projects were supported over consecutive years by a single or even multiple institutions. For example, for eastern gorilla populations, we found that only 3 of the 26 institutions provided support, and only 5 of the 37 projects received support, for the entire nine-year period. Second, support is not coordinated in any way and, as a result, is unevenly distributed across subspecies or projects. For example, the most threatened subspecies, *diehli*, received the smallest percentage of support, and 20% of support for western lowland gorillas went to a single project. Part of this inequality probably reflects variation between the study sites themselves; more established sites with a greater international presence are more likely to be known by zoos and thus supported. Third, support is very unevenly distributed between institutions; although 26 institutions contributed to eastern gorilla conservation, two organizations provided the support for over 40% of the projects. Finally, there was unequal support for the various types of projects, with research-based projects receiving much more support than park infrastructure or local communities.

We do not have any direct measures of how effective the above contributions are to the *in situ* conservation of gorillas. We can, however, examine how zoo involvement overlaps with current conservation priorities for gorillas, detailed in a number of recent papers (Tutin, *et al.*, 2005; Butynski, 2001; Kalpers *et al.*, 2003; Stokes, Chapter 15, this volume). The recurring message is that what is needed includes the establishment and long-term, consistent, sustainable support of protected areas and protected area personnel; capacity building and institutional strengthening in habitat countries; enforcement of wildlife laws; and close monitoring of populations. As noted above, a significant portion of zoo contributions has been dedicated to population monitoring, either through applied census work or basic studies. However, zoos have focused significantly less on the other areas—support of protected areas, capacity building and law enforcement—and, with a few exceptions, have not provided long-term/consistent support. Thus, zoos need to rework their approach to ensure their conservation activities are maximally effective. And this reworking must occur immediately if zoos are to meaningfully contribute to the conservation of great apes; current estimates are that gorilla populations could be functionally extinct from western equatorial Africa within the next few decades (Walsh *et al.*, 2003). Thus, the options and opportunities to save gorillas decrease daily and zoos should act responsibly and promptly.

1.4. A New Approach to Zoo-Based Conservation

In order for zoos to play a significant role in the conservation of gorillas, a new strategy needs to be developed that is both long term and focused. We would argue that this approach should also be collective; working together, zoos can raise significant funds and leverage considerable outside support for gorilla conservation. The current piecemeal approach of individual zoos providing small sums of money to projects, although certainly a valuable contribution, is simply an insufficient strategy given the conservation challenges that lie ahead. This is not to say that zoos should stop this practice; support by individual zoos can and does play an important role in increasing resources available to conservationists, often during times of crisis. Because of this, we have provided a summary of priority actions for gorilla conservation (Table 16.3) and how zoos could contribute to these efforts, either individually or collectively, for example, by joining together to adopt a park. We would suggest that institutions that continue in a small grant approach develop strategies that enable them to support projects over the long term (five years or more), which will enable recipients to focus on implementing their conservation activities rather than immediately trying to search for the next year's funding.

It would appear, however, that additional, coordinated strategies are needed to ensure a significant contribution by zoos to gorilla conservation. The AZA Ape TAG is currently formulating a more strategic approach to conservation and research that would take advantage of the collective strengths of AZA members. The basic concept is that the TAG and associated Species Survival Plans (SSP®) would select a portfolio of priority projects for support by the AZA and the roughly 120 institutions that house apes in their collections. The goals of the action plan would be to strengthen zoos' connection to *in situ* conservation.

In formulating such a plan, it is critical that the projects selected are of strategic importance for conservation and have a high probability of success; it is, therefore, also important that an expert peer review system be used to decide which projects merit support. In the current uncoordinated situation, individual zoo directors receive requests for financial and logistical support from numerous individuals and organizations during all times of the year. Often, there is no way to evaluate the need and quality of these proposals, to compare them with other potential projects that might be supported, or to time the requests for financial assistance so that they coincide with annual budgeting processes. The end result is that many excellent projects may not be funded at all, and, in addition, there is a higher probability that limited zoo conservation funds may not be used efficiently or for the highest priority purposes.

The TAG strategy for resolving this challenge and encouraging more institutions to get involved is to develop a collective action plan that directly links with priorities established by the field community (e.g., Kormos and Boesch, 2003; Tutin, *et al.*, 2005). This action plan will then be sent to all

TABLE 16.3. Priority actions for gorilla conservation and the role of zoos (adapted from www.western-gorilla.org).

Priority	Why?	What is needed?	What can zoos do?
Support of parks	Insufficient funding and lack of financial security are the biggest obstacles to effective management of protected areas	Evaluate existing parks for ability to protect viable populations of gorillas and genetic and ecological variation	Provide funds for park surveys; make long-term commitment to park infrastructure (e.g., adopt a park program)
Law enforcement	Major decline in many gorilla populations is from illegal activities; numerous examples that effective law enforcement is foundation for conservation	Train, equip, and provide competitive salaries for wildlife law enforcement personnel; regulate and control bushmeat trade; improve understanding of laws	Provide funds for training and salary of and equipment for guards; educate U.S. public and government about international legislation regarding wildlife; provide educational materials for habitat countries
Capacity building/training	Training of in-country personnel to monitor and manage wildlife resources is lacking	Creation of in-country capacity and international collaboration for evaluation of research and conservation activities; improved communication/access to conservation/research activities	Establish grants/scholarships for students; sponsor internships for park staff; donate expertise
Long-term, basic research projects	Baseline data on basics such as distribution, numbers, basic demographic parameters, as well as monitoring of changes is needed; presence of researchers deters illegal activities	Population survey of all gorilla species; regular censuses of small, critically endangered populations; baseline information on demographics, etc.	Long-term funding for repeated census work, basic research studies; dedicate staff time for surveys
Support for legislation/government funding	Legislation already in place to provide support for gorillas, but need to make sure funds are appropriated and effectively used	Continued pressure on Congress to appropriate funds to the Great Ape Conservation Fund; appropriate use of other governmental funds	Create an informed constituency by educating public about these issues; hold a letter-writing campaign; join the Bushmeat Crisis Task Force

AZA institutions housing apes with a cover letter explaining what the TAG intends to accomplish and membership levels for participation. Member benefits will vary depending on the level of funding; for example, zoos contributing at the lowest membership level will be able to say they support a limited number of projects in the action plan whereas those at higher levels will be able to show support for a larger number of projects. Through such a plan the cost to individual institutions may be comparatively low, but the total sum of funds sent to the field and to specific projects will be significantly larger and longer term than the funds currently provided by zoos. This general approach has been successfully implemented by several other AZA Conservation and Science Programs (e.g. Madagascar Conservation Action Program, www.madagascarfaunagroup.org), and the conservation subcommittee of the Ape TAG is now developing the specifics for its program implementation.

The benefits of a collaborative approach could be numerous. For example, this approach could place zoos in a much more favorable position in a highly competitive fundraising environment. Having significant funds in hand indicates that the community is serious about implementing the plan. Having a portfolio of well-designed projects that match priorities set by field scientists will also help to reassure donors that the projects are important and have a high probability of success. Planning and working cooperatively indicates a willingness to share resources—human and financial—and to work together for the greater good. In these difficult financial times, donors want to know that their limited funds will be spent in the most effective and efficient way possible, that they will have the desired effect, and that recipient organizations are working together to reach the stated goals.

What would be AZA'S role? As a member services organization, AZA staff's role is to facilitate, support, and promote its members' work in conservation and science, and to help build members' capacity to respond to conservation challenges. It does this by assisting with cooperative planning, partnership-building, national-level fundraising and fund management, training, and by generating positive media coverage. In terms of its fundraising and management role, AZA would assist with national level fundraising and serve as a pass-through for funds going to members and partners. AZA member institutions and partners would do the actual work and receive the recognition they deserve for their support and participation.

It is too early to tell whether a cooperative approach to planning and fundraising will be successful. Certainly, it will have to overcome intrinsic organizational biases that favor the individual institution over the collective— in this case, the AZA. Certainly, involvement in a collective effort could mean loss of individual organizational identity, and this is an issue that needs to be specifically addressed. If this approach is going to work, participating institutions will need to derive increased benefits in terms of funds raised, conservation action accomplished, improved staff morale and retention, and increased public recognition. But all this is possible. For example, AZA could provide template press releases that are easily modified for local consumption. Local newspapers and television would run stories on what their institution was doing for ape conservation, both individually and collectively. Field workers, representing priority projects, could be made available to speak at participating institutions, thus providing further acknowledgment of the institution's role in supporting conservation. This increased visibility can hopefully be leveraged into improved fundraising at the local level, not only for ape conservation, but for other institutional priorities as well.

Many challenges lie ahead for zoos and their individual and collective efforts to assist great apes. However, given the many threats to wild ape populations, it is critical for zoos to become more engaged in ape conservation, and it is also essential that they do it now. Time is of the essence. Options for conserving our closest living relatives in nature will certainly be fewer with each passing year. In order to have a significant effect on global conservation efforts, accredited zoos must advance beyond a piecemeal approach to conservation of great apes and other endangered creatures, and they must also do this in partnership with other relevant agencies and organizations.

References

Butynski, T.M. (2001). Africa's great apes. In: Beck, B.B., Stoinski, T.S., Hutchins, M., Maple, T.L., Norton, B., Rowan, A., Stevens, E.F., and Arluke, A. (eds.), *Great Apes and Humans: The Ethics of Coexistence*, Smithsonian Institute Press, Washington, DC, pp. 3–56.

Kalpers, J., Williamson, E.A., Robbins, M.M., McNeilage, A., Nzamurambaho, N.L., and Mugiri, G. (2003). Gorillas in the crossfire: population dynamics of the Virunga mountain gorillas over the past three decades. *Oryx* 37(3):326–337.

Kormos, R., and Boesch, C. (2003). Regional Action Plan for the Conservation of Chimpanzees in West Africa. Washington, DC: Conservation International.

Tutun, C., Stokes, E., Boesch, C., Morgan, D., Sanz, C., Reed, T., Blom, A., Walsh, P., Blake, S., and Kormos, R. (2005). Regional Action Plan for the Conservation of Chimpanzees and Gorillas in Western Equatorial Africa Washington, DC: Conservation International.

Walsh, P.D., Abernethy, K.A., Bermefo, M., Beyers, R., De Wachter, P., Akou, M.E., Huijbregts, B., Mambounga, D.I., Toham, A.K., Kilbourn, A.M., Lahm, S.A., Latour, S., Maisels, F., Mbina, C., Mihindou, Y., Obiang, S.N., Effa, E.N., Starkey, M.O., Telfer, P., Thibault, M., Tutin, C.E.G., White, L.J.T., and Wilkie, D.S. (2003). Catastrophic ape decline in western equatorial Africa. *Nature* 422(6932):611–614.

Chapter 17
The Bushmeat Crisis Task Force (BCTF)

Heather E. Eves, Michael Hutchins, and Natalie D. Bailey

1. Bushmeat and BCTF

There is an immediate and rapidly growing threat to the future of Africa's irreplaceable wildlife: the unsustainable, often illegal, unregulated commercial harvesting of wildlife for meat, also known as the bushmeat trade.[1] A multi-billion dollar industry, bushmeat is now the most significant threat to wildlife populations, including great apes, in Africa today (Bennett *et al.*, 2002b). Complex interactions between extractive industries (logging, mining, oil), transportation systems (roads and railroads), human population growth, absence of dietary alternatives, lack of governmental infrastructure, and widespread poverty have resulted in a rapid increase in the commercial trade in wildlife for meat. The commercial bushmeat trade results in the depletion of a wide range of wildlife, ranging from large- to small-bodied species. Due to their relatively low productivity, tropical forests are particularly vulnerable to the impacts of the commercial wildlife trade. The result is that, while forests may remain intact, they may be nearly devoid of wildlife—a phenomenon that has been termed the "empty forest syndrome" (Bennett *et al.*, 2002a).

The African bushmeat crisis affects a broad range of taxa, including endangered and threatened species. Hunting targets include great apes, elephants, duikers (forest antelope), other primates, rodents, reptiles, and birds. Dramatic reductions in common prey species also affect a number of carnivores (Ray *et al.*, 2002).

While Africans have hunted wildlife for many millennia, it is only recently that the crisis has arisen as growing urban populations have commercialized the trade. The massive off-take of wildlife has subsequently become unsustainable. In the Congo Basin alone, harvest estimates for all species combined

---

[1] *Wild meat* refers to meat legally and sustainably harvested for personal consumption, whereas *game meat* refers to meat that is sustainably, legally harvested through a regulated commercial trade.

range from 1 to 3.4 million metric tons per year; data from virtually all areas studied suggest that this intensity of hunting simply cannot be sustained (Wilkie and Carpenter, 1999; Robinson and Bennett, 2000; Bennett *et al.*, 2002a; Fa *et al.*, 2002).

The bushmeat trade is not only a wildlife and habitat conservation crisis, it is also a looming human tragedy (Eves and Hutchins, 2000). Wild animals are a significant source of protein for rural African communities, often making up 40–60% of protein needs (Auzel and Wilkie, 2000; Wilkie *et al.*, 2002). Loss of these important resources is a food security issue and could result in serious nutritional deficiencies, or even starvation, if appropriate dietary substitutes cannot be identified.

The commercial bushmeat trade is also a serious threat to human health. Hunting and meat preparation exposes people to emerging infectious diseases, such as Ebola, that are carried by wildlife hosts (Hahn, 2000; Wolfe, 2000). The epidemiology of Ebola is not well understood. However, outbreaks in Central Africa in 1994, 1996, 2001, and 2003 have all been associated with close contact with or consumption of nonhuman primates (Anon., 2001, 2002; Hearn, 2001). Governments and researchers are now working to prevent further outbreaks of this highly contagious and usually fatal disease (Lawson, 2002).

Similarly, some of the world's leading scientists (e.g., Ziff, 2002) have linked the bushmeat trade with the emergence of HIV/AIDS. Simian immunodeficiency virus (SIV) may have entered human bloodstreams during the butchering of nonhuman primates, such as chimpanzees, for meat and mutated into HIV in human hosts, eventually spreading around the globe (Gao *et al.*, 1999; Hahn *et al.*, 2000; Keele *et al.*, 2006).

Bushmeat is a short-term "band-aid" that cannot resolve the long-term pressures by which it is driven: poverty, lack of dietary alternatives, and rapidly growing human populations. When large-scale commercial exploitation causes wildlife populations to decline to near or total extinction, humans will still face these underlying problems. Unless we address the bushmeat crisis' root causes, it is unlikely that we can solve this highly complex conservation and human welfare issue.

Some conservationists have promoted sustainable harvest of wildlife to address the bushmeat problem (e.g., Bowen-Jones *et al.*, 2002). A recent global review of wildlife exploitation identified a complex set of conditions necessary for sustainable hunting of wildlife to occur: 1) harvest rates must not exceed production; 2) wildlife management goals must be clearly specified; and 3) biological, social, and political conditions must be in place to allow effective management to occur (Robinson and Bennett, 2000). Many of these conditions are not in place in the majority of the Congo Basin, or elsewhere in Africa (Robinson and Bennett, 2000). For such conditions to exist, it is essential that massive capacity building and public awareness efforts are supported through the collaborative efforts of many stakeholders (Eves and Stefan, 2002).

Collaboration among multiple stakeholders with divergent approaches to an issue is often necessary to resolve complex problems that require simultaneous implementation of many interrelated actions. The most successful collaborations are those in which partners share responsibility, authority and accountability for achieving the desired results (Chrislip and Larson, 1994). Collaborators must share a vision that is greater than that of any one participant and must believe that the other partners are essential for achieving success. With these aims in mind, numerous multi-organizational conservation coalitions have been formed in the past decade, including the Biodiversity Support Project (BSP), International Gorilla Conservation Program (IGCP), Central African Regional Program for the Environment (CARPE), and the Africa Biodiversity Collaborative Group (ABCG).

The bushmeat crisis erupted in Africa in the late 1980s and early 1990s as dramatic evidence of the scale of the wildlife trade in the Congo Basin became available (Hart, 1978; Anadu et al., 1988; Fa et al., 1995). Some have suggested that this shift was due to the sudden development of logging activity in the region, following the over-exploitation of forest resources elsewhere in the world (Wilkie et al., 1992). The many agencies and individuals observing and documenting the dramatic growth in the trade recognized that this issue demanded immediate attention. In February 1999, 34 experts representing 28 organizations and agencies were invited to a meeting organized by the Conservation and Science Department of the Association of Zoos and Aquariums (AZA). The purpose of the meeting was to discuss the causes of and potential solutions to the bushmeat crisis. This group agreed to form the Bushmeat Crisis Task Force (BCTF) to unify the then many independent and uncoordinated efforts. The group agreed that solutions must engage African partners, respect the authority of each affected nation's government, and take the needs of people, as well as wildlife, into account (Eves and Hutchins, 2001).

The BCTF mission is to build a public, professional, and government constituency aimed at identifying and supporting effective solutions to the bushmeat crisis in Africa and around the world. BCTF's primary goals are to: 1) work with BCTF members and partners to focus attention on and raise awareness of the bushmeat crisis among both the public and key decision-makers; 2) establish an information database and mechanisms for information sharing; 3) facilitate engagement of African partners and stakeholders; and 4) facilitate collaborative decision-making, fund-raising, and actions among BCTF members and partners.

The BCTF operates under the direction of a seven-member elected Executive Committee appointed by a Steering Committee, which includes representation from the 36 member organizations (as of 2006) that provide financial and other support for the organization. Additional funding for projects comes from grants and donations. The BCTF network includes a diverse set of nongovernmental conservation organizations (those engaged in ecosystem conservation, species conservation, animal protection, and accredited zoological parks and aquaria), and cooperates extensively with African and U.S. government agencies. In addition, BCTF works with more than 20 other bushmeat networks around the globe to coordinate actions with African partners. The network is constantly expanding to take advantage of newly identified resources and expertise.

In May 2001, BCTF organized and hosted a meeting of bushmeat experts from more than 20 countries to reach a common understanding of the emerging crisis, to identify priority solution areas and actions, and to establish the basis for a global action network. The meeting resulted in the BCTF Action Plan, a set of priority solutions to address the bushmeat crisis in Africa.

BCTF's long-term strategy includes national and international policy development, educational outreach and capacity building, and improved protected area management and monitoring. Short-term solutions include formation of public-private partnerships, development of protein and economic alternatives, formation of bushmeat hunter and trader associations, and development of awareness campaigns for the public and key decision-makers. No single organization or agency can effectively achieve all of these goals. BCTF works with its Supporting Members and partners to develop collaborative efforts that employ the strengths of each to address these priority actions. BCTF therefore plays a critically important facilitation role.

2. Bushmeat and Great Apes

Left unresolved, the commercial bushmeat trade could potentially result in the extinction of African great ape populations within a single generation (Goodall, 2000). Gorillas (*Gorilla gorilla* ssp.), chimpanzees (*Pan troglodytes* ssp.), and bonobos (*Pan paniscus*) are estimated to comprise only 1–4% of the total bushmeat trade (Bowen-Jones and Pendry, 1999). However, given their low fecundity and long period of infant dependence, and combined with other factors, such as habitat loss and disease, these losses are significant. Though the biodiversity concept affirms the value of all wildlife, great apes are of special concern to numerous stakeholders around the globe. Indeed, there are many ethical, cultural, economic, eco-logical, political, and scientific arguments for giving special attention to great apes and their conservation (Beck *et al.*, 2001; Eves *et al.*, 2002). These arguments are summarized below.

2.1. Ethical Perspectives

Scientific studies have revealed a close genetic relationship between humans and great apes, and there are striking similarities between apes and humans in behavior, anatomy, and physiology, which all point to a common ancestry (Moore, 1996; Corbey and Roebroeks, 2001). Arguments for great ape conservation are sometimes based on our kinship with and perceived ethical obligations to these sentient and highly intelligent nonhuman animals (Goodall,

1990; Fouts, 1997; Warren, 2001). Some have gone so far as to suggest that great apes be accorded legal rights similar to those of humans (Cavalieri and Singer, 1993; Wise, 2001). Similarly, Rose (1998) has recommended that religious missionaries and educators in Africa create "wildlife missions . . . to foster moral and humanistic concerns for living wildlife." Access to wildlife resources by future genrations of Africans is also an important ethical consideration (Eves and Wolf, 2005).

While funding campaigns based on kinship with and ethical obligations to great apes have been enormously successful in the United States and Europe, such efforts must be carefully examined for their appropriateness and effectiveness in Africa. In the award-winning BBC film *Ape Hunters* (2002), numerous interviews with bushmeat hunters and market sellers revealed their disagreement with the notion that apes are like humans. Even a participant in a "poacher to protector" project suggested that there is nothing wrong with eating apes, as they are an important source of protein.

While emotionally compelling to many Westerners, value-based concepts like "animal rights" raise questions about their ethical appropriateness and practical application in developing countries (Hutchins and Wemmer, 1986; Redmond, 1998; Hutchins, 2002). Do comparatively wealthy, well-fed developed societies have the right to impose their moral and cultural values on others, especially indigenous peoples living in developing countries? Will they be effective in reaching conservation goals? Appeals for the conservation of endangered species, particularly those based on philosophical arguments, are unlikely to be successful with impoverished people more concerned about finding their next meal (Myers, 1979). In addition, traditional African concepts of nature and the human-animal relationship are no less sophisticated, multi-dimensional, or paradoxical than those found in Western societies (see Kellert, 1996; Morris, 1998, 2000).

2.2. Cultural Perspectives

Arguments for great ape conservation may be more effective if based on their cultural importance to African peoples. Numerous references to apes in the mythology and traditions of African societies illustrate their societal importance (Richards, 1993). Animal totems have been identified with different cultural/kin groups (clans), and these totems reflect an intimate relationship with and dependency on nature (Morris, 1998, 2000). Until recently, some regional and local cultural taboos prohibited human consumption of apes. However, the importance of these prohibitions has diminished as the region's economic and social collapse has gradually eroded cultural traditions (Tashiro, 1995; Won wa Musiti, 1999). Cultural pride may be an effective means to promote wildlife conservation in African countries, but this requires an awareness of the scarcity of wildlife and a sense of personal responsibility for what is happening to it (Mordi, 1991). Conservation International's bushmeat awareness campaign in Ghana is based on local peoples' traditional connection to wildlife through their cultures' totemic system and stresses the endangered status of many traditional totemic animals (Conservation International, 2002).

It is important to note that in some cultures the traditional view of apes is negative. For example, the Mende people of West Africa believe there is a link between chimpanzees and the practice of witchcraft (and therefore political power), and that, as long as chimpanzees survive, "conservation objectives will be achieved at the risk of human degradation" (Richards, 1995). The effects of traditional perspectives and attitudes can therefore be counterproductive, suggesting that a thorough understanding of local beliefs is necessary before launching tradition-based conservation education or awareness initiatives in Africa.

2.3. Economic Perspectives

Some conservationists have used economic arguments to advance the cause of conservation, highlighting the economic value of wildlife and nature as a reason for their preservation (Myers, 1979; Mann and Plummer, 1995). Great apes can generate enormous amounts of foreign currency for conservation efforts, particularly through ecotourism and related industries. Mountain gorillas in Rwanda and Uganda are one of the sole sources of income for many local people in the region (Wilkie and Carpenter, 1999). Tourism is a relatively benign economic activity when compared with other alternatives. However, there are many potential pitfalls with this approach. For example, contact between humans and habituated gorilla troops increases the chances of disease transmission and can alter behavior in ways that are detrimental to apes and their environments (Litchfield 1997; and Chapter 4, this volume). In addition, financial profits from tourism do not always benefit local communities, thus making it less likely that local people will support conservation efforts. Furthermore, the infrastructure to support large numbers of tourists (e.g., hotels, roads, waste disposal, etc.) can be environmentally destructive, making it essential that ecotourism programs are appropriately regulated (Ecotourism Society, 1993). Economies that are dependent on tourism are at great risk during times of social instability, and this can be a drawback (Plumptre et al., 1997). Civil war and regional conflicts are still far too common in many contemporary African nations.

Westerners who never travel to Africa can also provide economic support for wildlife conservation by donating to conservation organizations that work in the region. Appeals for public support to address the bushmeat crisis may be more effective when great apes are at the forefront of awareness campaigns. That is, the public may be more willing to contribute when they become aware that gorillas and chimpanzees are threatened with extinction due to human hunting and consumption. For better or worse, public interest in supporting conservation appears to decrease when the focus is on landscapes or other less popular, or well-known species (Stoinski et al., 2002).

2.4. Ecological Perspectives

It has been argued that certain, so-called "keystone" species should be conserved because of the critical functional roles they play in their ecosystems (Wilson, 1992). Great apes occupy an important ecological niche in the forests of Central and West Africa (Cowlishaw and Dunbar, 2000). They contribute to habitat creation and seed dispersal (Terborgh, 1999) and have cooperative feeding strategies with other species (Ruggiero and Eves, 1998). Due to important seed dispersal roles of terrestrial mammals, the habitats of gorillas and other mega-herbivores can be significantly impacted by local extinctions (Maisels *et al.*, 2001). The loss of such keystone species from over-hunting could eventually alter the composition of tropical forests and permanently change the natural ecology of the region. While current hunting levels suggest that only 1–4% of the bushmeat trade is composed of great apes, populations may be far more heavily impacted than once thought. Ape populations in Gabon were reduced by 50% in just 17 years (1983–2000) as a direct result of bushmeat hunting (Walsh *et al.*, 2003). It is further predicted that ape populations will likely decline by 80% in less than 20 years (Walsh *et al.*, 2003) over much of the region. Such catastrophic predictions catapult apes into critically endangered status. With ape meat continuing to appear in bushmeat markets, the prognosis is dire.

Great apes have the potential to be important "flagship species" for promoting conservation of apes and the many animal and plant species that share their forest habitat (AWF, 2001). When conservation dollars are used to create and maintain protected areas, employ ecoguards, and conduct public awareness campaigns on behalf of great apes, many other species of wildlife and plants also benefit.

2.5. Political Perspectives

National and international law protects great apes in all of the countries where they exist. All African ape range states are signatories of the Convention on International Trade in Endangered Species of Wild Fauna and Flora (CITES) and are therefore bound to assure strict control of international trade in these species. In 2000, CITES signatories and interested NGOs, including BCTF, committed to the formation and support of the CITES Bushmeat Working Group (BWG), a committee composed of the directors of wildlife and protected areas from affected central and western African nations. The CITES BWG works together to monitor and control the illegal bushmeat trade in the region, including the trade in great apes.

Both the U.S. government and the co-led Great Ape Survival Project (GRASP) initiative of the United Nations Environment Programme (UNEP) and United Nations Educational, Scientific and Cultural Organisation (UNESCO) have established funds to support great ape conservation. The U.S. Great Ape Conservation Fund has the authority to allocate US $5 million in support of ape conservation efforts per year, but the current congressional allocation is for approximately US $1 million per year. GRASP has received US $2.9 million to support specific projects and to develop ape conservation action plans in each range state. Thus, great apes are generating international political will and the necessary human and financial resources to undertake focused conservation action.

Finally, efforts are underway to establish a new category of global recognition for the great apes as being species of "outstanding universal value from a scientific, education and cultural perspective, warranting a special conservation effort" through UNESCO's World Heritage Program (van Hooff, 2001). This would result in great apes being declared "World Heritage Species" and, as the supporters of this effort suggest, result in further resources and actions to support their conservation (Wrangham *et al.*, Chapter 14, this volume).

2.6. Scientific and Health Perspectives

Great apes are of enormous scientific value. Due to their shared ancestry with humans, great apes are a rich source of information on the early origins of human culture and behavior (Goodall, 1971; Stanford, 2001). The study of great apes contributes significantly to our understanding of basic animal behavior, ecology, and biology (Goodall, 1986).

As mentioned previously, great apes and bushmeat are currently at the center of one of the most significant health issues facing the globe: HIV/AIDS. Leading experts reported their findings on linkages between bushmeat and HIV/AIDS at a U.S. congressional briefing in February 2002. More than 13 species of primates commonly consumed as bushmeat are infected with some form of SIV. The briefing's dramatic conclusion by researcher Dr. Beatrice Hahn and her colleagues was that "the bushmeat trade is not only driving chimpanzees to extinction, not only exposing humans to other SIVs and, likely, other pathogens, but it's wiping out the very species that could lead to a fuller understanding of HIV/AIDS" (Ziff, 2002). Recently, Dr. Hahn and colleagues traced the origins of pandemic and nonpandemic HIV-1 to distinct, genetically isolated chimpanzee communities in southern Cameroon, where SIVcpz antibodies and nucleic acids in fecal samples demonstrated prevalence rates of up to 35% (Keele *et al.*, 2006). While some conservationists fear that drawing attention to the connection between great apes and human disease could ultimately lead to their demise, Hahn and her colleagues have presented a cogent argument for conserving great apes in their natural habitats. Indeed, critical biomedical research depends upon fecal samples from wild chimpanzees. Finding a cure for HIV/AIDS and preventing other emerging diseases from entering the global human population necessitates the preservation of ecosystems in which these viruses evolved.

The recent outbreak of Ebola in northwestern Republic of Congo underscores the urgency of addressing the bushmeat trade. Reports of Ebola-related great ape deaths in and around the Lossi Gorilla Conservation Area began to emerge in late 2002. A wave of human deaths was reported in early 2003. By February 2003 more than 150 gorillas and chimpanzees and nearly 60 humans were reported dead as a result of the disease (WHO, 2003). Given that the Congo Basin contains a major concentration of apes, these losses could have long-term effects on conservation efforts.

3. Bushmeat, Apes, and BCTF

Clearly, great apes have the potential to play a pivotal role in how the conservation and development communities address the bushmeat crisis. As a nexus for bushmeat information and action, BCTF brings together many organizations focused on apes and ape conservation, including the American Society of Primatologists (ASP), the Jane Goodall Institute (JGI), the Dian Fossey Gorilla Fund International (DFGFI), the Wildlife Conservation Society, and many zoological institutions that care for apes in their collections. While BCTF maintains a focus on all wildlife species affected by the bushmeat trade, it recognizes that great apes are especially important. BCTF has worked with scientific and conservation organizations, including the Great Ape Survival Project (GRASP), ASP, and AZA's Ape Advisory Group to raise the profile of the bushmeat issue. Some of BCTF's most important ape-focused activities are summarized below.

3.1. Information Sharing

BCTF gathers, evaluates, and provides information related to the bushmeat crisis through its research archive and projects database, which catalogues hundreds of scientific and media articles and dozens of relevant projects throughout Africa related to the bushmeat crisis. A collaborative effort to combine these databases into a comprehensive, web-based Bushmeat Information Management and Analysis Project (IMAP) was initiated in 2002 and formally launched in September 2004. This geo-referenced set of databases provide users—including conservation groups, African governments, and other key decision-makers—with scientific information that can be used to evaluate current activities and threats to great apes and their habitats and identify priority areas for action. There is a detailed section including species ranges and fact pages dedicated to the African great apes, which can be integrated with other landscape level layers including logging concession activity, road development and human population presence.

BCTF has summarized information concerning the impact of bushmeat on great apes, which it has widely distributed through its website, www.bushmeat.org (Bailey and BCTF, 2001; Bailey et al., 2001). In a review of BCTF's research archive, Bailey and Stein (2001) found that 28% of bushmeat projects, 30% of media articles, and 16% of peer-reviewed articles listed were focused on great apes. "To put these figures in perspective, the numbers of scientific studies and field projects reviewed from the BCTF Research Archive on the great apes alone are greater than the total number of entries for duikers, elephants, carnivores, and rodents combined" (Bailey and Stein, 2001). This document also provides a list of conclusions and proposed solutions, many of which are being actively pursued by both BCTF and its members.

3.2. Protected Areas

Protected areas have been identified as a key to the future of great ape populations across their range. Effectively managed protected areas may provide apes a haven from the bushmeat trade. Unfortunately, many protected areas are "paper parks" that exist on paper, but have not secured the necessary infrastructure (e.g., trained guards and administrators, vehicles and other equipment, guard quarters, fences, etc.) to ensure protection of the habitat and the wildlife contained within (Mittermeier et al., 2000). Many BCTF members are working actively with the U.S.-government funded Congo Basin Forest Partnership, which provided US $53 million in aid during its first three-year phase (2003–2005) to help protect key forest areas throughout the region (Anon, 2003). Several members also have specific initiatives and projects directly related to the development and maintenance of protected areas and park personnel training. BCTF also supported its members in their efforts to build awareness of the bushmeat issue at the World Parks Congress in South Africa in September 2003.

3.3. Linkages with Private Industry

Logging companies—mostly European and Malaysian—have built roads deep into tropical forest and provided a means by which hunters can more readily transport wildlife into central and western African urban centers. In addition, some logging companies do not provision their employees, thus increasing hunting pressures in logging concessions. As approximately 80% of Central African forests are contained in logging concessions (GFW, 2000), it is critical that effective wildlife management programs be implemented on these lands. BCTF was invited to participate in the World Bank sponsored CEO Tropical Africa Working Group, which was established to bring together conservation groups and logging company executives to discuss the development of effective wildlife management within logging concessions of Central Africa. It remains to be seen how effective such cooperative approaches will ultimately be. However, a public-private partnership between the Republic of Congo, the Wildlife Conservation Society, and the logging company Congolaise Industrielle de Bois has shown promising results in protecting endangered species, such as gorillas and chimpanzees, in areas where controlled hunting exists (Elkan and Elkan, 2002).

BCTF is also working with its members and the World Bank to support efforts to monitor and control the bushmeat trade along Cameroon's national railway. Similarly, BCTF is working with Supporting Members that have field programs in Cameroon to gather information regarding the potential impacts of the Chad-Cameroon pipeline constructed in 2002.

In early September 2005, BCTF was invited to participate in the International Petroleum Industry Environmental Conservation Association (IPIECA) and International Association of Oil and Gas Producers (OGP) workshop titled: "Biodiversity and the Oil and Gas Industry: Central and West Africa." This was an important gathering of oil and gas industry executives in Angola to discuss issues of importance regarding the environment and the oil and gas industry. A presentation provided executives with important detailed information regarding the overall bushmeat trade as well as the linkages with developing infrastructure and private industry campaigns.

3.4. International Policy Development

BCTF has been a primary supporter of the CITES Bushmeat Working Group, which has been mandated to coordinate activities regarding the illegal international bushmeat trade for Central African countries (Cameroon, Republic of Congo, Democratic Republic of Congo, Central African Republic, Gabon, and Equatorial Guinea). This group commissioned an important policy and legislation review regarding the bushmeat issue for all member countries.

Senior-level members of the U.S. Congress, Department of State, Department of Agriculture, and Department of Interior regularly contact BCTF for up-to-date information and resources related to the bushmeat crisis. BCTF has actively engaged in awareness raising efforts among U.S. key decision makers on other occasions, including BCTF's Capitol Hill event in May 2000 and National Press Club Event in May 2001. The former provided background on the bushmeat issue and the plight of the great apes to hundreds of key decision makers and their staff prior to the passing of the Great Ape Conservation Act. The latter brought the bushmeat issue to scores of media professionals in the Washington DC area. In addition, BCTF participated in and provided logistical support for a July 2002 Congressional Hearing on Bushmeat in the U.S. House of Representatives Resources Subcommittee on Fisheries Conservation, Wildlife, and Oceans.

BCTF has provided presentations to and maintains contact with the International Conservation Caucus (ICC)—a Congressional caucus now present in both the House of Representatives and Senate dedicated to highlighting important international conservation issues to our nation's leading legislators. In addition, BCTF works with the U.S. Department of State and other agencies regarding the importance of the bushmeat issue and the need to assure a collaborative effort of U.S. government agencies focused on illegal wildlife trade.

3.5. Professional Training and Capacity Building

BCTF also works extensively with Africa's three regional wildlife colleges (École pour la Formation des Spécialistes de la Faune de Garoua, Cameroon; College of African Wildlife Management, Mweka, Tanzania; Southern African Wildlife College, South Africa) to assist in the development of bushmeat-related curricula for mid-career wildlife professionals throughout sub-Saharan Africa. As these three institutions train many of the wildlife managers in Africa, including bushmeat in each curriculum will be an enormous benefit to increasing understanding of the challenges and solutions among African wildlife professionals. As of 2006, the bushmeat curriculum at École de Faune is fully implemented among all students at the wildlife college, and efforts are now underway to share the curriculum with other wildlife, forestry, and natural resources training institutions in francophone Africa.

3.6. Public Awareness

In recent years, African conservation efforts have focused largely at high policy levels or at local community levels with mixed success. However, a stronger focus on public awareness in African urban centers could help educate the primary consumers of bushmeat. African societies have become more and more urbanized and thus are further removed from the realities of rural life. The 2002 BBC film *Ape Hunters* showed a bushmeat seller who stated that for every animal she purchased there were 20 being born in the forest every day. This is reminiscent of American attitudes at the beginning of the last century, where a perception of unlimited natural resources and lack of regulations, led to rampant over-exploitation and extinction or near extinction of species, such as the passenger pigeon, bison, waterfowl, and even the white-tailed deer (Hornaday, 1913; Lund, 1980; Warren, 1997).

Using media to reach out to urban consumers is a compelling strategy for two reasons. First, it enables conservationists to target their efforts on areas of highest human population density and commercial activity. Second, it targets an audience that has other dietary and economic options, and last, the potential for successful intervention is high. Regulating urban bushmeat markets could result in local communities retaining some level of legal and potentially sustainable wildlife harvest to supplement local income and meet both cultural and protein needs (Eves *et al.*, 2002). In 2003, the CITES BWG supported a review effort to identify opportunities for a massive Central African public awareness campaign targeted at urban consumers. An NGO, the International Conservation and Education Fund was later established to provide capacity building and training to Central African media professionals, facilitating the production of public awareness campaigns on key environmental issues for local and national African publics.

Another partner in public awareness and education efforts is the Pan Africa Sanctuary Alliance (PASA), the umbrella organization representing primate sanctuaries throughout Africa. Because most primate sanctuaries developed out of a need to care for the orphans of the bushmeat trade, PASA approached BCTF in 2000 to request assistance in coordinating support for public education on the bushmeat issue in Africa. From 1993 to 2003, the number of primate sanctuaries increased from just 2 to 19. There is a great need to reach out to rural and urban communities regarding the bushmeat crisis, to reduce the trade in apes and other species and, therefore, to eliminate the continued need for sanctuaries. BCTF and its members provided supporting funds for an important bushmeat awareness campaign targeted at law enforcement officers led by PASA, CITES, and the Dian Fossey Gorilla Fund International in eastern Democratic Republic of Congo in 2004.

Raising awareness in the United States, especially among the public and key decision-makers, is essential to mobilizing resources for implementing solutions to the bushmeat crisis in Africa and around the world. As the wealthiest and most powerful country in the world, the United States has a special obligation to assist disadvantaged countries in their efforts to resolve difficult conservation and human welfare issues. In cooperation with AZA's Conservation Education Committee, BCTF and many volunteers have developed a comprehensive *Bushmeat Education Resource Guide* (BERG) to provide training, graphics, programs, activities and evaluation tools to U.S. and international institutions that wish to provide education on the bushmeat issue. The BERG concept was created in response to the numerous requests for bushmeat information, particularly regarding great apes. While the BERG has been developed to provide a comprehensive perspective on the many species that are targeted in the bushmeat trade, many of our colleagues have indicated their intention to pair BERG signage with zoo ape exhibits or ape-related events.

BCTF has found that the majority of concerned individuals who ask for information on bushmeat are motivated by a concern for great apes. Therefore, many articles to which BCTF has contributed have focused on ape conservation and welfare, including *Newsweek, US News and World Report, National Geographic Adventure, People, USA Today, The Washington Post, Scientific American, Christian Science Monitor,* and programs on National Public Radio, Voice of America, BBC News, Discovery Channel Canada and many others. Scientific articles by BCTF, our members, and partners have appeared in *Endangered Species Update, Conservation Biology in Practice, Oryx, Trends in Ecology and Evolution,* and others. These efforts have been extremely effective in raising the general awareness and support for this critical conservation crisis among the public, conservationists, and university scientists.

4. Conclusions

A coordinated effort by multiple stakeholders to identify and implement solutions to the bushmeat crisis is the only way that wildlife populations will survive for future generations of Africans and concerned individuals around the world. The threat of this unsustainable, illegal, commercial trade has the potential to extinguish populations of many unique species, including the African great apes, within a single generation. Failure to resolve this crisis will be a tragedy of global proportions. With input and action from a vast array of partners in Africa and around the world, BCTF hopes to shape a future for wildlife and people beyond this immediate crisis. As the concerned public and media calls for the protection of great apes, it is our responsibility to provide resources for their conservation in African range countries. The massive harvest of wildlife for meat compels us to act quickly, as our options are likely to become more limited with each passing year. While we believe that conservationists should not cry wolf, we are convinced that the bushmeat issue deserves to be called a crisis. BCTF and its members are working to mobilize immediate, massive, and collective action from the global community to assure a common future for African wildlife, including great apes, and human communities.

Acknowledgments. Julie Stein, Liz Gordon, Tim Clark, Andrew Tobiason, BCTF Supporting and Contributing Members, and the BCTF Steering Committee deserve thanks for their continuing support and work on behalf of BCTF.

References

African Wildlife Foundation. (2001). Virunga heartland is biologically rich—but embattled. Washington, DC (20 February 2001). http://www.awf.org/news/1029

Anadu, P.A., Elaman, P.O., and Oates, J.F. (1988). The bushmeat trade in southwestern Nigeria: A case study. *Human Ecology* 16(2):199–208.

Anonymous. (2001). Government seeks help to fight Ebola outbreak. All Africa News Agency. (12 December 2001). http://allafrica.com/stories/200112120718.html

Anonymous. (2002). WHO blames gorilla for Ebola cases. *Washington Post.*

Anonymous. (2003). Congo Basin Forest Partnership: U.S. Contribution Fact Sheet. U.S. Department of State Bureau of Oceans and International Environmental and Scientific Affairs. (17 February 2003), www.state.gov/g/oes/rls/fs/2002/15617.htm

Auzel, P., and Wilkie, D.S. (2000). Wildlife use in Northern Congo: Hunting in a commercial logging concession. In: Robinson, J.G., and Bennett E.L. (Eds.) *Hunting for Sustainability in Tropical Forests*, New York, Columbia University Press, pp. 413–426.

Bailey, N.D., and BCTF. (2001). Effects of Bushmeat Hunting on Populations of African Great Apes. Summary Document. Silver Spring, MD, Bushmeat Crisis Task Force. http://www.bushmeat.org/html/apesbushmeat.html

Bailey, N.D., and Stein, J.T. (2001). BCTF Species Summary: African Great Apes and the Bushmeat Trade. Silver Spring, MD, Bushmeat Crisis Task Force.

Bailey, N.D., Stein, J.T., and BCTF. (2001). BCTF Fact Sheet: African Great Apes and the Bushmeat Trade. Silver Spring, MD, Bushmeat Crisis Task Force.

BBC Cymru Wales (Producer). (2002). *Ape Hunters* (video). Wales, U.K., BBC Wales/U.K.

Beck, B.B., Stoinski, T.S., Hutchins, M., Maple, T.L., Norton, B., Rowan, A., Stevens, E.F., and Arluke, A. (2001). *Great Apes and Humans: The Ethics of Coexistence.* Washington, DC, and London, Smithsonian University Press.

Bennett, E.L., Eves, H.E., Robinson, J.G., and Wilkie, D.S. (2002a). Why is eating bushmeat a biodiversity crisis? *Conservation in Practice* 3(2):28–29.

Bennett, E.L., Milner-Gulland, E.J., Bakarr, M.I., Eves, H.E., Robinson, J.G., and Wilkie, D.S. (2002b). Hunting the world's wildlife to extinction. *Oryx* 36(4):1–2.

Boo, E. (1990). Ecotourism: The Potentials and Pitfalls. Volume 1. Washington, DC, World Wildlife Fund.

Bowen-Jones, E., and Pendry, S. (1999). The threat to primates and other mammals from the bushmeat trade in Africa, and how this threat could be diminished. *Oryx* 33(3):233–246.

Bowen-Jones, E., Brown, D., and Robinson, E. (2002). Assessment of the solution-oriented research needed to promote a more sustainable bushmeat trade in Central and West Africa. Department for Environment Food and Rural Affairs, Wildlife & Countryside Directorate. pp. 1–125.

Bushmeat Crisis Task Force. (2000). Bushmeat Crisis Task Force Information Packet. Silver Spring, MD, Bushmeat Crisis Task Force.

Cavalieri, P., and Singer, P. (1993). *The Great Ape Project: Equity Beyond Humanity,* New York, St. Martin's Press.

Chrislip, D., and Larson, C. (1994). *Collaborative Leadership: How Citizens and Civic Leaders Can Make a Difference,* San Francisco, Jossey-Bass.

Conservation International. (2002). Ghana's old hunting trade generates new problems. Press release. http://www.conservation.org/xp/CIWEB/newsroom/press_releases/090502.xml

Corbey, R., and Roebroeks, W. (2001). *Studying Human Origins: Disciplinary History and Epistemology.* Amsterdam, Amsterdam University Press.

Cowlishaw, G., and Dunbar, R. (2000). *Primate Conservation Biology.* Chicago: University of Chicago Press.

Ecotourism Society. (1993). *Ecotourism Guidelines for Nature Tour Operators.* North Bennington, Vermont, The Ecotourism Society.

Elkan, P., and Elkan, S. (2002). Engaging the private sector: a case study of the WCS-CIB-Republic of Congo project to reduce commercial bushmeat hunting, trading and consumption inside a logging concession. *Communiqué* November 2002, pp. 40–42.

Eves, H.E., Gordon, E.A., Stein, J.T., and Clark, T.W. (2002). Great ape conservation in Central Africa: Addressing the bushmeat crisis. *Endangered Species Update* 19(4):171–178.

Eves, H.E., and Hutchins, M. (2001). The Bushmeat Crisis Task Force: Cooperative U.S. efforts to curb the illegal commercial bushmeat trade in Africa. In: Conway, W.G., Hutchins, M., Souza, M., Kapetanakos, Y., and Paul, E. (Eds.) *The AZA Field Conservation Resource Guide.* Atlanta: Zoo Atlanta. pp. 181–186.

Eves, H.E., and Stefan, A. (2002). Bushmeat Crisis Task Force: collaborative action toward addressing the bushmeat crisis in Africa. In: Mainka, S. and Trivedi, M. (Eds.) *Links Between Biodiversity Conservation, Livelihoods and Food Security: The Sustainable Use of Wild Species for Meat.* Gland, Switzerland and Cambridge, UK, International Union for Conservation of Nature and Natural Resources.

Eves, H.E., and Wolf, C.M. (2005). Forum: Ethics of hunting: perspectives on the bushmeat trade in Africa. *Frontiers in Ecology and the Environment* 7(3):394–395.

Fa, J.E., Juste, J., Del Val, J.P., and Castroviejo, J. (1995). Market hunting on mammal species in Equatorial Guinea. *Conservation Biology* 9(5):1107–1115.

Fa, J.E., Peres, C.A., and Meeuwig, J. (2002). Bushmeat exploitation in tropical forests: an intercontinental comparison. *Conservation Biology* 16(1):232–237.

Fouts, R. (1997). *Next of Kin: What Chimpanzees Have Taught Me about Who We Are*. New York, William Morrow and Company, Inc.

Gao, F., Bailes, E., Robertson, D.L., Chen, Y., Rodenburg, C.M., Michael, S.F., Cummins, L.B., Arthur, L.O., Peeters, M., Shaw, G.M., Sharp, P.M., and Hahn, B.H. (1999). Origin of HIV-1 in the chimpanzee *Pan troglodytes troglodytes*. *Nature* 397:436–441.

Global Forest Watch Cameroon Report. (2000). *An Overview of Logging in Cameroon*. Washington, D.C., World Resources Institute.

Goodall, J. (1990). *Through a Window: My Thirty Years with the Chimpanzees of Gombe*, Boston, Houghton Mifflin Company.

Goodall. J. (2000). At-Risk Primates. *The Washington Post*, 8 April 2000.

Hahn, B.H., Shaw, G.M., DeCock, K.M., and Sharp, P.M. (2000). AIDS as a zoonosis: scientific and public health implications. *Science* 287(5453):607–614.

Hart, J.A. (1978). From subsistence to market: a case study of the Mbuti net hunters. *Human Ecology* 6(3):325–353.

Hearn, J. (2001). *Scientific American* June, 2001. http://www.sciam.com/article.cfm?articleID=000709E4-03F5-1C70-84A9809EC588EF21

Hornaday, W.T. (1913). *Our Vanishing Wildlife*. New York: Charles Scribner's Sons. In: Warren, L.S. 1997 (Ed.) *The Hunter's Game: Poachers and Conservationists in Twentieth-Century America*, New Haven, Yale University Press.

Hutchins, M. (2002). Rattling the Cage: Toward Legal Rights for Animals (book review). *Animal Behaviour* 61:855–858.

Hutchins, M., and Wemmer, C. (1986). Wildlife conservation and animal rights: Are they compatible? In: Fox, M., and Mickley, L.D. (Eds.) *Advances in Animal Welfare Science 1986/87*. Washington, DC: Humane Society of the United States, pp. 111–137.

Kellert, S.R. (1998). *The Value of Life: Biological Diversity and Human Society*. Washington, DC, Island Press.

Lawson, A. (2002). Monkey brains off the menu in Central Africa. Reuters. http://abcnews.go.com/wire/SciTech/reuters20020101_21.html

Keele, B.F., Van Heuverswyn, F., Li, Y., Bailes, E., Takehisa, J., Santiago, M.L., Bibollet-Ruche, F., Chen, Y., Wain, L.V., Liegeois, F., Loul, S., Mpoudi Ngole, E., Bienvenue, Y., Delaporte, E., Brookefield, J.F., Sharp, P.M., Shaw, G.M., Peeters, M., and Hahn, B.H. (2006). Chimpanzee Reservoirs of Pandemic and Nonpandemic HIV-1. *Science*. Published online 25 May 2006. http://www.sciencemag.org/cgi/content/abstract/1126531v1

Litchfield, D. (1997). *Treading Lightly: Responsible Tourism with the African Great Apes*. Adelaide, Australia, Traveller's Medical and Vaccination Centre Pty Ltd.

Lund, T.A. (1980). *American Wildlife Law*. Berkeley, University of California Press.

Maisels, F., Keming, E., Kemei, M., and Toh, C. (2001). The extirpation of large mammals and implications for montane forest conservation: the case of the Kilum-Ijim Forest, North-west Province, Cameroon. *Oryx* 35(4):322–331.

Mann, C.C., and Plummer, M.L. (1995). *Noah's Choice: The Future of Endangered Species*. New York, Alfred A. Knopf.

Masland, T. (2003). Gorillas in the midst of an outbreak. Newsweek (20 January 2003). www.msnbc.com/news/858229.asp

Mittermeier, R.A., Robles Gil, P., and Mittermeier, C.G. (1997). *Megadiversity. Earth's Biologically Wealthiest Nations*. Monterrey, Mexico, CEMEX.

Moore, J. (1996). Savanna chimpanzees, referential models and the last common ancestor. In: McGrew, W.C., Marchant, L., and Nishida, T. (Eds.). *Great Ape Societies*. Cambridge, Cambridge University Press, pp. 275–292.

Mordi, A. (1991). Attitudes towards wildlife in Botswana. New York, Garland.

Morris, B. (1998). *The Power of Animals: An Ethnography*. Oxford, U.K., Berg.

Morris, B. (2000). *Animals and Ancestors: An Ethnography*. Oxford, U.K., Berg.

Myers, N. (1979). *The Sinking Ark: A New Look at the Problem of Disappearing Species*. New York, Pergamon Press.

Noss, A.J. (1998). Cables and snares in a central African forest. *Environmental Conservation* 25(3):228–233.

Plumptre, A.J., Bizumuremyi, J.B., Uwimana, F., and Ndaruhebeye, J.D. (1997). The effects of the Rwandan civil war on poaching of ungulates in the Parc National des Volcans. *Oryx* 31(4):265–273.

Ray, J.C., Stein, J.T., and BCTF. (2002). BCTF Fact Sheet: Forest Carnivores and the African Bushmeat Trade. Bushmeat Crisis Task Force. Silver Spring, MD. http://www.bushmeat.org/docs.html

Redmond, I. (1998). The Ape Alliance: European logging companies fuel trade in ape meat. Press release. http://bushmeat.net/pr298.htm

Richards, P. (1993). Natural symbols and natural history: chimpanzees, elephants and experiments in Mende thought. In: Milton, K. (Ed.) *Environmentalism: The View from Anthropology*. London and New York: Routledge, pp. 144–159.

Richards, P. (1995). Local understandings of primates and evolution: Some Mende beliefs concerning chimpanzees. In: Corbey, R. and Theunissen, B. (Eds.) *Ape, Man, Apeman: Changing Views Since 1600*. Leiden, Germany, Leiden University Department of Prehistory.

Robinson, J.G., and Bennett, E.L. (2000). *Hunting for Sustainability in Tropical Forests*. New York, Columbia University Press.

Rose, A.L. (1998). Growing commerce in bushmeat destroys great apes and threatens humanity. *African Primates* 3:1–2.

Ruggiero, R.G., and Eves, H.E. (1998). Bird-mammal associations in forest openings of northern Congo (Brazzaville). *African Journal of Ecology* 36:183–193.

Stoinski, T.S., Allen, M.T., Bloomsmith, M.A., Forthman, D.L., and Maple, T.L. (2002). Educating the public about complex environment issues: Should we do it and how? *Curator* 45:125–139.

Tashiro, Y. (1995). Economic difficulties in Zaïre and the disappearing taboo against hunting bonobos in the Wamba Area. *PanAfrica News,* 2(2). http://jinrui.zool. kyoto-u.ac.jp/PAN/2(2)/tashiro.html

Terborgh, J. (1999). *Requiem for Nature*. Washington, DC, Island Press.

Van Hooff, J.A.R.A.M. (2001). Conflict, reconciliation and negotiation in non-human primates: the value of long-term relationships. In: Noë, R., van Hooff, J.A.R.A.M., and Hammerstein, P. (eds.) *Economics in Nature: Social Dilemmas, Mate Choice, Biological Markets*. Cambridge, Cambridge University Press, pp. 67–90.

Warren, L.S. (1997). *The Hunter's Game: Poachers and Conservationists in Twentieth-Century America*. New Haven, Yale University Press.

Warren, M.A. (2001). The moral status of great apes. In: Beck, B.B., Stoinski, T.S., Hutchins, M., Maple, T.L., Norton, B., Stevens, E.F., and Arluke, A., (Eds.) *Great Apes and Humans: Ethics of Coexistence*, Washington, DC, Smithsonian Institution Press, pp. 313–328.

Walsh, P.D., Abernethy, K.A., Bermejo, M., Beyers, R., De Wachter, P., Akou, M.E., Huijbregts, B., Mambounga, D.I., Tohem, A.K., Kilbourn, A.M., Lahm, S.A., Latour, S., Maisels, F., Mbina, C., Mihindou, Y., Obiang, S.N., Effa, E.N., Starkey, M.P., Telfer, P., Thibault, M., Tutin, C.E.G., White, L.J.T., and Wilkie, D.S. (2003). Catastrophic ape decline in western equatorial Africa. *Nature* 6 April 2003.

Wilkie, D.S. (2001). Bushmeat hunting in the Congo Basin—a brief overview. In: Bakarr, M.I., Fonseca, G., Mittermeier, R., Rylands, A.B., and Painemilla, K.W. (Eds.) *Hunting and Bushmeat Utilization in the African Rain Forest: Perspectives Toward a Blueprint for Conservation Action*, Washington, DC, Conservation International.

Wilkie, D.S., and Carpenter, J.F. (1999). Bushmeat hunting in the Congo Basin: An assessment of impacts and options for mitigation. *Biodiversity and Conservation* 8(7):927–955.

Wilkie, D.S., Sidle, J.G., and Boundzanga, G.C. (1992). Mechanized logging, market hunting, and a bank loan in Congo. *Conservation Biology* 6(4):570–580.

Wilkie, D.S., Bennett, E.L., Eves, H.E., Hutchins, M., and Wolf, C. (2002). Roots of the bushmeat crisis: Eating the world's wildlife into extinction. *AZA Communiqué*, November 2002, pp. 6–7.

Wilson, E.O. (1992). *The Diversity of Life*. New York, W.W. Norton & Company.

Wise, S.M. (2001). *Rattling the Cage: Toward Legal Rights for Animals*. Cambridge, Massachusetts, Perseus Publishing.

Wolfe, N.D., Eitel, M.N., Gockowski, J., Muchaal, P.K., Nolte, C., Prosser, A.T., Torimiro, J.N., Weise, S.F., and Burke, D.S. (2000). Deforestation, hunting and the ecology of microbial emergence. *Global Change and Human Health* 1(1):10–25.

Won wa Musiti, B. (1999). The silent forest. *World Conservation* (3–4):14–15.

World Heath Organization. (2003). Ebola haemorrhagic fever in the Republic of Congo—Update 4. (18 February 2003). www.who.int/csr/don/2003_02_18/en/

Ziff, S. (2002). "Bushmeat" and the Origin of HIV/AIDS: A Case Study of Biodiversity, Populations Pressures, and Human Health. Summary Document. Environmental and Energy Study Institute, Washington, D.C. http://www.eesi.org/publications/02.19.02bushmeat.pdf

Author Index

Subject Index

Printed in the United States of America